Herbal Medicine Trials

This book comprehensively provides the research and validation of herbal medicines within modern clinical trials. The book covers the basics of herbal remedies, their historical and cultural contexts, and delves into specific methodologies such as traditional evidence-based medicine, reverse pharmacology, and integrative medicine. It examines pharmacokinetics, pharmacodynamics, and herb–drug interactions and outlines the roles of various stakeholders in herbal medicine trials. Detailed attention is given to quality control, standardization, and safety evaluation, alongside methods for conducting randomized clinical trials. Ethical and regulatory challenges are addressed extensively, with separate focus on regulations in the United States, the European Union (EU), and non-EU countries, including specific challenges in the Indian system of medicines. The text also highlights the use of artificial intelligence and machine learning in clinical trials and presents recent advancements, best practices, and the future trends in herbal medicine research.

Key Features:

- Provides a thorough overview of herbal medicines, detailing their historical and cultural backgrounds alongside modern scientific research
- Explores specific research methodologies such as traditional evidence-based medicine, reverse pharmacology, and integrative medicine approaches tailored to herbal medicines
- Includes in-depth examinations of how herbal medicines interact within the body and with other drugs
- Emphasizes the importance of standardization, quality control, and safety evaluation strategies in conducting clinical trials for herbal medicines
- Discusses ethical and regulatory challenges, with specific sections on regulations in the United States, EU, and non-EU countries

This book is an essential resource for regulatory professionals, researchers, and clinicians seeking to bridge traditional herbal practices with rigorous scientific research.

Herbal Medicine Trials

Methodologies, Regulatory Challenges, and Future Directions

Edited by
Anoop Kumar and Devesh Tewari

CRC Press
Taylor & Francis Group
Boca Raton London New York

CRC Press is an imprint of the
Taylor & Francis Group, an **informa** business

Designed cover image: Shutterstock

First edition published 2026
by CRC Press
2385 NW Executive Center Drive, Suite 320, Boca Raton FL 33431

and by CRC Press
4 Park Square, Milton Park, Abingdon, Oxon, OX14 4RN

CRC Press is an imprint of Taylor & Francis Group, LLC

ISBN: 978-1-032-86259-0 (hbk)
ISBN: 978-1-032-86262-0 (pbk)
ISBN: 978-1-003-52206-5 (ebk)

DOI: 10.1201/9781003522065

Typeset in Times
by Apex CoVantage, LLC

Contents

Foreword ... vii
Preface ... ix
About the Editors ... xi

Chapter 1 Herbal Medicinal Trials: Methodologies, Regulatory Agencies, and Future Directions ... 1
Neeraj Choudhary

Chapter 2 Introduction to Clinical Trials with Special Focus on Botanicals ... 15
Neeraj K. Sethiya, Mahendra Singh, Sonu Sharma, Vishal Kumar, and Vrinda Poply

Chapter 3 Competitive Landscape Analysis and Market Dynamics of Herbal Medicines ... 25
Kartikeya Joshi, Ayush Tyagi, Naresh Suyunari, Dhruv Ghoshal, and Dhruv Bhardwaj

Chapter 4 Integrating Traditional Wisdom and Modern Science: Evidence-Based Approaches and Reverse Pharmacology in Herbal Medicine ... 36
Anjana Sathyan, Anu Ruhila, and Galib Ruknuddin

Chapter 5 Stakeholders and Their Roles in the Clinical Trials of Herbal Medicines: Navigating Collaboration and Compliance ... 49
Ramneek Khaira, Paranjeet Kaur, and Roqia Bashery

Chapter 6 Quality Control, Standardization, and Safety Evaluation Strategies of Herbal Medicines for Clinical Trials ... 56
Dhanashree Sharma, Navratan Soni, Mukul Yadav, Meena Chandran, and David Paul

Chapter 7 Ethical Aspects of Clinical Trials of Herbal Medicines Across the Globe ... 75
Tanya Ralli, Nitish Singh Jangwan, Swati Dobhal, Shradha Bisht, and Mamta F. Singh

Chapter 8 Regulatory Aspects of Clinical Trials of Herbal Medicines in the United States ... 91
Pankaj Dagur, Naresh Rangra, Motseoa Grace Molefi, Ankita Jadhav, and Manik Ghosh

Chapter 9 Regulatory Aspects of Clinical Trials of Herbal Medicines in Non-European Countries ... 105
Uma Ranjan Lal, Megha Soni, and Snigdha Lal

Chapter 10 Good Clinical Practices for Clinical Trials of Herbal Medicine ... 131
Sachin Karmakar, Tanzeem Nigar, Anwesha Das, Arijit Nandi, and Wei-Yu Lin

Chapter 11 Regulatory Aspects, Challenges, and Opportunities in Clinical Trials of the Indian System of Medicines ... 147
Maya Elizabeth Thomas, Anoop Kumar, and Devesh Tewari

Chapter 12 Latest Advancements in Clinical Trials of Herbal Medicines ... 155
Chandrakanta, Archana Negi Sah, and Sweta Bawari

Chapter 13 Clinical Trials of the Indian System of Medicine Based on Case Studies 167

Khushi Gupta, Avadh Biharee, Aatif Subhan, and Suresh Thareja

Chapter 14 Clinical Trials of the Traditional Chinese Medicine System Based on Case Studies 182

Mithun Singh Rajput and Gopal Natesan

Chapter 15 Challenges and Future Perspectives of Herbal Medicines 194

Raja Chakraverty, Prasanth Thumpati, Swarnali Goswami, Abhik Paul, and Tatini Debnath

Chapter 16 Future Trends, Emerging Technologies, and Interdisciplinary Collaborations for Clinical Trials of Herbal Drugs 200

Richi Goel, Versha Chandila, and Sweta Bawari

Index 213

वैद्य राजेश कोटेचा
सचिव
Vaidya Rajesh Kotecha
Secretary

सत्यमेव जयते

आज़ादी का अमृत महोत्सव

भारत सरकार
आयुष मंत्रालय
आयुष भवन, 'बी' ब्लॉक, जी.पी.ओ. कॉम्प्लेक्स,
आई.एन.ए. नई दिल्ली–110023
Government of India
Ministry of Ayush
Ayush Bhawan, B-Block, GPO Complex,
INA, New Delhi-110023
Tel. : 011-24651950, Fax : 011-24651937
E-mail : secy-ayush@nic.in

Foreword

Herbal medicine clinical trials represent a vital convergence of traditional systems of medicine and modern biomedical science. Across cultures and centuries, medicinal plants have served as the cornerstone of healthcare, yet their transition from time-honored wisdom to evidence-based medicine continues to present unique challenges. The book, *Herbal Medicine Trials: Methodologies, Regulatory Challenges, and Future Directions*, edited by Anoop Kumar and Devesh Tewari is a commendable and timely effort that elucidates the design and conduct of herbal medicine trials using advanced scientific approaches and illustrative case studies.

This book provides a comprehensive overview of the methodologies, regulatory frameworks, and emerging tools that define the evolving landscape of herbal medicine research. It bridges traditional knowledge with modern scientific rigor, highlighting the potential of systems biology, omics technologies, and artificial intelligence to advance the credibility and global acceptance of herbal medicines.

For researchers, clinicians, and regulators, this book offers both guidance and inspiration. I am confident that the insights presented here will inspire new thinking, advance scientific collaboration, and guide the next generation of clinical research in herbal medicine.

(Rajesh Kotecha)

Preface

Herbal medicines have been an integral part of human healthcare for centuries, forming the backbone of traditional systems across diverse cultures. In recent decades, growing scientific curiosity, rising global demand, and increasing recognition by healthcare systems have brought herbal therapies into the spotlight of clinical research. Yet designing and conducting clinical trials for herbal medicines remains one of the most intricate challenges in modern biomedical science.

This book, *Herbal Medicine Trials: Methodologies, Regulatory Challenges, and Future Directions*, was conceived to address this very challenge. It aims to provide a comprehensive framework for understanding how clinical trials of herbal interventions differ from those of conventional pharmaceuticals and why specialized approaches are necessary.

The chapters within explore the nuances of trial methodologies, including standardization of multicomponent herbal preparations, placebo design, participant selection, outcome assessment, and statistical considerations. They also critically examine the regulatory environments governing herbal clinical trials in different regions of the world—highlighting both gaps and opportunities for harmonization. Recognizing that the credibility of herbal medicine research depends on both scientific rigor and cultural sensitivity, the book emphasizes ethical conduct and transparency throughout.

Equally important is the book's forward-looking orientation. It discusses emerging tools such as omics technologies, systems biology, artificial intelligence, and digital health platforms that can revolutionize the design and interpretation of herbal clinical trials. By weaving traditional knowledge with modern methodologies, this volume envisions a future where herbal medicines are evaluated and integrated into healthcare with the same rigor as any other therapeutic intervention.

This work is intended for researchers, clinicians, regulators, students, and practitioners who seek to engage with herbal medicine at the interface of tradition and modernity. It is our hope that the insights shared here will foster innovation, strengthen regulatory confidence, and contribute to safer and more effective global healthcare solutions.

About the Editors

Anoop Kumar is currently working as Assistant Professor in the Department of Pharmacology, Delhi Institute of Pharmaceutical Education and Research (DIPSAR), Delhi Pharmaceutical Sciences & Research University (DPSRU), Government of NCT of Delhi. Recently, he has also been included in the top 2% list of scientists released by Stanford University. Earlier, he worked as Assistant Professor in the Department of Pharmacology and Toxicology, National Institute of Pharmaceutical Education and Research (NIPER), Raebareli, and as Head and Associate Professor in the Department of Pharmacology, Indo-Soviet Friendship (ISF) College of Pharmacy, Moga, Punjab. He has also worked as a research scientist in the Medical Affairs and Clinical Research Department of Sun Pharmaceutical Industries Limited, Gurugram, and at the Translational Health Science Institute (THSTI), Faridabad. He has also worked as an expert in the Drug Discovery Hackathon 2020, a joint initiative of the All India Council for Technical Education (AICTE), the Council of Scientific and Industrial Research (CSIR), and the Government of India. He has also received various fellowships, including the Graduate Aptitude Test in Engineering (GATE), the University Grants Commission Basic Scientific Research Fellowship (UGC BSR), and the Department of Science and Technology Innovation in Science Pursuit for Inspired Research (DST INSPIRE).

He has authored more than 120 research and review articles and 30 book chapters in international journals and with publishers of repute. He has also edited five international books for reputed publishers such as Elsevier, Springer, and Taylor and Francis. He is an associate editor and an editorial board member of various international journals published by Elsevier, Wiley, and Springer Nature. He has delivered more than 50 expert talks at international and national conferences. He has worked as member secretary of the Institutional Animal Ethics Committee (IAEC) and as animal house in-charge. Currently, he is working as a member secretary of the Human Ethics Committee approved by the Central Drugs Standard Control Organization (CDSCO) and treasurer of the International Society for Pharmacoeconomics and Outcomes Research (ISPOR) India Chapter. His laboratory also receives funding from the Department of Science and Technology – Science and Engineering Research Board (DST SERB) (Start-up Research Grant (SRG) and Core Research Grant (CRG)), the Indian Council of Medical Research (ICMR), the National Innovation Foundation (NIF)–India, and the Indian Institute of Technology Delhi (IIT Delhi). His research interests are in the following areas: (1) drug repurposing, (2) meta-analysis of clinical trials, (3) network pharmacology, (4) pharmacoeconomic studies, (5) signal analysis in pharmacovigilance, and (6) animal experimental studies.

Devesh Tewari is currently serving as Assistant Professor in the Department of Pharmacognosy and Phytochemistry and as in-charge of the Ultra Performance Liquid Chromatography (UPLC) facility at Delhi Pharmaceutical Sciences & Research University (DPSRU), New Delhi. He also served as a visiting scientist at the Institute of Genetics and Animal Biotechnology of the prestigious Polish Academy of Sciences, Poland. He is among the youngest pharmacognosy teachers in the world featured in the list of the "World Ranking of Top 2% Scientists" by Stanford University and Elsevier for the last 4 years, including the career-wise list. He is a renowned researcher in the field of natural product drug discovery and formulation development, with a main focus on the exploration of medicinal plants of the Himalayan region through ethnopharmacology. Dr. Devesh is a leading expert in the field of natural product drug discovery. In addition, he has also served on high-level committees formed to clinically validate some natural product–based drugs derived from grassroots innovations/ethnopharmacology. Having significant research achievements in the field of pharmacological validation and phytochemistry of Western Himalayan plants, he received several important awards for plant drug research, including the Young Scientist Award by the Uttarakhand State Council for Science and Technology (UCOST), Government of Uttarakhand; the Best Paper Award by the Ministry of Ayurveda, Yoga and Naturopathy, Unani, Siddha and Homoeopathy (AYUSH), Government of India; the Best Concept Note Award by the Pharmacy Council of India (PCI), Government of India; the Research Appreciation Award for three consecutive years from Lovely Professional University (LPU); first, second, and third prize for the presentation of research work, including monetary awards for research and Hindi writing. He is also a member of the Pharmacopoeia Committee (PCIM&H), an expert member for specific publications related to COVID-19 of the Ministry of AYUSH, and an expert reviewer for grant proposals of international research councils. Dr. Devesh has received various research grants from the Department of Science and Technology (DST), Government of India. He has served as a resource person/expert and presenting author at prestigious international and national conferences and as an editorial board member for several prestigious journals.

1 Herbal Medicinal Trials

Methodologies, Regulatory Agencies, and Future Directions

Neeraj Choudhary
Department of Pharmacognosy, GNA School of Pharmacy, GNA University, Phagwara, Punjab, India

1.1 HERBAL MEDICINE TRIALS

Herbal clinical trials are crucial for validating the safety, efficacy, and quality of plant-based medicines within an evidence-based framework. They address key challenges such as variability in plant composition, standardization of active compounds, and the design of suitable placebo controls. Recent research emphasizes their growing role in modern healthcare. For instance, a 2025 scoping review of randomized controlled trials (RCTs) on herbal medicine for obesity highlighted the urgent need for methodological standardization to enhance the reliability of outcomes [1]. A 2023 cross-sectional study on herbal trials for rare diseases further demonstrated that inadequate placebo reporting and ethical approval disclosures continue to undermine reproducibility and scientific credibility [2]. Additionally, assessment of randomized trials on traditional Chinese patent medicines revealed persistent weaknesses in blinding, registration, and ethics oversight, despite gradual improvements in trial quality. Together, these findings underscore the importance of well-designed herbal clinical trials in strengthening global confidence, ensuring patient safety, and supporting the integration of herbal therapies into modern medical practice. Here are some key reasons why clinical trials are important for herbal medicines. They can cause adverse reactions or interact with conventional drugs, and clinical trials help identify these risks.

- Trials assess the potential toxicity of herbal products, which is essential for public health safety.
- Clinical trials provide the necessary evidence to validate the therapeutic benefits of herbal medicines, which is crucial for their integration into mainstream healthcare.
- Trials help standardize herbal products by ensuring consistent formulations and dosages, which is vital for reproducible results.
- Many countries require clinical trials to demonstrate safety and efficacy before granting regulatory approval for herbal products.
- Trials enforce quality control measures, ensuring that herbal products meet certain standards for purity and potency.
- Clinical trials provide scientific validation, which enhances public confidence in herbal medicines and supports their use alongside conventional treatments.
- Rigorous clinical trials are necessary for herbal products to gain acceptance in the global market and to be recognized by international health authorities.
- Clinical trials facilitate the advancement of herbal medicine by identifying effective compounds and optimizing formulations.
- Evidence from trials can lead to the integration of herbal medicines into national healthcare systems, promoting a more holistic approach to healthcare [3, 4].

1.2 ETHICAL CONSIDERATIONS IN CLINICAL TRIALS WITH HERBAL PRODUCTS

According to the WHO guidelines set by the World Health Organization (WHO), the basic ethical norms governing human participation in research equally apply to herbal treatments and research involving these substances. Obtaining consent is necessary, as well as ensuring fairness in subject selection, evaluating risks and benefits to ensure they are favorable for possible participants, and ensuring a good experimental design [5]. Specific concerns particularly relevant to clinical trials involving herbal products include the risk of product adulteration, which has been documented in several studies. For instance, research on raw herbal trade samples in South India found that about 12% of market samples were adulterated with other species or substances, highlighting the risk of compromised product quality and safety. Additionally, unintentional or intentional adulteration with synthetic chemicals has been reported and poses a significant health risk. Another major concern is the potential interactions between herbal remedies and other pharmaceuticals or substances, which are often poorly understood, leading to unpredictable effects during clinical use. Furthermore, there is limited availability of robust data on reproductive toxicity and organ-specific toxicity for

DOI: 10.1201/9781003522065-1

many herbal products. This gap in toxicological data creates challenges for establishing safe dosage ranges. Finally, clinical trials often suffer from incomplete prior dose-finding studies, which are vital for defining effective and safe dose parameters in herbal medicine research [6, 7]. Addressing these concerns is critical for ensuring the safety, efficacy, and reliability of herbal medicinal trials. The widespread conviction in various parts of the globe that herbal remedies are advantageous and harmless can introduce bias, which can be mitigated by meticulously considering the study's design, including the inclusion of suitable control groups [8]. Hence, it is imperative to select proficient clinicians as investigators to ensure timely identification and suitable management of any noted adverse event or deterioration of a preexisting ailment. Herbal studies should be subjected to the same level of scrutiny by ethics boards as conventional therapy regimens.

1.3 METHODOLOGY OF CLINICAL TRIALS

For a new herbal medicine, a new use for an existing herbal medicine, or a different way of taking the medicine, the principles and requirements for a clinical trial should be similar to those for conventional drugs. These guidelines can be found in Annex VI of the WHO's Good Clinical Practice (GCP) protocols [9]. Randomized controlled clinical studies that are well established are considered to provide the most reliable and conclusive evidence of effectiveness. These studies promote the adoption of herbal treatments in many places and among individuals with diverse cultural backgrounds. Nevertheless, the implementation of techniques such as randomization and the administration of a placebo may be hindered by ethical concerns and technical limitations. For instance, it may be impractical to include a placebo control in a study if the herbal treatment has a potent or noticeable aroma or flavor, such as items containing certain essential oils. Furthermore, patients who have previously received treatment with the herbal drug being studied, which possesses a distinct sensory attribute, cannot be randomly assigned to control groups. For herbal treatments that have a strong taste, placebos with the same taste may have a comparable effect. Under such circumstances, it is recommended to employ a minimal dose of the identical herbal remedy as a control. Alternatively, a well-established treatment can be utilized as a positive control [10].

Large-scale observational studies can serve as a highly beneficial method for assessing the effectiveness of herbal treatments, and traditional medical ideas and principles suggest that the prevention, diagnosis, improvement, and treatment of sickness are tailored to the individual patient's distinctive needs [11]. Hence, it is imperative not to disregard single-case studies when assessing the effectiveness of herbal therapy. Various methods for conducting clinical trials on herbal drugs, including descriptive and explanatory studies, are included later.

1.4 DESCRIPTIVE STUDIES

Descriptive studies document atypical or novel occurrences such as the presence of sudden infant death syndrome in several siblings within one family or the prevalence of albinism within a single family. The researcher records the findings and establishes correlations between the observed events and potential causes. These studies are not randomized or predesigned. There are often two categories, namely case reports and case series. A subset of individuals with unique clinical features is included in case reports. All the fundamental characteristics are documented, and each patient is regarded as a distinct case with full control over all the factors. The patient is being monitored and assessed to determine the potential result. The results are compared to the initial values or are indicated as either successful or unsuccessful outcomes of the treatment administered. If the therapy is successful, a hypothesis is formulated for a comprehensive study to investigate the correlation between the treatment and the observed outcome [12].

In a case series, regular observations are recorded from patients who have been exposed to a specific medicine or a group of drugs. In addition, they may also investigate the medical records of previous patients who experienced the same event to identify any potential cause-and-effect relationship, if one exists. These are valuable for forecasting the occurrence of a negative outcome associated with a recently introduced medication when information is scarce about such occurrences. Case reports and case series are retrospective studies that offer comprehensive accounts of individual instances or a group of cases with comparable features. Typically, they provide an overview of the subject area, along with a detailed account of the specific instances and their distinguishing characteristics. Nevertheless, it is important to note that case reports and case series cannot draw statistical inferences or make generalizations. Additionally, they do not include data from comparative groups of patients or individuals. Therefore, case reports and case series can bring awareness to a novel phenomenon, clinical manifestation, or possibly an emerging disease, thereby stimulating additional research. Nevertheless, the study design is not suitable for identifying connections, correlations, or causal linkages [13].

1.4.1 Explanatory Studies

1.4.1.1 Experimental Studies

Experimental studies, also known as interventional study designs, represent the pinnacle of clinical studies concerning scientific evidentiary quality and have the most power to suggest or prove causal relationships. These study designs test hypotheses, and the investigators control the exposure and assess the outcome. Experimental studies can be conducted both with and without comparison groups. Comparative studies are more convincing in demonstrating causal relationships, as the intervention being studied is distributed to one group and the other represents a control to confirm that the outcome is related to the intervention and not due to time or change. These study designs aim to test hypotheses, with investigators having control over the exposure and assessing the outcome. Experimental investigations can be undertaken with or without comparison groups. Since the intervention under study is given to one group while the other serves as a control, comparative studies are more effective in establishing causal relationships. This is because they ensure that the outcome is attributable to the intervention and not to chance, time, or other unrelated factors. There are generally two types of studies: non-randomized and randomized studies [14].

1.4.1.1.1 Non-randomized Clinical Trials

Non-randomized studies refer to research investigations that do not involve random assignment of participants to different groups or conditions. Patients are chosen based on specific criteria for selection. The participants are not assigned randomly to specific treatment(s) but rather receive a treatment based on the progression of their sickness. Typically, phase IV of a clinical trial proceeds in the following manner. Moreover, randomization is often unfeasible in numerous human experimental trials. Several surgical experiments have been developed with distinct indications and applications. Due to their specific patient groups, randomization is either not feasible or considered unethical. For example, dual kidney failure requires kidney transplantation. While dialysis is a viable choice, it cannot be equated with renal transplantation. Therefore, patients cannot be randomly assigned to these alternatives [15].

1.4.1.1.2 Randomized Controlled Trials

Patients are followed prospectively after being randomly assigned to either the treatment or control group in this type of study. The study therapy provides the most persuasive evidence of its projected effect on human health, and the process of randomization reduces any bias in study selection. RCTs are considered the most dependable and yield the most important results among several experimental study types. The significance of randomization lies in its ability to ensure that the treatment groups are evenly distributed in terms of both known and unknown prognostic factors. It is crucial to ensure that the intervention group and control group are studied simultaneously so that the observation phase takes place within the same period for both groups. The combination of blinding and placebo effectively decreases the possibility of bias. Double blinding refers to a research method in which both the investigator and the subject are kept unaware of the treatment being administered to the subject. Blinding the subjects may be unattainable in certain cases, such as when the therapy involves surgery or an educational procedure. However, it is frequently feasible to ensure that the evaluators who assess the results are uninformed of the treatment [16].

While the scientific rigor and statistical validity of the results may be strong, the practical significance in real-world applications may be restricted. To guarantee relevance, it is crucial that the endpoints are well stated, capable of being replicated, and have significant clinical relevance. Nevertheless, if a new medicine is planned for use in a patient population with significant biological diversity, the value of a clinical study may be diminished by the use of highly selected subject groups. Homogeneous study groups can help identify statistically significant associations that may not have practical benefits in clinical practice. To guarantee the practical applicability of clinical trials, it is crucial that the group of participants accurately reflects the broader patient population being studied. This includes considering factors such as biological diversity, age, gender, race, and the range of disease severity and symptoms. When conducted using rigorous scientific methods, RCTs with a prospective design, controlled intervention, balanced comparison groups, and measures to eliminate bias, such as randomization, placebo controls, and double blinding, offer robust scientific evidence for establishing a cause-and-effect relationship between the intervention and the outcome [17]. The comparison of RCTs and observational studies is mentioned in Table 1.1 and Figure 1.1.

For instance, if you wish to ascertain whether your new toothpaste is more effective in preventing cavities compared to your previous toothpaste, you employ a random allocation method to assign a substantial number of patients to either an intervention group, which uses the new toothpaste, or to a control group, which uses the old toothpaste. Subsequently, you would quantify the number of cavities that occur within the different groups over time.

1.4.2 Observational Studies

Observational studies, often known as epidemiological studies, can be categorized into descriptive and analytical investigations. Descriptive studies focus solely on describing the inherent relationship between components such as the exposure and/or the outcome. On the other hand, analytical studies try to identify an association or quantify a correlation between the exposure and the outcome [13].

1.4.2.1 Cross-sectional Studies

It is a type of study that examines a population at a specific point in time to get information on a particular characteristic or phenomenon. An epidemiological study investigates the correlation between diseases (or other health-related traits) and other factors of interest within a certain population at a given moment. In this type of study, both the exposure and results are examined simultaneously. Most suitable for measuring the extent of an illness or risk factor, as well as for measuring the precision of a diagnostic test, cross-sectional studies, while also retrospective, can be both descriptive and analytical. An analytical cross-sectional study assesses the possible connection between an exposure and an outcome. Due to the simultaneous measurement of both the exposure and the outcome, a cross-sectional study cannot evaluate temporality, a crucial factor in establishing causal relationships [13].

TABLE 1.1
Comparison of Randomized Controlled Trials and Observational Studies[13, 19]

S. No	Study design	Strengths	Weaknesses
1	**Randomized controlled trials**	• Equitable distribution of confounding factors without bias. • Increased likelihood of blinding. • Randomization simplifies statistical analysis. • Best design to reduce bias. • Only designed to account for unknown variables. • Able to determine the size of the treatment effect. • Able to determine treatment-predictive variable interactions.	• There may be instances where ethical issues arise. • Provides aggregated outcomes for a population rather than personalized ones. • Applicable to patients who match specific predetermined criteria for inclusion, and • criteria for excluding certain individuals or elements from a particular group or study. • Regarding the inclusion and exclusion criteria for randomized controlled trials • Challenging for uncommon medical conditions and infrequent groups of patients.
2	**Observational studies**	• Applicable to every demographic for which data is accessible. • Capable of identifying both risk variables and prognostic factors. • Capable of investigating detrimental encounters. • Capable of examining existing methodologies. • Capable of recognizing long-term patterns that are not influenced by religious or spiritual factors. • Capable of investigating uncommon exposures and uncommon outcomes. • Capable of creating risk models. • Capable of enabling the creation of hypotheses. • Capable of enabling the assessment of long-term outcomes. • Enable the investigation of the clinical course of patients who do not satisfy the criteria for randomized controlled trials.	• The outcome may provide a general overview of a population rather than personalized results, depending on the design. • Unable to account for unobserved variables that may influence the results. • Insufficient ability to effectively manage various biases, such as selection bias, and limited capacity to account for confounding. • Prospective observational studies might also incur significant costs for data collection processes.
2.1	Cross-sectional	• Prompt and precise data, tailored to each individual, with the capability to account for various confounding factors. • Has the ability to evaluate multiple results. • Economically affordable and uncomplicated. • Ethically secure.	• Not suitable for uncommon medical conditions. • Inadequate for acute illnesses. • Establishes correlation at most, not causation. • Vulnerable to recall bias. • Potential unbalanced distribution of confounding factors. • Possible unequal group sizes.
2.2	Case-control	• Rapid and cost-effective. • The only viable approach for extremely rare conditions or those with a significant time delay between exposure and outcome. • Requires a smaller sample size compared to cross-sectional research. • Capable of managing and accounting for many variables that may influence the outcome. • Effective for treating uncommon medical conditions. • Can evaluate various exposures.	• Unable to compute prevalence. • Can only evaluate a single result. • Inadequate choice of controls can induce bias. • It may be a challenge to accurately identify a sufficient number of cases. • Dependence on memory or documented evidence to ascertain exposure status. • Challenging selection of control groups. • Possible recall and selection bias.
2.3	Cohort studies	Ethically sound. Ability to match subjects. Ability to determine the timing and direction of events. Standardization of eligibility criteria and result assessments. Administrative simplicity and cost-effectiveness compared to randomized controlled trials.	Identification of controls may provide challenges. Exposure may be associated with an undisclosed confounding factor. Implementing blinding is challenging. Randomization is absent.

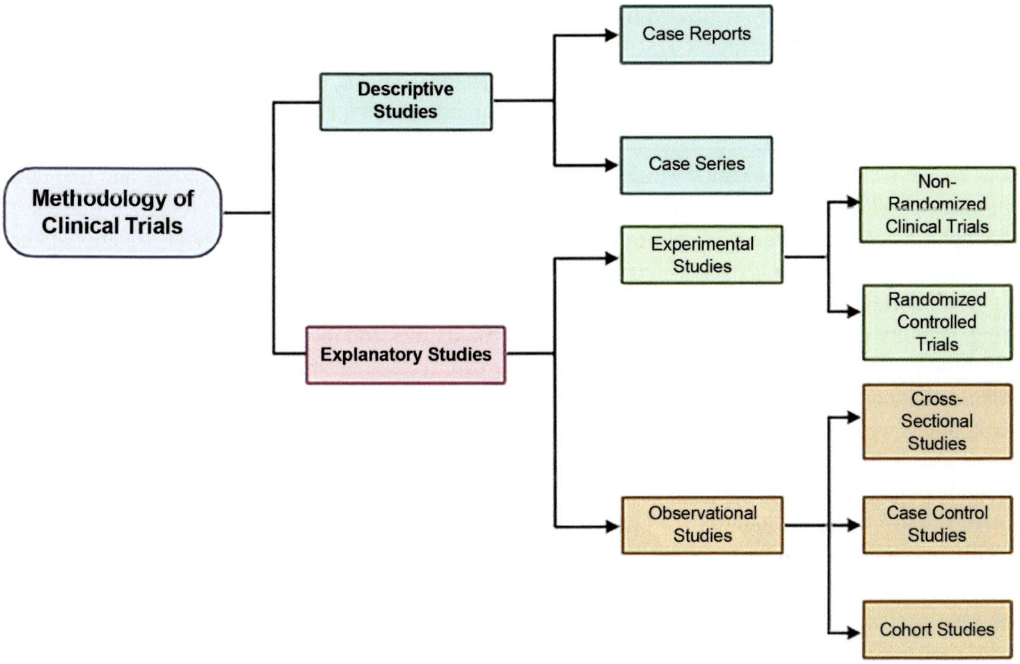

FIGURE 1.1 Methodology of clinical trials.

1.4.2.2 Case-Control Studies

Case-control studies are utilized to retrospectively analyze many groups of subjects and determine characteristics that are correlated with a specific result. For instance, the research design is employed to identify potential risk factors for the development of a disease by comparing a group that has the ailment with a group that does not. The study design is capable of testing hypotheses by discovering associations and correlations, but it cannot establish causal linkages. This study involves the selection of patients who have a specific outcome or disease, as well as a suitable group of controls who do not have the outcome or disease. The selection process takes into account factors such as matching and careful consideration of appropriate controls. Subsequently, information is gathered to determine whether the subjects have been exposed to the factor being investigated. The objective is to investigate the underlying factors that led to the result in one group while not affecting the other. To minimize confounding variables, it is essential to recruit all participants from a homogeneous community. This study design is susceptible to recollection bias, where events may be recalled differently between the two groups. Nevertheless, this approach is cost-effective, can be accomplished swiftly, and can be executed by individuals or small teams of researchers [18]. Case-control studies are frequently employed to investigate uncommon illnesses, as a prospective cohort study would necessitate an exceedingly large sample size to gather an adequate number of cases, an extended study duration, or both. Case-control studies are more cost-effective and require less time compared to prospective cohort studies, for example, to identify whether smoking is a risk factor for the onset of lung cancer. A comparison between the smoking history of persons who have been diagnosed with lung cancer (cases) and those who have not (controls) should be carried out [13].

1.4.2.3 Cohort Studies

Cohort studies are a type of research method used to investigate the relationship between certain factors and outcomes over a period. A non-randomized study design in which patient groups are divided based on exposure or treatment, either before (retrospective) or after (prospective) the start of the investigation. Cohort studies involve the selection of two or more groups that have distinct exposure or risk variables, which are then compared to determine any variations in the outcome, such as the occurrence of a disease. Therefore, this represents a contrasting approach to the design of case-control studies, where the selection of groups is based on the outcome itself. Cohort studies can be either prospective or retrospective. Retrospective study designs are susceptible to selection and recall biases, although they are significantly less time-consuming and cost-effective. To mitigate selection bias, it is crucial that the comparison groups are made as comparable as feasible, except for the specific exposures or potential risk factors being investigated. The investigators can regulate this to some degree [13, 19].

1.5 REGULATORY REQUIREMENTS IN INDIA

India follows a dual regulatory framework for herbal medicines. On one side are phytopharmaceuticals, defined under the New Drugs and Clinical Trials Rules (NDCTR 2019) as purified and standardized fractions of medicinal plants intended for therapeutic use. These are regulated by the Central Drugs Standard Control Organization (CDSCO) as new drugs and must undergo the full pathway of dossier submission, preclinical studies, and phased clinical trials, similar to synthetic drugs. On the other side are classical AYUSH/ASU medicines (Ayurveda, Siddha, Unani), which are regulated by the Ministry of AYUSH under AYUSH-specific GCP guidelines. Phytopharmaceuticals, however, require complete chemistry, manufacturing, and controls data, including botanical identity, source traceability, chemical fingerprinting, contaminant testing, stability data, and toxicology studies. ASU medicines mainly require authenticated raw materials, documented formulation details, and justification of dose based on traditional texts, along with bridging safety or clinical data where necessary [20]. Clinical trials for phytopharmaceuticals under NDCTR require a detailed application to CDSCO, including protocols with defined endpoints, an investigator's brochure; chemistry, manufacturing, and controls data; safety data; a statistical analysis plan; informed consent forms; adverse event reporting systems; and compensation plans, with mandatory registration on the Clinical Trials Registry-India [21]. For ASU products, GCP provides guidance such as traditional diagnostic criteria, pragmatic comparators, and informed consent tailored to the AYUSH context. Regardless of category, all herbal clinical trials in India must comply with GCP standards and the Indian Council of Medical Research (ICMR) National Ethical Guidelines (2017), which emphasize participant safeguards, informed consent, protection of vulnerable groups, and serious adverse effect reporting. Together, this dual pathway balances the rigor of modern drug regulation with the recognition of traditional systems, ensuring that both phytopharmaceuticals and ASU medicines are tested ethically and scientifically before wider therapeutic claims are made. In India, ethical oversight for herbal clinical trials is guided by the ICMR National Ethical Guidelines for Biomedical and Health Research (2017). These guidelines apply equally to both phytopharmaceuticals and ASU trials, ensuring that participants give proper informed consent, that vulnerable groups are protected, and that ethics committees work in a standardized and transparent way to support these ethical rules; the Pharmacopoeia Commission for Indian Medicine and Homoeopathy and the National Medicinal Plants Board set quality standards, prepare pharmacopeial monographs, and promote good agricultural and collection practices. This ensures that herbal raw materials are consistent and that clinical trials can be reproduced reliably [22]. Together, these systems create a balanced framework that respects traditional knowledge while also meeting modern scientific and regulatory standards, helping to build evidence on the safety and effectiveness of herbal medicines and increasing their acceptance worldwide.

1.6 REGULATORY AGENCIES IN OTHER COUNTRIES

Regulatory agencies in different countries play a crucial role in ensuring the safety, quality, and efficacy of herbal medicines and clinical trials. In the United States, the Food and Drug Administration; in the European Union, the European Medicines Agency, along with the Committee on Herbal Medicinal Products; and in China, the National Medical Products Administration regulate traditional Chinese medicines (TCM), while the Chinese Pharmacopoeia Commission sets official standards for herbal formulations. Japan regulates herbal medicines through the Ministry of Health, Labour and Welfare and ensures safety and efficacy via the Pharmaceuticals and Medical Devices Agency. Similarly, the Therapeutic Goods Administration in Australia regulates complementary medicines, including herbal products, under strict quality and safety standards (Figure 1.2). These agencies, though differing in their approaches, share the common objective of protecting public health while promoting the safe integration of herbal medicines into modern healthcare systems.

1.6.1 Australia

The Therapeutic Goods Administration (TGA) plays a central role in regulating clinical trials of herbal medicines

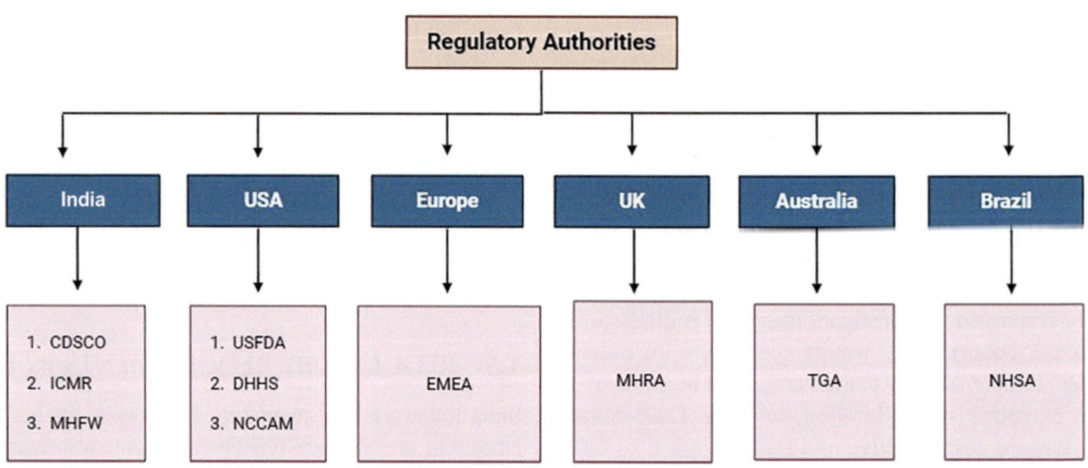

FIGURE 1.2 Various regulatory authorities for clinical trials.

in Australia. While herbal medicines fall under the complementary medicine framework, their clinical testing is subject to the same scientific and ethical rigor as pharmaceutical drugs, depending on risk level. The TGA's primary role is risk stratification, oversight, and quality assurance. For herbal clinical trials, the TGA ensures that trial sponsors provide evidence of botanical identity, standardization of active phytochemicals, quality assurance through Good Manufacturing Practices, and nonclinical safety data where applicable. The Clinical Trial Notification scheme allows most herbal trials to proceed after ethics approval, with the TGA acting as a regulatory recipient of trial details. This pathway emphasizes research ethics committees as the primary step in ensuring scientific and ethical standards. Conversely, for novel or higher-risk herbal products (such as concentrated extracts or complex polyherbal formulations with limited prior human use), the Clinical Trial Exemption scheme requires the TGA to directly evaluate preclinical and clinical data before approving trials, thereby adding a stricter layer of regulatory scrutiny. The TGA's role extends to monitoring safety during herbal trials. Sponsors must report all serious adverse events, and the TGA maintains authority to suspend or terminate trials if safety concerns arise. Furthermore, herbal products that progress from trials to market as "registered medicines" must demonstrate efficacy through well-designed RCTs, whereas "listed medicines" can make only limited claims based on traditional evidence. This creates a tiered evidentiary standard that balances Australia's recognition of traditional herbal knowledge with the demand for modern scientific validation. By shaping protocol standards, risk classification, trial oversight, and post-market vigilance, the TGA ensures that herbal medicine trials in Australia are scientifically credible, ethically sound, and aligned with international norms such as ICH-GCP. This dual focus on innovation and patient protection has positioned Australia as a leader in regulating herbal and complementary medicine trials. Scholarly reviews highlight that the TGA's model serves as a pragmatic template for integrating traditional medicines into evidence-based healthcare [23, 24].

1.6.2 Canada

Health Canada's Natural and Non-prescription Health Products Directorate plays a central role in ensuring that herbal medicines tested in clinical trials meet international standards of safety, efficacy, and quality. Within the Natural Health Products Regulations, Health Canada's oversight ensures that herbal products used in research are not only standardized and authenticated but also clinically validated through well-designed studies. For herbal clinical trials, sponsors must submit a Clinical Trial Application that demonstrates the product's quality (botanical authentication, good agricultural and collection practices, marker-compound quantification, batch-to-batch reproducibility, impurity, and contaminant testing) alongside preclinical or historical evidence supporting its safety profile. Importantly, Health Canada requires that herbal trials follow ICH-GCP guidelines, thereby protecting participants, ensuring ethical informed consent, and mandating independent Research Ethics Board review before study initiation. In terms of trial design, it encourages sponsors to build on traditional-use evidence for initial safety and dose justification but emphasizes that RCTs are needed to substantiate therapeutic claims, particularly for novel extracts, new indications, or higher-risk populations. The agency also requires adverse-event reporting, data-monitoring systems, and pharmacovigilance planning, highlighting the importance of ongoing risk management given the multicomponent nature of botanicals. By demanding these safeguards, Health Canada ensures that herbal medicines are tested with the same rigor as conventional drugs when claims go beyond traditional use. This role makes Canada's regulatory framework a bridge between traditional medicine and evidence-based therapeutics, ensuring both scientific credibility and public trust in herbal clinical research [25, 26].

1.6.3 China

The National Medical Products Administration (NMPA, formerly CFDA) is China's primary regulatory body for pharmaceuticals, including TCM and modern herbal drugs. For herbal medicines intended for therapeutic claims, the NMPA requires sponsors to submit clinical trial applications, including pharmacological and toxicological data, pharmacokinetics where applicable, and phase-wise clinical protocols that comply with the GCP guidelines harmonized with ICH standards. Importantly, the NMPA differentiates between classical TCM prescriptions, which may be approved with reduced clinical requirements if they are included in authoritative TCM references with long histories of safe use, and new herbal formulations or modified products, which must undergo rigorous nonclinical and clinical evaluation similar to chemical drugs. The Chinese Pharmacopoeia Commission plays a complementary role by issuing the *Chinese Pharmacopoeia*, which provides detailed monographs on herbal drugs, including authentication, quality standards, marker compounds, safety thresholds for heavy metals and pesticides, and analytical methods. These pharmacopeial standards serve as the scientific basis for trial material characterization and regulatory review. Together, the NMPA and the Chinese Pharmacopoeia Commission ensure that clinical trials on herbal drugs in China are grounded in both modern regulatory science and TCM traditions, balancing innovation with patient safety [27, 28].

1.6.4 EMEA

In the European Union, the regulation of herbal medicines is managed by the European Medicines Agency (EMA) through the Committee on Herbal Medicinal Products, which provides scientific opinions, herbal monographs, and a "Community list" to harmonize the safe and effective use of botanicals across EU member states. The regulatory framework is primarily guided by the Traditional Herbal Medicinal Products Directive. Under this directive, herbal products can follow two major pathways: (1) traditional-use registration, which requires at least 30 years of continuous medicinal use (15 of which must be within the EU) and allows bibliographic or pharmacological data to establish safety and plausible efficacy, thereby reducing the need for new trials; and (2) well-established use registration, which applies to herbal products with at least 10 years of use within the EU and sufficient published scientific evidence, enabling reliance on bibliographic clinical data rather than newly conducted studies. However, when herbal medicines are developed with new therapeutic claims, novel combinations, or indications beyond traditional or well-established use, they must undergo the same rigorous clinical trial requirements as conventional drugs, including RCTs conducted under GCP standards. To support developers, the agency has published critical guidelines such as the "Guideline on Non-clinical and Clinical Documentation of Herbal Medicinal Products" (EMEA/HMPC/32116/2005), which outlines expectations for designing herbal clinical trials, including patient selection, comparator choice, phytochemical standardization, pharmacokinetics, and adverse-event monitoring. The Herbal Medicinal Products Committee also provides monographs that consolidate safety and efficacy data, serving as authoritative references for clinical trial justification. In this way, the EMA and the Herbal Medicinal Products Committee play a dual role: They facilitate streamlined approvals for traditional herbal medicines while simultaneously demanding evidence-based clinical trials for herbal products making modern drug-like claims. This ensures both the protection of public health and the integration of traditional knowledge into regulated medicine. Scholarly analyses highlight how this regulatory structure balances respect for ethnopharmacology with the scientific rigor of modern clinical research [27, 29].

1.6.5 Japan

In Japan, herbal medicines, particularly Kampo medicines, are regulated under the authority of the Ministry of Health, Labour and Welfare (MHLW), with scientific and technical evaluations conducted by the Pharmaceuticals and Medical Devices Agency (PMDA). Kampo medicines occupy a unique position: Many formulas are recognized under traditional use, but when claims are expanded or new formulations are developed, clinical evidence is required. The MHLW provides the regulatory framework, which includes requirements for GCP, safety monitoring, and quality control of herbal raw materials. The PMDA evaluates investigational new drug applications, clinical trial protocols, and data submissions to ensure that herbal formulations are standardized, safe, and effective. Importantly, Japanese regulators emphasize standardization and batch-to-batch reproducibility, as well as the integration of Kampo into modern medical practice. Clinical trials must document not only efficacy and safety but also diagnostic frameworks when Kampo-specific approaches (e.g., *Sho* diagnosis) are employed. Thus, Japan's regulatory system supports the use of traditional herbal medicines within a modern evidence-based medicine context, ensuring both cultural relevance and compliance with international clinical trial standards [30, 31].

1.6.6 South Korea

South Korea's Ministry of Food and Drug Safety (MFDS) regulates herbal medicines through an integrated framework that both respects Traditional Korean Medicine (TKM) heritage and enforces modern drug development standards: Products marketed solely on long-standing traditional use may follow simplified/traditional pathways, but any herbal product pursuing novel therapeutic claims must obtain formal marketing approval and support its claims with GCP–compliant RCTs. MFDS requires chemistry, manufacturing, and controls quality documentation for multicomponent botanicals, botanical authentication (voucher specimens), marker-compound profiling, batch comparability, impurity and microbial testing, and stability data because upstream variability directly affects clinical reproducibility. Where appropriate, documented traditional human use can inform safety/dose selection and may justify limited nonclinical testing, but for modified extracts, new combinations, or novel indications, comprehensive toxicology and clinical pharmacology are normally required. Trial protocols must obtain Institutional Review Board approval, define clinically meaningful endpoints, include robust safety-monitoring plans and a risk management/post-marketing surveillance strategy, and demonstrate statistically adequate power for efficacy claims. This dual pathway—traditional use versus modern drug route—allows TKM products to be developed for both local traditional markets and international regulatory acceptance, but crucially, MFDS demands RCT evidence and standardized quality when therapeutic claims go beyond historical practice [32, 33].

1.6.7 The United States

In the United States, the Food and Drug Administration (FDA), through its Center for Drug Evaluation and Research, regulates herbal medicines intended for therapeutic use as botanical drugs under the Federal Food, Drug, and Cosmetic Act. To initiate human studies, sponsors must submit an investigational new drug application supported by well-controlled clinical trials. FDA's "Botanical Drug Development: Guidance for Industry" (2016) provides detailed instructions for conducting herbal clinical trials, emphasizing the need for RCTs, despite allowance for some flexibility in preclinical requirements if sufficient historical safety data exist. The guidance places strong weight on chemistry, manufacturing, and controls to ensure batch-to-batch consistency and quality, which are critical when the active components of a botanical are not fully characterized. All trials must comply with ICH-GCP standards, including requirements for trial design, ethics committee oversight, participant protection, and adverse-event reporting. The FDA plays a dual role in herbal clinical trials: it enforces stringent scientific and ethical standards while also enabling botanical drug development by accommodating traditional evidence where appropriate. This pathway has already led to successful approvals such as Veregen® (*sinecatechins* ointment from green tea extract) and Fulyzaq® (crofelemer from Croton lechleri sap). Scholarly reviews highlight the FDA's leadership in this domain, noting that while the framework ensures safety and efficacy, it also creates opportunities for botanicals to enter modern drug markets [34, 35].

1.6.8 WHO

The WHO plays a pivotal role in shaping global standards for clinical trials involving herbal medicines by providing methodological, regulatory, and pharmacovigilance frameworks that many national authorities adopt or adapt. The General Guidelines for Methodologies on Research and Evaluation of Traditional Medicine outline trial designs suitable for herbal medicines, emphasizing appropriate outcome measures, standardization of plant materials, and culturally sensitive ethical practices. To ensure upstream quality, the WHO guidelines on Good Agricultural and Collection Practices for Medicinal Plants provide procedures for cultivation, collection, and traceability of medicinal plants, thereby safeguarding reproducibility in clinical investigations. Safety monitoring during and after clinical trials is reinforced through the WHO guidelines on Safety Monitoring of Herbal Medicines in Pharmacovigilance Systems, which integrate herbal products into existing pharmacovigilance systems and adapt causality assessments to complex, multi-constituent formulations. In addition, the WHO Monographs on Selected Medicinal Plants supply detailed botanical, pharmacological, and quality standards that inform the selection and authentication of herbal trial material [9, 30]. Together, these WHO guidelines ensure that herbal medicine clinical trials are scientifically rigorous, ethically grounded, and globally harmonized.

1.7 REGULATORY BODIES, FRAMEWORKS, AND PROCEDURES IN INDIA

The CDSCO in India, which operates under the Ministry of Health and Family Welfare, is the primary entity responsible for formulating regulatory protocols and benchmarks for pharmaceuticals, cosmetics, diagnostics, and devices. It provides regulatory direction by making changes to statutes and rules and oversees the process of approving new drugs. The primary goal of this initiative is to establish uniformity in clinical research practices and facilitate the introduction of safer pharmaceutical products into the Indian market [36]. The Drugs Controller General of India (DCGI) is tasked with granting regulatory authorizations for the execution of clinical studies and is responsible for granting marketing licenses for pharmaceuticals in India. In addition to the DCGI office, several governmental entities are participating in the regulatory oversight of new pharmaceuticals. Upon submission of an application, the office of the Deputy Controller General (Intellectual Property) examines it. The duration for approval is contingent upon the regulatory state of the trial in other nations. To accelerate the approval procedure for studies that have already been approved by regulatory agencies in other countries, the DCG(I) office has categorized all applications into two groups: A and B. Category A comprises clinical studies with protocols that have received approval from EMEA or regulatory agencies in the United States, the United Kingdom, Switzerland, Australia, Canada, Germany, South Africa, or Japan. Permission is granted for conducting such studies, provided that the protocol is approved by the respective governments. The estimated timeframe for reviewing and approving Category A applications is two to four weeks. The Category B clinical trial applications undergo evaluation through the standard approval process by an expert committee. The approval of the application typically requires a period of 8 to 12 weeks. This review period does not account for any possible delays caused by incomplete applications or the time it takes for the sponsor to respond to queries made by the regulatory authority. As part of the application, a concise summary of information is provided in the form of an investigators' brochure. This brochure contains comprehensive data on pharmacology, toxicity, and any relevant clinical experience. The DCGI occasionally seeks advice from other autonomous government organizations, such as ICMR or the Department of Biotechnology (for biotech products), on a case-by-case basis, which leads to an extension of the review timeframe. Obtaining DCGI permission is necessary to obtain a test license for importing trial supplies, a process that typically takes two weeks. The aforementioned processes, when executed simultaneously, necessitate a cumulative duration of 14 weeks. Once the DCGI grants approval for the clinical study, it takes an additional 2–4 weeks for the Director General of Foreign Trade to approve the shipment of blood samples. However, the timescales for the procedure may be further extended if the regulatory body raises concerns or requests additional evidence, and the modifications are not handled or received promptly. In addition to applying to the DCGI office, it is possible to submit it simultaneously to the ethics committees of the prospective sites, typically hospitals and/or clinics. Ethics committees often adhere to the ICH-GCP principles and Schedule Y of the Drugs and Cosmetics Act of 1945. Typically, these ethics committees are associated with study locations, although they can also operate as autonomous ethics bodies [37, 38].

1.7.1 Drugs and Cosmetics Act (1940) and Drugs and Cosmetics Rules (1945)

The purpose of this law, which has been in place since 1940, is to guarantee that all pharmaceuticals and personal care items marketed within the nation meet minimum quality requirements for safety and efficacy. The document contains chapters, rules, and schedules and is regularly revised to enhance safety, effectiveness, and drug quality. Schedule Y, in conjunction with rules 122A, 122B, 122D, 122DA, 122DAC, and 122E, is the primary document that regulates clinical research in the country. According to the law, it is obligatory for all clinical research that falls under the scope of Schedule Y to adhere to the relevant conditions. The document includes 12 appendices, which consist of formats for clinical trial protocols, informed consent forms, ethics committee approval templates, and a framework for serious adverse-event reporting [39].

1.7.2 Ethical Guidelines of the Indian Council of Medical Research

The revised ICMR guidelines released in 2006 are called the "Ethical Guidelines for Biomedical Research on Human Participants." These guidelines encompass two overarching facets of clinical research—the fundamental principles that must be adhered to, and recommendations about specific areas of study, such as research involving children or herbal research. Researchers are required to know both these documents, comply with the regulations in the former, and follow the instructions in the latter [40].

1.7.3 Indian Good Clinical Practice Guideline

In 2001, the CDSCO produced a clinical practice guideline known as the GCP guideline. This guideline was intended to be exclusive to India. However, unlike the ICH-GCP guideline, it has not been updated since its inception [41].

However, there are different phases of clinical trials, which are summarized in Table 1.2.

1.8 CURRENT CHALLENGES FOR HERBAL MEDICINES

Regulatory difficulties arise when it comes to the research applications and commercialization of traditional herbal medicines. The various challenges associated with herbal medicine clinical trials are mentioned in the following section.

1.8.1 Standardization of Herbal Drugs

Consistency in composition and biological activity is a crucial prerequisite for the safe and effective use of herbal medications. Nevertheless, herbal medications often do not meet this criterion due to challenges such as (1) complexities in plant identification, (2) genetic differences, (3) variations in cultivation conditions, (4) diverse harvesting techniques and extract processing, and (5) insufficient

TABLE 1.2
Different Phases of Clinical Trials[34, 37]

Parameters	Preclinical trial (Usually done on animals to determine if the drug is safe enough for human testing)	Phase I (Determine pharmacological actions and tolerability)	Phase II (Evaluate safety and efficacy)	Phase III (Evaluate effectiveness and risk-benefit ratio)	Phase IV (Monitor long-term effects and effectiveness)
Duration	3–6 years since drug discovery	Months	Months–years	Years	Ongoing following FDA approval
Sample size	Not specific	Small	Large	Larger	Impacting larger populations beyond
Population	*In vitro* and *in vivo* studies	A healthy population/ maybe with a targeted disease, e.g., cancer and TB	Population with the target disease	Diverse population with the target disease	Diverse population with the target disease and new age groups, gender
Types of studies	Not specific	Unblinded and uncontrolled	Maybe placebo (an inactive substance) and an active, controlled	Randomized and controlled	Expanded safety comparison
Factors to be identified	Mechanism of action, efficacy (the ability of the drug to act against the pathogen/ disease), safety (concerns the medical risk to the patient)	Pharmacodynamics (side effects/desired effects, mechanism of action of the drug), pharmacokinetics (absorption, distribution, metabolism, and elimination of the drug), tolerated dose	Drug-drug and drug-disease interactions, efficacy at various doses, patient safety	Dosage intervals, risk benefit information, efficacy and safety for subgroups	Epidemiological data, efficacy, and safety within large, diverse populations, pharmacoeconomics (comparison of the value of one pharmaceutical drug or drug therapy to another)

knowledge about the active pharmacological components. Employing chromatographic techniques and marker molecules to standardize herbal products can guarantee uniformity from one batch to another. Nevertheless, this does not guarantee consistent pharmacological effectiveness or stability. The absence of standardization of herbal treatments poses a significant challenge for researchers, as they cannot depend on commercially accessible herbal products for their research projects [42].

1.8.2 Quality of Herbal Preparations

For an herbal medicine to be deemed effective, it is imperative to implement quality assurance measures to guarantee that the product delivers the anticipated benefits. Even without statistics on effectiveness, it is crucial to prioritize quality assurance because the quality of a product or service is a key factor in ensuring safety. The adulteration of plants is a significant issue. Common adulterants include botanicals, poisonous metals, bacteria, microbial toxins, insecticides, and fumigation agents. The medicines that were most commonly detected included ephedrine, chlorpheniramine, methyltestosterone, and phenacetin. Approximately 10% to 15% of these drugs were discovered to include lead, mercury, or arsenic [8, 43]. The prevalence of heavy metal contamination is uncertain, while a single study revealed that 64% of samples gathered in India had substantial levels of lead (64% for mercury, 41% for arsenic, and 9% for cadmium). Administering such therapies to patients can result in significant harm and may complicate the evaluation of safety in a clinical trial [44]. Quality must be ensured at every level, including the sourcing of herbal raw materials, the processing of herbals, and the production of final herbal medications.

1.8.3 Evidence of Clinical Efficacy

The scientific data derived from randomized clinical studies are robust only for numerous applications of acupuncture, certain herbal medicines, and a subset of manual therapies. Only a minute proportion of the numerous therapeutic plants utilized globally have undergone thorough testing in randomized, controlled trials. Although animal studies and anecdotal clinical experiences may show promise and indicate widespread use of a plant, it is important to note that such findings cannot accurately predict the outcomes of well-designed randomized, controlled trials. The majority of trials do not provide detailed information on the effectiveness of the treatment, and the length of time for which the patients are observed is typically brief. The practical significance of the observed impacts is not always evident. For example, the majority of Indian trials testing hepatoprotective drugs are conducted in an open and uncontrolled manner. Due to the inherent tendency of most acute liver problems to naturally resolve, establishing a direct correlation between amelioration and herbal medicine is challenging [45]. An inherent challenge in all clinical research on herbal medications is determining if several products, extracts, or even different batches of the same extract may be considered similar and equivalent. For instance, Echinacea products may include other botanical extracts, utilize various plant species such as *E. purpurea*, *pallida*, or *angustifolia*, incorporate different plant components such as the herb or root, and may have been manufactured using distinct extraction methods such as hydrophilic or lipophilic extraction. The herbal business is exempt from conducting clinical studies, and industry experts contend that the substantial research expenses cannot be recouped due to the difficulty in patenting herbal goods compared to novel chemical compounds. However, randomized, controlled studies are the most effective methods for demonstrating the effectiveness of any therapy [8, 45, 46].

1.9 SAFETY CONCERNS—DRUG INTERACTIONS

Herbal medicines are often seen as being relatively safer than synthetic drugs. Although this assertion may be accurate, case reports demonstrate that serious adverse effects and significant drug interactions can arise. For example, the herb Ephedra, which was sold as a dietary supplement, resulted in a minimum of 12 fatalities, as well as heart attacks and strokes. Additional notable safety concerns include the hepatotoxicity of kava and the renal consequences of aristolochic acid. In addition, the potential interactions between herbal treatments and other medications are a significant cause for concern. Hypericum extracts can reduce the levels of certain medicines by inducing enzymes. The absence of regulatory criteria for the effectiveness and safety of herbal products did not generate significant apprehension in the past, as these items were commonly seen as being so safe that even if they were ineffective, they caused minimal toxicity. Nevertheless, the current scenario is transforming, leading to a growing corpus of literature that delves into the adverse effects and interplay of herbal treatments. In addition to the immediate dangers of negative side effects and potential drug interactions, there is also a secondary risk that an herbal therapy lacking proven effectiveness may undermine, prolong, or substitute for a reliable type of traditional treatment [47, 48].

1.10 CHALLENGES IN CLINICAL TRIALS

It is highly challenging, unfeasible, or occasionally unattainable to create active and control groups that possess the same morphology as the herbal drug. Additionally, the utilization of a placebo presents comparable challenges, as herbal research medicine may include a potent fragrance or a distinct taste that cannot be replicated during the production of a placebo. An integrated approach of multiple herbal systems does not make a distinction between the ailment and the patient. This strategy poses challenges for establishing the inclusion and exclusion criteria in a clinical trial. Administering a study drug to a population with diverse constitutions may result in inconsistent outcomes. Quality control of herbal medications is frequently intricate and challenging. Conventional therapy methods suggest several therapies at different illness stages in the same patient, adding another factor to consider in a clinical trial. To minimize costs and expenses, it is important to create relevant and acceptable requirements for assessing the safety and efficacy of different categories of herbal medications [39].

1.11 CONCLUSION AND FUTURE DIRECTIONS

The complex nature of the extract presents challenges in quantification, and there have been several issues noted during clinical trials, such as batch-to-batch variation, the use of a placebo instead of the innovative product, a lack of a sufficient quality control system, and inadequate safety and efficacy assessment requirements for various herbal medicines. These hurdles can be surmounted by implementing the latest methodology and recommendations for clinical trials. Effective planning of the scientific clinical study design is crucial for the success of any clinical trial in this context [49]. To ensure accurate clinical trial findings for herbal remedies, it is essential to conduct double-blind studies with a sufficient number of carefully chosen patients, ideally following the same standards as those used in clinical trials for new drug development. Herbal medicine clinical trials should only be conducted when the standardization and identification of markers have been accomplished to ensure consistent examination of substances. The CDSCO in India, functioning under the Ministry of Health and Family Welfare, is the main authority tasked with developing and regulating norms and standards for pharmaceuticals, cosmetics, diagnostics, and devices. It offers regulatory guidance by modifying statutes and rules and supervises the process of authorizing new medications. The main objective of this project is to standardize clinical research processes and streamline the entry of safer pharmaceutical goods into the Indian market. The DCGI is responsible for giving regulatory authorizations for the implementation of clinical studies and for issuing marketing licenses for pharmaceuticals in India. Ensuring consistency in both the content and biological activity of herbal drugs is essential for their safe and successful usage [34]. However, herbal medications frequently fail to meet this requirement due to difficulties such as intricate plant identification, genetic disparities, variations in cultivation conditions, diverse harvesting techniques and extract processing, and inadequate understanding of the active pharmacological constituents. Creating active and control groups with identical morphology to the herbal medication is extremely difficult, impractical, or often impossible. Creating active and control groups with the same color, smell, and taste as the herbal treatment, which cannot be replicated when producing a placebo, is extremely challenging. These problems can be mitigated or surmounted by implementing the latest procedures and recommendations for clinical studies [8, 50]. In this situation, where traditional medicine and knowledge are on the brink of extinction, it is imperative to take decisive and rigorous action to save and conserve our cultural heritage. It is now the responsibility of regulatory organizations to oversee the controlled and high-quality distribution of herbal products and to support their advancement to the clinical trial phase. The achievement of the aim would be within reach if regulatory authorities collaborate closely with academic institutions, research institutes, research clinics and centers, hospitals, industry, and pharmacy faculties.

REFERENCES

[1] Chan, J. S. W., Lau, M. S., Muhamad Rosli, S. H., Mustapha Kamal, S. K., Wan Seman, W. N. N., Ali, N. H., Tan, T. Y. C., Ahmad, I. F., Krishnan, P., Borhan, M. K., Syed Mohamed, A. F., & Lim, X. Y. (2025). Methodological insights into intervention and outcomes in randomised controlled trials of herbal medicine for obesity: A scoping review. *Therapeutic Advances in Chronic Disease*, 16, 1–44. https://doi.org/10.1177/20406223241310206

[2] Li, Y., Du, P., Zhang, X., Ren, C., Shi, X., Dong, X., & Zhang, C. (2023). Qualified placebo for trials of herbal medicine treatment in rare diseases? A cross-sectional analysis. *Orphanet Journal of Rare Diseases*, 18(1), 373. https://doi.org/10.1186/s13023-023-02987-w

[3] Koonrungsesomboon, N., Sakuludomkan, C., Na Takuathung, M., Klinjan, P., Sawong, S., & Perera, P. K. (2024). Study design of herbal medicine clinical trials: A descriptive analysis of published studies investigating the effects of herbal medicinal products on human participants. *BMC Complementary Medicine and Therapies*, 24(1), 391. https://doi.org/10.1186/s12906-024-04697-7

[4] Gamil, N. M., Elsayed, H. A., Hamed, R. M., Salah, E. T., Ahmed, A. M., Mostafa, H. A., El-Shiekh, R. A., & Abou-Hussein, D. (2025). Insights from herb interactions studies: A foundational report for integrative medicine. *Future Journal of Pharmaceutical Sciences*, 11(1), 46. https://doi.org/10.1186/s43094-025-00791-7

[5] Koonrungsesomboon, N., Morakote, N., & Karbwang, J. (2021). Ethical considerations and challenges in herbal drug trials with the focus on scientific validity and risk assessment. *Phytotherapy Research*, 35(5), 2396–2402. https://doi.org/10.1002/ptr.6962

[6] Hassen, G., Belete, G., Carrera, K. G., Iriowen, R. O., Araya, H., Alemu, T., Solomon, N., Bam, D. S., Nicola, S. M., Araya, M. E., Debele, T., Zouetr, M., & Jain, N. (2022). Clinical implications of herbal supplements in conventional medical practice: A US perspective. *Cureus*, 14(7), e26893 https://doi.org/10.7759/cureus.26893

[7] Wolf, S., Gelis, L., Dörrich, S., Hatt, H., & Kraft, P. (2017). Evidence for a shape-based recognition of odorants *in vivo* in the human nose from an analysis of the molecular mechanism of lily-of-the-valley odorants detection in the Lilial and Bourgeonal family using the C/Si/Ge/Sn switch strategy. *PLoS One*, 12(8), e0182147. https://doi.org/10.1371/journal.pone.0182147

[8] Wang, H., Chen, Y., Wang, L., Liu, Q., Yang, S., & Wang, C. (2023). Advancing herbal medicine: Enhancing product quality and safety through robust quality control practices.

Frontiers in Pharmacology, 14, 1265178. https://doi.org/10.3389/fphar.2023.1265178

[9] Parveen, A., Parveen, B., Parveen, R., & Ahmad, S. (2015). Challenges and guidelines for clinical trial of herbal drugs. *Journal of Pharmacy and Bioallied Sciences*, 7(4), 329–333. https://doi.org/10.4103/0975-7406.168035

[10] Razavy, S., Lee, L., & Zaslawski, C. (2020). A pre-trial evaluation of blinding for a Chinese herbal medicine trial. *Contemporary Clinical Trials Communications*, 19, 100632. https://doi.org/10.1016/j.conctc.2020.100632

[11] Langham, S., Langham, J., Goertz, H. P., & Ratcliffe, M. (2011). Large-scale, prospective, observational studies in patients with psoriasis and psoriatic arthritis: A systematic and critical review. *BMC Medical Research Methodology*, 11, 32. https://doi.org/10.1186/1471-2288-11-32

[12] Kiani, A. K., Naureen, Z., Pheby, D., Henehan, G., Brown, R., Sieving, P., Sykora, P., Marks, R., Falsini, B., Capodicasa, N., Miertus, S., Lorusso, L., Dondossola, D., Tartaglia, G. M., Ergoren, M. C., Dundar, M., Michelini, S., Malacarne, D., Bonetti, G., . . . Bertelli, M. (2022). Methodology for clinical research. *Journal of Preventive Medicine and Hygiene*, 63(2), E267–E278. https://doi.org/10.15167/2421-4248/jpmh2022.63.2S3.2769

[13] Song, J. W., & Chung, K. C. (2010). Observational studies: Cohort and case-control studies. *Plastic and Reconstructive Surgery*, 126(6), 2234–2242. https://doi.org/10.1097/PRS.0b013e3181f44abc

[14] Thiese, M. S. (2014). Observational and interventional study design types: An overview. *Biochemia Medica*, 24(2), 199–210. https://doi.org/10.11613/BM.2014.022

[15] Conway, C. (2023). Nonrandomized controlled trials. *Translational Sports Medicine*, 2, 251–254. https://doi.org/10.1016/B978-0-323-91259-4.00033-3

[16] Jeong, J. H., Lee, J. Y., Kim, J. Y., Seo, Y. K., Kang, W. C., Kang, H. W., Park, S. J., Jang, H. K., Park, Y. C., & Jung, I. C. (2020). Safety and efficacy of Jujadokseo-hwan for memory deficit (amnesia) in mild neurocognitive disorder: A protocol for randomized, double-blind, placebo-controlled, parallel-group, multicenter clinical trial. *Medicine (United States)*, 99(8), e19231. https://doi.org/10.1097/MD.0000000000019231

[17] Bhide, A., Shah, P. S., & Acharya, G. (2018). A simplified guide to randomized controlled trials. *Acta Obstetricia et Gynecologica Scandinavica*, 97(4), 380–387. https://doi.org/10.1111/aogs.13309

[18] Drukker, M., Weltens, I., van Hooijdonk, C. F. M., Vandenberk, E., & Bak, M. (2021). Development of a methodological quality criteria list for observational studies: The observational study quality evaluation. *Frontiers in Research Metrics and Analytics*, 6, 675071. https://doi.org/10.3389/frma.2021.675071

[19] Chidambaram, A. G., & Josephson, M. (2019). Clinical research study designs: The essentials. *Pediatric Investigation*, 3(4), 245–252. https://doi.org/10.1002/ped4.12166

[20] Bhatt, A. (2016). Phytopharmaceuticals: A new drug class regulated in India. *Perspectives in Clinical Research*, 7(2), 59–61. https://doi.org/10.4103/2229-3485.179435

[21] Vennu, V., & Saini, P. P. (2020). India's clinical trial regulatory changes, indian researchers' awareness of recently changed regulations, and the impact of the new drugs and clinical trial rules: A review. *Indian Journal of Pharmaceutical Sciences*, 82(5), 726–740. https://doi.org/10.36468/pharmaceutical-sciences.702

[22] Mathur, R., & Swaminathan, S. (2018). National ethical guidelines for biomedical and health research involving human participants, 2017: A commentary. *Indian Journal of Medical Ethics*, 3(3), 1–16. https://doi.org/10.20529/IJME.2018.065

[23] Masters, C. S. (2008). Commercialism, choice and consumer protection: Regulation of complementary medicines in Australia. *Medical Journal of Australia*, 188(1), 21–50. https://doi.org/10.5694/j.1326-5377.2008.tb01905.x

[24] Dixit, S. K., & Sambasivan, M. (2018). A review of the Australian healthcare system: A policy perspective. *SAGE Open Medicine*, 6, 2050312118769211. https://doi.org/10.1177/2050312118769211

[25] Laeeque, H., Boon, H., Kachan, N., Cohen, J. C., & D'Cruz, J. (2006). The Canadian Natural Health Products (NHP) regulations: Industry perceptions and compliance factors. *BMC Health Services Research*, 6, 63. https://doi.org/10.1186/1472-6963-6-63

[26] Ng, J. Y., & Luong, M. (2020). Evaluation of the Canadian natural health product regulatory framework in academic research: A scoping review. *European Journal of Integrative Medicine*, 37, 101159. https://doi.org/10.1016/j.eujim.2020.101159

[27] Chauhan, M. K., & Malik, S. (2016). Regulatory guidelines for approval of biosimilars in India, Europe, Brazil and China: A comprehensive overview. *International Journal of Pharmacy and Pharmaceutical Sciences*, 8(10), 7–11. https://doi.org/10.22159/ijpps.2016v8i10.11753

[28] Liang, Z., Lai, Y., Li, M., Shi, J., Lei, C. I., Hu, H., & Ung, C. O. L. (2021). Applying regulatory science in traditional Chinese medicines for improving public safety and facilitating innovation in China: A scoping review and regulatory implications. *Chinese Medicine (United Kingdom)*, 16(1), 23. https://doi.org/10.1186/s13020-021-00433-2

[29] Anquez-Traxler, C. (2011). The legal and regulatory framework of herbal medicinal products in the European Union: A focus on the traditional herbal medicines category. *Therapeutic Innovation & Regulatory Science*, 45(1), 15–23. https://doi.org/10.1177/009286151104500102

[30] Goda, Y. (2022). Regulatory science of natural products. *Journal of Natural Medicines*, 76(4), 732–747. https://doi.org/10.1007/s11418-022-01639-w

[31] Nakamura, K., & Shibata, T. (2020). Regulatory changes after the enforcement of the new Clinical Trials Act in Japan. *Japanese Journal of Clinical Oncology*, 50(4), 399–404. https://doi.org/10.1093/jjco/hyaa028

[32] Lee, B., Choi, Y., Kim, P. W., Yang, C., & Lee, M. S. (2021). Regulation and status of herbal medicine clinical trials in Korea: A narrative review. *Integrative Medicine Research*, 10(2), 100688. https://doi.org/10.1016/j.imr.2020.100688

[33] Chee, D. H. (2019). Korean clinical trials: Its current status, future prospects, and enabling environment. *Translational and Clinical Pharmacology*, 27(4), 115–118. https://doi.org/10.12793/tcp.2019.27.4.115

[34] Ali, F., Khasimbi, S., Sharma, K., Trivedi, M., Ali, A., & Ahmad, J. (2020). Regulatory perspectives of herbal medicinal products. *Bioactive Phytochemicals: Drug Discovery to Product Development*, 18, 158–171. https://doi.org/10.2174/9789811464485120010012

[35] Wu, C., Lee, S. L., Taylor, C., Li, J., Chan, Y. M., Agarwal, R., Temple, R., Throckmorton, D., & Tyner, K. (2020). Scientific and regulatory approach to botanical drug development:

A U.S. FDA perspective. *Journal of Natural Products*, 83(2), 552–563. https://doi.org/10.1021/acs.jnatprod.9b00949

[36] Gogtay, N. J., Ravi, R., & Thatte, U. M. (2017). Regulatory requirements for clinical trials in India: What academicians need to know. *Indian Journal of Anaesthesia*, 61(3), 192–199. https://doi.org/10.4103/ija.IJA_143_17

[37] Thakkar, S., Anklam, E., Xu, A., Ulberth, F., Li, J., Li, B., Hugas, M., Sarma, N., Crerar, S., Swift, S., Hakamatsuka, T., Curtui, V., Yan, W., Geng, X., Slikker, W., & Tong, W. (2020). Regulatory landscape of dietary supplements and herbal medicines from a global perspective. *Regulatory Toxicology and Pharmacology*, 114, 104647. https://doi.org/10.1016/j.yrtph.2020.104647

[38] Dubale, S., Usure, R. E., Mekasha, Y. T., Hasen, G., Hafiz, F., Kebebe, D., & Suleman, S. (2025). Traditional herbal medicine legislative and regulatory framework: A cross-sectional quantitative study and archival review perspectives. *Frontiers in Pharmacology*, 16, 1475297. https://doi.org/10.3389/fphar.2025.1475297

[39] Parveen, A., Parveen, B., Parveen, R., & Ahmad, S. (2015). Challenges and guidelines for clinical trial of herbal drugs. *Journal of Pharmacy and Bioallied Sciences*, 7(4), 329–333. https://doi.org/10.4103/0975-7406.168035

[40] Tandon, N., & Yadav, S. S. (2017). Contributions of Indian Council of Medical Research (ICMR) in the area of medicinal plants/traditional medicine. *Journal of Ethnopharmacology* 197, 39–45. https://doi.org/10.1016/j.jep.2016.07.064

[41] Castelino, L. J., Anoop Narayanan, V., Fernandes, S. D., Kumar, P., & Sandeep, D. S. (2018). Good clinical practices: An Indian perspective. *Research Journal of Pharmacy and Technology*, 11(7), 3209. https://doi.org/10.5958/0974-360X.2018.00590.5

[42] Indrayanto, G. (2024). Regulation and standardization of herbal drugs: Current status, limitation, challenge's and future prospective. *Profiles of Drug Substances, Excipients and Related Methodology*, 49, 153–199. https://doi.org/10.1016/bs.podrm.2023.11.003

[43] Bhalla, A., & Pannu, A. K. (2022). Are Ayurvedic medications store house of heavy metals? *Toxicology Research*, 11(1), 179–183. https://doi.org/10.1093/toxres/tfab124

[44] Abdelmonem, B. H., Kamal, L. T., Elbaz, R. M., Khalifa, M. R., & Abdelnaser, A. (2025). From contamination to detection: The growing threat of heavy metals. *Heliyon*, 11(1), e41713. Elsevier Ltd. https://doi.org/10.1016/j.heliyon.2025.e41713

[45] Mukherjee, P. K., Banerjee, S., Das Gupta, B., & Kar, A. (2022). Evidence-based validation of herbal medicine: Translational approach. *Evidence-Based Validation of Herbal Medicine: Translational Research on Botanicals*, 1–41. https://doi.org/10.1016/B978-0-323-85542-6.00025-1

[46] Leisegang, K., Jimenez, M., Durairajanayagam, D., Finelli, R., Majzoub, A., Henkel, R., & Agarwal, A. (2022). A systematic review of herbal medicine in the clinical treatment of benign prostatic hyperplasia. *Phytomedicine Plus*, 2(1), 100153. https://doi.org/10.1016/j.phyplu.2021.100153

[47] Dores, A. R., Peixoto, M., Castro, M., Sá, C., Carvalho, I. P., Martins, A., Maia, E., Praça, I., & Marques, A. (2023). Knowledge and Beliefs about Herb/Supplement consumption and herb/supplement–drug interactions among the general population, including healthcare professionals and pharmacists: A systematic review and guidelines for a smart decision system. *Nutrients*, 15(10), 2298. https://doi.org/10.3390/nu15102298

[48] Tamboli, F. A., Jogi, P. R., Waghmare, R., Bhosale, V., Jogi, P., Gavhane, D., & Rathod, M. (2025). Herbal medicines and drug interactions: A growing concern in healthcare. *International Journal of Pharmaceutical Chemistry and Analysis*, 12(1), 8–15. https://doi.org/10.18231/j.ijpca.2025.002

[49] Jadhav, R., Das, J., Rajyaguru, S., Kyada, S., Kale, V., Teja, P. K., Bhutia, D. O., & Chauthe, S. K. (2025). Regulations, current development, and future prospects of phytopharmaceuticals, a new class of herbal medicines in India. *Discover Pharmaceutical Sciences*, 6(1), 1–25. https://doi.org/10.1007/s44395-025-00006-4

[50] Salm, S., Rutz, J., van den Akker, M., Blaheta, R. A., & Bachmeier, B. E. (2023). Current state of research on the clinical benefits of herbal medicines for non-life-threatening ailments. *Frontiers in Pharmacology*, 14, 1234701. https://doi.org/10.3389/fphar.2023.1234701

2 Introduction to Clinical Trials with Special Focus on Botanicals

Neeraj K. Sethiya, Mahendra Singh, Sonu Sharma, Vishal Kumar, and Vrinda Poply
Faculty of Pharmacy, School of Pharmaceutical and Populations Health Informatics, DIT University, Mussoorie Diversion Road, Dehradun, Uttarakhand, India

2.1 INTRODUCTION

Clinical trials are investigations into the safety, effectiveness, and possible advantages of novel medical treatments, such as medications, technologies, and therapies, that are carried out on human subjects [1]. Further, clinical trials provide vital information on the effects, dosage, and adverse responses of medications by adhering to standardized methodology for testing safety and efficacy in the broader population. Small-scale safety testing is usually the first of four stages that clinical trials go through, followed by large-scale studies to verify effectiveness and track long-term consequences, as per the National Institutes of Health (NIH). Furthermore, clinical trials play a crucial role in advancing medical knowledge, allowing regulatory agencies to make informed decisions about the approval of new treatments, as per European Medicines Agency (EMA) requirements [2]. Finding a drug's safety, dose range, and adverse effects among a small sample of healthy patients or volunteers (20–100 participants) is the main goal of phase I studies. Phase II studies assess the medication's effectiveness and monitor its safety in a larger sample size (100–300 people). These studies are frequently controlled and randomized. A larger patient population (1,000–3,000 participants) is used in phase III studies to verify the drug's efficacy, track adverse effects, and compare it to current or conventional therapies. Important data for regulatory clearance are provided by this step. Phase IV studies, often referred to as post-marketing monitoring, take place following a drug's approval for general use. They seek to identify any long-term adverse effects and evaluate the medication's efficacy in a larger demographic [1–4]. An outline of the basic requirements for product development using botanicals was depicted in Figure 2.1.

2.2 ETHICAL CONSIDERATIONS

2.2.1 Informed Consent and Participant Rights

A fundamental component of ethical clinical research is informed consent, which mandates that participants be given enough details about the study's goals, methods, risks, and rewards to enable them to make an educated decision about taking part [5]. Respecting participant rights means granting them the freedom to withdraw without facing consequences, in addition to giving them clear and thorough information [6]. More than just a formality, consent is a continuous process that guarantees participants' autonomy and knowledge throughout the study [7].

2.2.2 The Role of Institutional Review Boards

Institutional Review Boards (IRBs) oversee participant protection by examining clinical trial protocols to make sure participant risks are kept to a minimum and ethical requirements are fulfilled. By concentrating on participant welfare and making sure that consent processes and participant rights are given priority, this oversight committee determines whether trials adhere to ethical standards [8]. Additionally, IRBs keep an eye on ongoing studies, making sure that any negative outcomes are recorded and that trials are modified if there are substantial hazards to participants [9–10].

2.2.3 Good Clinical Practices

Good Clinical Practices (GCP) is a global scientific and ethical norm that promotes reliable findings while protecting the rights, safety, and welfare of clinical trial participants. Following GCP rules necessitates adherence to procedures, thorough data collection, and the upholding of ethical norms at every stage of the research process [11, 12]. The International Council for Harmonization (ICH) created GCP criteria, which place a strong emphasis on participant safety, thorough documentation, and openness in the conduct of clinical trials [13, 14]. Integrating GCP guidelines, IRB supervision, and informed consent guarantees that clinical trials are carried out ethically, increasing scientific understanding while honoring and safeguarding participants.

DOI: 10.1201/9781003522065-2

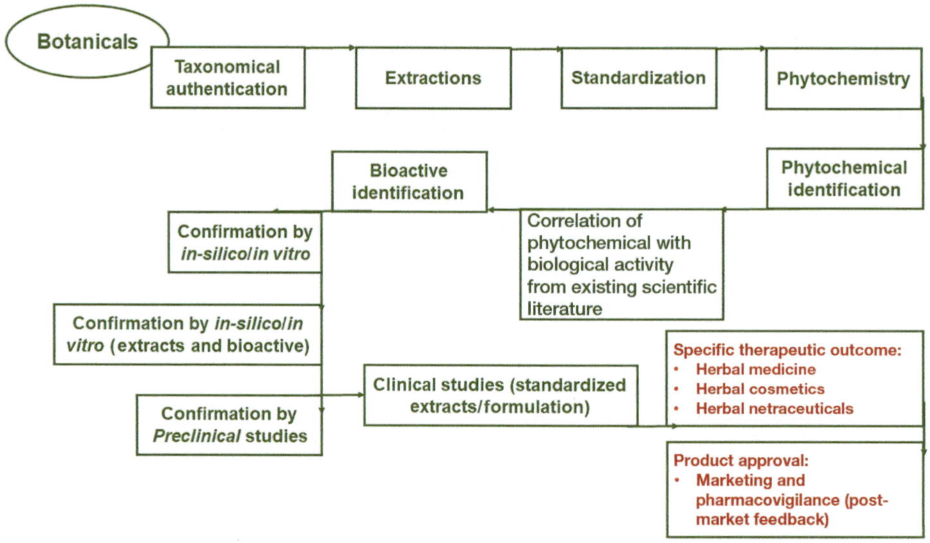

FIGURE 2.1 Outline of the product development process using botanicals via clinical trials.

2.3 REGULATORY FRAMEWORK

The development, evaluation, and approval of new medications and therapies depend heavily on international regulatory agencies such as the EMA, the ICH of Technical Requirements for Pharmaceuticals for Human Use (ICH), and the U.S. Food and Drug Administration (FDA). Before novel treatments can be offered to the public, they must first prove that they are safe, effective, and fulfill strict quality criteria.

2.3.1 FDA, EMA, and ICH: An Overview

Before new medications and treatments are made available to the public, the FDA, which oversees the drug approval process in the United States, makes sure they pass stringent testing and adhere to legal standards. Preclinical research, clinical trials (phases I, II, and III), and a thorough examination of the data provided by the drug sponsor are all part of this process. In a similar vein, the EMA oversees medication regulation in the EU, implementing protocols to evaluate the efficacy and safety of pharmaceuticals in all EU member states. More broadly, the ICH seeks to standardize and expedite the drug development process by coordinating technical norms across the United States, EU, and Japan. The ICH recommendations are essential for maintaining uniformity in regulatory standards across the globe, which minimizes testing duplication and delays in medication availability in various geographical areas [15–20].

2.3.2 Role in Ensuring Safety Standards

Because they establish stringent safety requirements, regulatory bodies are essential to preserving public health. These organizations require that pharmaceutical companies carry out thorough clinical trials to determine the possible dangers and side effects of novel treatments. To make well-informed judgments on risk-benefit ratios, they continue to track safety data throughout the drug's life cycle, even after it has been approved for sale. Because regulatory agencies can swiftly amend or revoke a drug's clearance if new safety concerns emerge, this vigilance is essential to protecting the public's health [21–25].

2.3.3 Transparency in Clinical Trials

Clinical trial transparency is crucial for both ethical and scientific reasons. To provide other researchers and stakeholders with thorough information on safety and efficacy, regulatory organizations frequently mandate the public registration of clinical studies and the publication of findings. Transparency improves research repeatability, reduces the possibility of selective reporting, and increases confidence in regulatory choices. As comprehensive information about a drug's effects becomes accessible, a transparent system enables patients and healthcare professionals to make more educated treatment decisions [26–29]. Making sure that new medications are safe and effective for human use is largely dependent on the strict regulatory scrutiny provided by organizations such as the FDA, EMA, and ICH. These organizations safeguard the public against potentially dangerous therapies and provide a foundation of confidence in the approval process by imposing strict standards and demanding openness.

2.4 ROLE OF HERBAL BIOACTIVE MOLECULES AND PHYTOCHEMICALS IN MODERN MEDICINE

2.4.1 History and Traditional Use of Herbal Medicine

For ages, herbal medicine has played a crucial role in conventional healthcare systems. With its roots in ancient India more than 3,000 years ago, Ayurveda promotes a comprehensive approach to health by using herbal remedies like Ashwagandha and Triphala. Classical writings such as the Charaka Samhita and Sushrutha Samhita [30–33] described these treatments. According to the Huangdi Neijing, Traditional Chinese Medicine (TCM), which has been practiced for more than 2,000 years, uses herbal combinations, such as ginseng and ephedra, to balance the body's yin and yang [34–37]. Similarly, Unani medicine, which originated in Greco-Arabic civilizations, focuses on balancing the humors (*akhlaat*) by combining natural medicines like myrrh and black seed with Hippocratic principles [38–39]. The scientific confirmation of bioactive substances was a necessary step in the move from traditional therapy to contemporary pharmacology. This change is best shown by the discovery of artemisinin in the TCM herb Artemisia annua, which has revolutionized the treatment of malaria [40–41]. Evidence-based medicine increasingly incorporates standardized phytopharmaceuticals and herbal extracts. Accurate measurement of active substances is made possible by methods like mass spectrometry and chromatography, which guarantee efficacy and safety [42, 43]. This development fosters advancements in drug discovery while conserving cultural heritage by bridging the gap between conventional thinking and contemporary research.

2.4.2 Importance of Herbal Bioactive Molecules

Herbal bioactive molecules are substances found naturally in plants that have medicinal properties. These compounds have a major impact on biological systems and add to the therapeutic qualities of plants. They are derived from main and secondary plant metabolites. Because of their significance in promoting health and preventing disease, they have attracted a lot of study attention.

2.4.3 Bioactive Compound Types

Alkaloids: Nitrogen-containing substances with analgesic, antimalarial, and antibacterial qualities include morphine, quinine, and berberine, respectively [44, 45]. Flavonoids: Polyphenolic substances with anti-inflammatory, antioxidant, and cardioprotective properties include quercetin and catechins [46]. Terpenoids: Compounds with antibacterial and antimalarial properties, such as menthol and artemisinin, are present in essential oils [47]. Glycosides: Digoxin and other cardiac glycosides are essential for the treatment of heart failure [48]. Polyphenols: Substances with anti-inflammatory, anticancer, and antioxidant qualities include curcumin and resveratrol [49, 50].

2.4.4 Biological Roles

These bioactive substances alter important biological processes: Anti-inflammatory: Alkaloids and flavonoids reduce inflammation by blocking pro-inflammatory enzymes and cytokines [51]. Antioxidant: Polyphenols reduce damage from oxidative stress by neutralizing free radicals [52]. Antimicrobial: Terpenoids and alkaloids have antimicrobial properties by rupturing bacterial membranes or preventing their metabolism [53]. Anticancer: Substances such as resveratrol and curcumin control angiogenesis, tumor growth, and apoptosis [54]. Cardioprotective: Flavonoids prevent atherosclerosis by enhancing endothelial function and lowering low-density lipoprotein (LDL) oxidation [55]. Herbal bioactive compounds' great diversity and effectiveness highlight their potential for drug discovery and development, bringing traditional medical practices into line with contemporary therapeutic demands.

2.5 CHALLENGES OF HERBAL BIOACTIVE MOLECULES IN DRUG DEVELOPMENT

Because of their varied pharmacological characteristics, herbal bioactive compounds have demonstrated promise in medicinal applications. However, several obstacles prevent their incorporation into contemporary medication research. The chemical complexity of herbal substances is a major problem. Because extracts can contain several different components, it can be difficult to isolate, identify, and standardize them. Reproducibility is further complicated by compositional variability brought on by variations in plant species, geographic location, and extraction techniques [56]. Poor solubility and bioavailability present another difficulty. The quick metabolism and poor water solubility of many herbal compounds, including flavonoids and alkaloids, restrict their systemic absorption and therapeutic effectiveness [57]. Significant obstacles are also presented by safety and toxicity issues. Despite being regarded as natural, certain herbal ingredients may have harmful consequences; hence, thorough toxicity profiles and regulatory approval are required [58]. Lastly, the absence of defined clinical assessment procedures and intellectual property rights prevents commercialization. The proprietary character of herbal medications is frequently less clear than that of synthetic pharmaceuticals, which lessens the incentives for pharmaceutical corporations [59–61]. Advances in extraction technologies, distribution systems based on nanotechnology, and cooperative efforts to create strong standards and regulatory norms are all necessary to meet these problems.

2.6 IMPORTANCE OF CLINICAL TRIALS FOR HERBAL BIOACTIVE MOLECULES

2.6.1 Preclinical Investigations of Herbal Extracts

Before being used in clinical settings, preclinical studies of herbal extracts are essential for assessing their possible therapeutic benefits, safety, and mode of action. The pharmacological characteristics of plant-based substances are usually investigated through *in vitro* (cell-based) and *in vivo* (animal models) research. Phytochemical profiling, mass spectrometry, and high-performance liquid chromatography (HPLC) are some of the methods used to identify and measure the active ingredients in herbal extracts. The biological actions seen in preclinical investigations are frequently attributed to these ingredients, which include terpenoids, alkaloids, flavonoids, and phenolic compounds [43]. The cytotoxicity, anti-inflammatory, antioxidant, and antibacterial properties of herbal extracts are frequently evaluated *in vitro*. These tests aid in locating viable subjects for additional research using animal models. For instance, *Curcuma longa* (turmeric) extract has been shown to have anti-inflammatory qualities by blocking pro-inflammatory cytokines in cell culture models [62]. Pharmacokinetics, toxicity, and therapeutic effects can be examined in a more complex biological system through *in vivo* research conducted on rats. Clinical trial development is therefore guided by the outcomes of animal models. Comprehensive safety evaluations must be carried out despite the encouraging results of preclinical research because herbal extracts may have negative side effects or interact with other medications [63]. Furthermore, to guarantee efficacy and reproducibility in therapeutic settings, standardization of herbal extracts is essential.

2.6.2 Clinical Trial Challenges Specific to Herbal Medicines

Due to their widespread use worldwide, herbal medications pose difficulties for clinical research. The absence of standardization is one of the main problems. Herbal medications, in contrast to synthetic pharmaceuticals, frequently differ in composition because of things like the plant's origin, growth techniques, and preparation procedures. Reliability in evaluating safety and efficacy may be hampered by this discrepancy [43, 58]. Furthermore, another level of difficulty is added by dosage complexity. Herbal medications frequently lack clear standards on ideal dosages, which causes inconsistency in clinical trial findings, in contrast to chemical pharmaceuticals that have specific dosages [64]. The safety profile of herbal products is another major obstacle. Although some herbs are thought to be harmless, some may have negative effects or interact poorly with prescription drugs, particularly if used in large quantities or over an extended period of time [65–67]. Clinical trials involving natural medications are further complicated by regulatory concerns. Preclinical data on safety and efficacy may be insufficient in many countries because herbal medicines are not subject to the same stringent regulatory procedures as pharmaceuticals [56]. Lastly, because of the strong cultural expectations and beliefs surrounding the use of natural treatments, placebo effects may be more noticeable in herbal trials. Trial results may be distorted by this psychological component, making it more difficult to discern between placebo and actual therapeutic effects [68, 69]. To guarantee dependable results, tackling these issues calls for meticulous planning, precise procedures, and suitable regulatory supervision.

2.6.3 Clinical Trial Design for Herbal Medicines

2.6.3.1 Key Considerations for Designing Trials Involving Phytochemicals

Because of their complex structure and varying bioavailability, phytochemicals must be carefully considered while designing clinical trials. To guarantee uniformity in the active ingredient content, which has a direct bearing on efficacy and safety assessment, phytochemical extracts must be standardized [70, 71]. Establishing mechanistic connections between phytochemicals and treatment outcomes requires the selection of suitable biomarkers [72, 73]. To address possible toxicity or interactions with other drugs, long-term safety studies and dose optimization are required [74]. Additionally, trial relevance is increased by comprehending pharmacokinetics and including a variety of patient populations [75]. The gap between evidence-based therapeutic applications and traditional phytomedicine can be closed by well-planned experiments that take these factors into account [56].

2.6.3.2 Standardization and Quality Control of Herbal Products in Trials

To guarantee the constant quality, safety, and effectiveness of botanical products, standardized extracts with clearly defined phytochemical composition are essential. Certain amounts of bioactive chemicals, which can change greatly across batches due to variables such as plant source, extraction techniques, and environmental conditions, are guaranteed to be present in the extract through standardization. Both clinical research and commercial application depend on the extract's therapeutic effects being repeatable, which is why this strategy is so important [76, 77]. By offering proof of the product's composition, standardized extracts also assist in fulfilling regulatory standards by lowering the possibility of contamination and adulteration [78]. To discover and measure the bioactive components that give plants their medicinal effects, it is essential to characterize the active ingredients in botanical extracts. In this process, analytical techniques, including nuclear magnetic resonance (NMR) spectroscopy, liquid chromatography-mass spectrometry (LC-MS), and HPLC, are crucial instruments. Individual substances in complex mixtures can be separated and quantified using HPLC [79]. LC-MS provides quantitative and qualitative information about a compound's concentration and molecular structure [80]. When it comes to verifying the authenticity and purity of phytochemicals, as well as the chemical structure of isolated molecules, NMR spectroscopy is incredibly useful [81]. A key factor in guaranteeing botanical validity is pharmacognosy. It entails examining materials obtained from plants to determine their provenance, quality, and possible adulterants. Plant species are accurately identified and free of contaminants or replacements when botanical authentication techniques, including morphological, microscopic, and molecular markers, are used. This procedure aids in avoiding plant species misidentification, which may result in hazardous or ineffective goods [82–84].

2.6.3.3 Dose Selection and Pharmacokinetics

To comprehend the therapeutic potential and safety of herbal medicines, it is essential to establish dose-response correlations for herbal bioactives. The link between a bioactive compound's dose and the strength of its biological action is measured by a dose-response relationship. Several variables, including the formulation, bioavailability, and particular phytochemical profile, can affect this association in the case of herbal bioactives. Studies on *Curcuma longa*, or turmeric, for instance, have demonstrated that the anti-inflammatory properties of curcumin are dose-dependent, meaning that greater dosages provide more noticeable effects but also raise the risk of side effects such as gastrointestinal distress [85]. Another crucial area of study is comprehending the pharmacokinetics of complex phytochemicals. Because phytochemicals are complex, the absorption, distribution, metabolism, and excretion (ADME) mechanisms involved in the pharmacokinetics of plant components vary greatly. These substances frequently go through phase I and phase II metabolic changes that affect how bioactive they are. For instance, the fast degradation of the grape polyphenol resveratrol in the liver and intestines limits its bioavailability, resulting in low systemic amounts even with considerable ingestion [86]. When assessing the efficacy of herbal bioactives, bioavailability and metabolic factors are crucial. Because of their enzymatic breakdown, high molecular size, or lipophilic nature, phytochemicals frequently have poor absorption rates. To increase their bioavailability, many approaches have been investigated, including the use of phospholipid complexes, nanoparticles, and co-administration with absorption enhancers [70]. Herb–drug interactions can also occur when herbal substances interact with drug-metabolizing enzymes such as cytochrome P450, which can impact the metabolism and therapeutic results of medications taken together [87].

2.6.3.4 Safety and Toxicity Considerations

An essential component of preclinical and clinical drug development is the use of dose-escalation studies to investigate possible toxicities. To find the maximum tolerated dose (MTD) and detect any dose-limiting toxicities (DLTs), these studies are intended to evaluate a drug's safety profile by progressively increasing its dosage. To gather crucial data for therapeutic dose selection in people, such studies frequently take a stepwise strategy, starting with low doses in healthy volunteers or animal models and increasing until notable side effects are seen [88]. Because herbal products are frequently used in conjunction with conventional medications, pharmacology must take herb–drug interactions into account. Compounds in some herbs can change how drugs are metabolized, frequently through the cytochrome P450 enzyme system, which can either stimulate or inhibit enzyme activity. As a result, co-administered medications may become less effective or more harmful [89]. The pharmacokinetic and pharmacodynamic interactions between medications and herbs require special consideration because, although they may not be obvious at first, they can have major clinical repercussions, such as bleeding risks when using anticoagulants [90, 91]. For any new medication or combination therapy to have a safety profile, adverse events must be reported. This involves recording both anticipated and surprising responses seen in post-marketing surveillance or clinical trials. Informed decision-making and the issuance of required safety warnings or recalls are made possible by thorough reporting, which guarantees that possible hazards are conveyed to regulatory agencies and healthcare practitioners [92, 93].

2.7 PROSPECTS AND EMERGING TRENDS IN HERBAL CLINICAL TRIALS

2.7.1 Personalized Medicine and Phytochemicals

The past several years have seen a marked slowdown in the approval of new medications. Novel ways of drug development are needed to speed up the creation of new molecules. By using pharmacogenetics and pharmacogenomics, translational medicine/research, a new field at the intersection of fundamental science and medicine, may accelerate and improve the efficacy of the drug development process. While pharmacogenomics focuses on the simultaneous effects of various mutations in the genome that may affect a patient's reaction to medication therapy, pharmacogenetics studies the genetic factors that contribute to individual differences in drug response. Therefore, by using these techniques in the medication development process, patient subpopulations that show better benefit/risk profiles and/or more effective responses may be identified [94]. In many fields of study, such as nutrition and phytochemical studies, mouse models are a vital resource. Conventional inbred mouse models have contributed to our understanding of health and illness by facilitating the identification of therapeutic targets and mechanisms of action. Nevertheless, the genetic heterogeneity often present in human populations is absent from these models, making it more difficult to extrapolate mouse results to human populations. In several disciplines, including cancer, immunology, and toxicology, the creation of genetically varied mouse models, such as the collaborative cross (CC) or diversity outbred (DO) models, has proven a helpful aid in overcoming this challenge. These instruments are still relatively new in the field of phytochemical research, nevertheless. Given the difficulties and several unsuccessful clinical trials in this sector, the application of CC and DO models has the potential to offer priceless insights for phytochemical translation from rodents to humans, as shown in other fields. These models may be useful for understanding the effects of genetic variability on individual responses to phytochemicals, predicting inter-individual variability in phytochemical bioavailability and efficacy, identifying genetic loci or genes governing responses to phytochemicals, identifying phytochemical mechanisms of action and therapeutic targets, and more. Such knowledge would be extremely helpful for the customized application of phytochemicals in humans [95].

2.7.2 Combination Therapy and Herbal Products

Combining synthetic medications with plant bioactives is a new approach that aims to reduce side effects and increase therapeutic efficacy. Because of their great potency and limited therapeutic windows, many synthetic medicines frequently have negative side effects. A supplementary strategy to attain greater results with fewer side effects is to incorporate herbal bioactives, which are recognized for their moderate action and variety of biological activities. This approach, referred to as "synergistic pharmacology," uses the interplay between synthetic and herbal substances to reduce toxicities and enhance therapeutic effects. When two or more drugs work together to produce a higher effect than the sum of their separate effects, this is known as synergy in multicomponent therapies. This interaction can alter metabolic pathways, improve absorption, and increase bioavailability. For instance, it has been demonstrated that mixing the natural ingredient curcumin with chemotherapy drugs can improve anticancer activity and decrease treatment resistance [96, 97]. Herbal bioactives provide a comprehensive strategy that can customize treatment to meet the needs of each patient, improving safety and effectiveness when added to multi-drug regimens [98].

2.7.3 Role of Nanotechnology in Herbal Medicine

An important problem in herbal medicine has been solved by the development of nanocarriers, which are efficient instruments for increasing the bioavailability of poorly soluble phytochemicals. The limited solubility and bioavailability of several phytochemicals, such as polyphenols, flavonoids, and alkaloids, restrict their potential for medicinal use [99]. The delivery of herbal bioactives by nanoparticles for targeted therapy has drawn considerable interest. Targeting ligands, like peptides or antibodies, can be functionalized onto nanoparticles to enable selective distribution to cells or tissues, reducing off-target effects and enhancing therapeutic results. Additionally, by shielding bioactives from deterioration, nanoparticles can prolong their half-life and improve their pharmacokinetics. According to recent research, nanocarrier systems can transport bioactives for neurological diseases, inflammatory conditions, and cancer [100].

2.8 CONCLUSION

Clinical investigations are essential for confirming the safety and effectiveness of herbal bioactive substances in modern medicine and for ensuring their scientific credibility. Herbal medicines are widely used, but they often lack the comprehensive clinical validation required to establish their therapeutic effectiveness. Clinical studies provide a methodical way to assess the pharmacokinetics, pharmacodynamics, safety, and efficacy of bioactive compounds derived from herbal sources. While phase I studies ensure the herbal compound's safety for human use, phase II trials evaluate the compound's efficacy in treating ailments. Phase III studies compare the herbal therapy against a placebo or conventional medications to confirm its safety profile and therapeutic effectiveness in a variety of groups. Understanding potential adverse effects, the optimal dosage, and drug interactions is made easier by this thorough testing. Because of their many therapeutic applications, phytochemicals—bioactive substances produced from plants—have enormous promise for use in contemporary clinical research. However, their clinical translation is hampered by issues such as poor water solubility, restricted bioavailability, compositional unpredictability, and possible toxicity. Along with the requirement for strong preclinical and clinical proof, standardization and large-scale production continue to be significant obstacles. These difficulties can be addressed by improving the stability, solubility, and targeted administration of drugs through developments in nanotechnology, analytical methods, and drug delivery systems. The therapeutic potential of phytochemicals can be unlocked by utilizing interdisciplinary techniques, opening the door for their incorporation into conventional medicine for the prevention and treatment of illness. It takes well-planned scientific trials to confirm safety and efficacy and to bridge the gap between traditional herbal use and contemporary medicine. Though they sometimes lack solid scientific proof, traditional medicines have enormous promise. We can discover the therapeutic potential of herbal medicines by using contemporary research approaches such as pharmacokinetic studies and randomized controlled trials. To guarantee financing, standardization, and ethical monitoring, cooperation between researchers, physicians, and legislators is crucial. In addition to protecting traditional knowledge, supporting such research guarantees its integration into mainstream medicine, promoting innovation and providing patients with safer, scientifically supported therapeutic alternatives that are based on nature's pharmacy.

REFERENCES

[1] Kandi, V., & Vadakedath, S. (2023). Clinical Trials and Clinical Research: A Comprehensive Review. *Cureus*, *15*(2), e35077. https://doi.org/10.7759/cureus.35077

[2] Umscheid, C.A., Margolis, D.J., & Grossman, C.E. (2011). Key Concepts of Clinical Trials: A Narrative Review. *Postgraduate Medicine*, *123*(5), 194–204. https://doi.org/10.3810/pgm.2011.09.2475

[3] Sun, D., Gao, W., Hu, H., & Zhou, S. (2022). Why 90% of Clinical Drug Development Fails and How to Improve It? *Acta Pharmaceutica Sinica B*, *12*(7), 3049–3062. https://doi.org/10.1016/j.apsb.2022.02.002

[4] Vrijens, B., Pironet, A., & Tousset, E. (2024). The Importance of Assessing Drug Exposure and Medication Adherence in Evaluating Investigational Medications: Ensuring Validity and Reliability of Clinical Trial Results. *Pharmaceutical Medicine*, *38*, 9–18. https://doi.org/10.1007/s40290-023-00503-w

[5] Emanuel, E.J., Wendler, D., & Grady, C. (2000). What Makes Clinical Research Ethical? *JAMA*, *283*(20), 2701–2711. https://doi.org/10.1001/jama.283.20.2701

[6] Varkey, B. (2020). Principles of Clinical Ethics and Their Application to Practice. *Medical Principles and Practice*, *30*(1), 17–28. https://doi.org/10.1159/000509119

[7] Joffe, S., Cook, E.F., Cleary, P.D., Clark, J.W., & Weeks, J.C. (2001). Quality of Informed Consent in Cancer Clinical Trials: A Cross-Sectional Survey. *Lancet*, *358*(9295), 1772–1777. https://doi.org/10.1016/S0140-6736(01)06805-2

[8] Kim, W.O. (2012). Institutional Review Board (IRB) and Ethical Issues in Clinical Research. *Korean Journal of Anesthesiology*, *62*(1), 3–12. https://doi.org/10.4097/kjae.2012.62.1.3

[9] Flynn, K.E., Hahn, C.L., Kramer, J.M., Check, D.K., Dombeck, C.B., Bang, S., Perlmutter, J., Khin-Maung-Gyi, F.A., & Weinfurt, K.P. (2013). Using Central IRBs for Multicenter Clinical Trials in the United States. *PLoS One*, *8*(1): e54999. https://doi.org/10.1371/journal.pone.0054999

[10] Nichols, C., Kunkel, L.E., Baker, R., Jelstrom, E., Addis, M., Hoffman, K.A., McCarty, D., & Korthuis, P.T. (2019). Use of Single IRBs for Multi-Site Studies: A Case Report and Commentary from a National Drug Abuse Treatment Clinical Trials Network study. *Contemporary Clinical Trials Communications*, *14*, 100319. https://doi.org/10.1016/j.conctc.2019.100319

[11] Anonymous. (2001). ICH Harmonised Tripartite Guideline: Guideline for Good Clinical Practice. *Journal of Post-graduate Medicine*, *47*(3), 199–203.

[12] Vijayananthan, A., & Nawawi, O. (2008). The Importance of Good Clinical Practice Guidelines and Its Role in Clinical Trials. *Biomedical Imaging and Intervention Journal*, *4*(1), e5. https://doi.org/10.2349/biij.4.1.e5

[13] Switula, D. (2000). Principles of Good Clinical Practice (GCP) in Clinical Research. *Science and Engineering Ethics*, *6*(1), 71–77. https://doi.org/10.1007/s11948-000-0025-z

[14] Mondal, H., & Mondal, S. (2024). A Brief Review on Good Clinical Practice and Its Training Methods. *Indian Dermatology Online Journal*, *15*(3), 377–382. https://doi.org/10.4103/idoj.idoj_603_23

[15] Langhof, H., Chin, W.W.L., Wieschowski, S., Federico, C., Kimmelman, J., & Strech, D. (2018). Preclinical Efficacy in Therapeutic Area Guidelines from the U.S. Food and Drug Administration and the European Medicines Agency: A Cross-Sectional Study. *British Journal of Pharmacology*, *175*(22), 4229–4238. https://doi.org/10.1111/bph.14485

[16] Rahi, S., & Rana, A. (2019). Role of ICH Guidelines in Registration of Pharmaceutical Products. *International Journal of Drug Regulatory Affairs*, *7*(4), 14–17. https://doi.org/10.22270/ijdra.v7i4.365

[17] Teixeira, T., Kweder, S.L., & Saint-Raymond, A. (2019). Are the European Medicines Agency, US Food and Drug Administration, and Other International Regulators Talking to Each Other? *Clinical Pharmacology & Therapeutics*, *107*(3), 507–513. https://doi.org/10.1002/cpt.1617

[18] Bode, G. (2020). Regulatory Guidance: ICH, EMA, FDA. In: Hock, F., Gralinski, M. (eds.), *Drug Discovery and Evaluation: Methods in Clinical Pharmacology*. Springer, Cham, 1, 1085–1138. https://doi.org/10.1007/978-3-319-68864-0_58

[19] Hufnagel, H.R., & Tennyson, S.D. (2024). Industry Perspective on Regulatory Authority (RA) Quality Reviews of Biosimilar Applications: An Evaluation of RA Guidance and Expectations for Chemical, Manufacturing, and Controls Information through In-Depth Query Analysis. *Expert Opinion on Biological Therapy*, *24*(7), 627–635. https://doi.org/10.1080/14712598.2024.2376197

[20] Alelayan, S., Al-Sheikh, L., Almotawa, R., Alkofide, H., & Alshammari, T. (2024). Are Drug Safety Measures Harmonized? A Real-World Evidence from Three Regulatory Authorities. *Saudi Pharmaceutical Journal*, *32*(3), 101963. https://doi.org/10.1016/j.jsps.2024.101963

[21] Greener, M. (2008). First Do No Harm: Improving Drug Safety through Legislation and Independent Research. *EMBO Reports*, *9*(3), 221–224. https://doi.org/10.1038/embor.2008.17

[22] Patel, M., & Miller, M.A. (2012). Impact of Regulatory Science on Global Public Health. *The Kaohsiung Journal of Medical Sciences*, *28*(7), S5–S9. https://doi.org/10.1016/j.kjms.2012.05.003

[23] Rägo, L., Sillo, H., Hoen, E., & Zweygarth, M. (2014). Regulatory Framework for Access to Safe, Effective Quality Medicines. *Antiviral Therapy*, *19*(3): 69–77. https://doi.org/10.3851/IMP2902

[24] Kandi, V., Vadakedath, S., Addanki, P.S., Godishala, V., & Pinnelli, V.B. (2023). Clinical Trials: The Role of Regulatory Agencies, Pharmacovigilance Laws, Guidelines, Risk Management, Patenting, and Publicizing Results. *Borneo Journal of Pharmacy*, *6*(1), 93–109. https://doi.org/10.33084/bjop.v6i1.3263

[25] Samara, C., Garcia, A., Henry, C., Vallotton, L., Cariolato, L., Desmeules, J., & Pinçon, A. (2023). Safety Surveillance During Drug Development: Comparative Evaluation of Existing Regulations. *Advances in Therapy*, *40*(5), 2147–2185. https://doi.org/10.1007/s12325-023-02492-3

[26] Alemayehu, D., Anziano, R.J., & Levenstein, M. (2014). Perspectives on Clinical Trial Data Transparency and Disclosure. *Contemporary Clinical Trials*, *39*(1), 28–33. https://doi.org/10.1016/j.cct.2014.07.002

[27] Mintzes, B., Lexchin, J., & Quintano, A.S. (2015). Clinical Trial Transparency: Many Gains But Access to Evidence for New Medicines Remains Imperfect. *British Medical Bulletin*, *116*, 43–53. https://doi.org/10.1093/bmb/ldv042

[28] Hudson, K.L., Lauer, M.S., & Collins, F.S. (2016). Toward a New Era of Trust and Transparency in Clinical Trials. *JAMA*, *316*(13), 1353–1354. https://doi.org/10.1001/jama.2016.14668

[29] Axson, S.A., Mello, M.M., Lincow, D., Yang, C., Gross, C.P., Ross, J.S., & Miller, J. (2021). Clinical Trial Transparency and Data Sharing Among Biopharmaceutical Companies and the Role of Company Size, Location and Product Type: A Cross-Sectional Descriptive Analysis. *BMJ Open*, *11*, e053248. https://doi.org/10.1136/bmjopen-2021-053248

[30] Singh, N., Bhalla, M., Jager, P., & Gilca, M. (2011). An Overview on Ashwagandha: A Rasayana (Rejuvenator) of Ayurveda. *African Journal of Traditional, Complementary and Alternative Medicines*, *8*(5 Suppl), 208–213. https://doi.org/10.4314/ajtcam.v8i5S.9

[31] Jaiswal, Y.S., & Williams, L.L. (2016). A Glimpse of Ayurveda: The Forgotten History and Principles of Indian Traditional Medicine. *Journal of Traditional and Complementary Medicine*, *7*(1), 50–53. https://doi.org/10.1016/j.jtcme.2016.02.002

[32] Peterson, C.T., Denniston, K., & Chopra, D. (2017). Therapeutic Uses of Triphala in Ayurvedic Medicine. *Journal of Alternative and Complementary Medicine*, *23*(8), 607–614. https://doi.org/10.1089/acm.2017.0083

[33] Pirintsos, S., Panagiotopoulos, A., Bariotakis, M., Daskalakis, V., Lionis, C., Sourvinos, G., Karakasiliotis, I., Kampa, M., & Castanas, E. (2022). From Traditional Ethnopharmacology to Modern Natural Drug Discovery: A Methodology Discussion and Specific Examples. *Molecules*, *27*(13), 4060. https://doi.org/10.3390/molecules27134060

[34] Cavalieri, S., & Rotoli, M. (1997). Huangdi Neijing: A Classic Book of Traditional Chinese Medicine. *Recenti Progressi in Medicina*, *88*(11), 541–546.

[35] Curran, J. (2008). The Yellow Emperor's Classic of Internal Medicine. *BMJ*, *336*(7647), 777. https://doi.org/10.1136/bmj.39527.472303.4E

[36] Che, C., Wang, Z.J., Chow, M.S.S., & Lam, C.W.K. (2013). Herb-Herb Combination for Therapeutic Enhancement and Advancement: Theory, Practice and Future Perspectives. *Molecules*, *18*(5), 5125–5141. https://doi.org/10.3390/molecules18055125

[37] Matos, L.C., Machado, J.P., Monteiro, F.J., & Greten, H.J. (2021). Understanding Traditional Chinese Medicine Therapeutics: An Overview of the Basics and Clinical Applications. *Healthcare*, *9*(3), 257. https://doi.org/10.3390/healthcare9030257

[38] Parveena, A., Ahmadc, M., Parveend, B., Parveene, R., & Iqbalf, M. (2022). The Traditional System of Unani Medicine, Its Origin, Evolution and Indianisation: A Critical Appraisal. *Indian Journal of Traditional Knowledge*, *21*(3), 511–521.

[39] Safadi, R. (2024). The Liver in Greco-Arabic and Islamic Medicine. *Clinical Liver Disease (Hoboken)*, *23*(1), e0137. https://doi.org/10.1097/CLD.0000000000000137

[40] Miller, L.H., & Su, X. (2011). Artemisinin: Discovery from the Chinese Herbal Garden. *Cell*, *146*(6), 855–858. https://doi.org/10.1016/j.cell.2011.08.024

[41] Su, X., & Miller, L.H. (2015). The Discovery of Artemisinin and Nobel Prize in Physiology or Medicine. *Science China Life Sciences*, *58*(11), 1175–1179. https://doi.org/10.1007/s11427-015-4948-7

[42] Noviana, E., Indrayanto, G., & Rohman, A. (2022). Advances in Fingerprint Analysis for Standardization and Quality Control of Herbal Medicines. *Frontiers in Pharmacology*, *13*, 853023. https://doi.org/10.3389/fphar.2022.853023

[43] Wang, H., Chen, Y., Wang, L., Liu, Q., Yang, S., & Wang, C. (2023). Advancing Herbal Medicine: Enhancing Product Quality and Safety through Robust Quality Control Practices. *Frontiers in Pharmacology*, *14*, 1265178. https://doi.org/10.3389/fphar.2023.1265178

[44] Heinrich, M., Mah, J., & Amirkia, V. (2021). Alkaloids Used as Medicines: Structural Phytochemistry Meets Biodiversity-An Update and Forward Look. *Molecules*, *26*(7), 1836. https://doi.org/10.3390/molecules26071836

[45] Chaachouay, N., & Zidane, L. (2024). Plant-Derived Natural Products: A Source for Drug Discovery and Development. *Drugs Drug Candidates*, *3*(1), 184–207. https://doi.org/10.3390/ddc3010011

[46] Panche, A.N., Diwan, A.D., Chandra, S.R. (2016). Flavonoids: An Overview. *Journal of Nutritional Science*, *5*, e47. https://doi.org/10.1017/jns.2016.41

[47] Siddiqui, T., Khan, M.U., Sharma, V., & Gupta, K. (2024). Terpenoids in Essential Oils: Chemistry, Classification, and Potential Impact on Human Health and Industry. *Phytomedicine Plus*, *4*(2), 100549. https://doi.org/10.1016/j.phyplu.2024.100549

[48] Fender, J., Klöcker, J., Boivin-Jahns, V., Ravens, U., Jahns, R., & Lorenz, K. (2024). Cardiac Glycosides—Quo Vaditis?—Past, Present, and Future? *Naunyn-Schmiedeberg's Archives of Pharmacology*, *397*, 9521–9531. https://doi.org/10.1007/s00210-024-03285-3

[49] Liu, W., Cui, X., Zhong, Y., Ma, R., Liu, B., & Xia, Y. (2023). Phenolic Metabolites as Therapeutic in Inflammation and Neoplasms: Molecular Pathways Explaining Their Efficacy. *Pharmacological Research*, *193*, 106812. https://doi.org/10.1016/j.phrs.2023.106812

[50] Rudrapal, M., Khairnar, S.J., Khan, J., Dukhyil, A.B., Ansari, M.A., Alomary, M.N., Alshabrmi, F.M., Palai, S., Deb, P.K., & Devi, R. (2022). Dietary Polyphenols and Their Role in Oxidative Stress-Induced Human Diseases: Insights into Protective Effects, Antioxidant Potentials and Mechanism(s) of Action. *Frontiers in Pharmacology*, *13*, 806470. https://doi.org/10.3389/fphar.2022.806470

[51] Gonfa, Y.H., Tessema, F.B., Bachheti, A., Rai, N., Tadesse, M.G., Singab, A.N., Chaubey, K.K., & Bachheti, R.K. (2023). Anti-Inflammatory Activity of Phytochemicals from Medicinal Plants and Their Nanoparticles: A Review. *Current Research in Biotechnology*, *6*, 100152. https://doi.org/10.1016/j.crbiot.2023.100152

[52] Losada-Barreiro, S., & Bravo-Díaz, C. (2017). Free Radicals and Polyphenols: The Redox Chemistry of Neurodegenerative Diseases. *European Journal of Medicinal Chemistry*, *133*, 379–402. https://doi.org/10.1016/j.ejmech.2017.03.061

[53] Huang, W., Wang, Y., Tian, W., Cui, X., Tu, P., Li, J., Shi, S., & Liu, X. (2022). Biosynthesis Investigations of Terpenoid, Alkaloid, and Flavonoid Antimicrobial Agents Derived from Medicinal Plants. *Antibiotics (Basel)*, *11*(10), 1380. https://doi.org/10.3390/antibiotics11101380

[54] Pavan, A.R., Silva, G.D.B., Jornada, D.H., Chiba, D.E., Fernandes, G.F.S., Chin, C.M., & Santos, J.L. (2016). Unraveling the Anticancer Effect of Curcumin and Resveratrol. *Nutrients*, *8*(11), 628. https://doi.org/10.3390/nu8110628

[55] Ciumărnean, L., Milaciu, M.V., Runcan, O., Vesa, S.C., Răchișan, A.L., Negrean, V., Perné, M., Donca, V.I., Alexescu, T., Para, I., & Dogaru, G. (2020). The Effects of Flavonoids in Cardiovascular Diseases. *Molecules*, *25*(18), 4320. https://doi.org/10.3390/molecules25184320

[56] Parveen, A., Parveen, B., Parveen, R., & Ahmad, S. (2015). Challenges and Guidelines for Clinical Trial of Herbal Drugs. *Journal of Pharmacy and Bioallied Sciences*, *7*(4), 329–333. https://doi.org/10.4103/0975-7406.168035

[57] Nasim, N., Sandeep, I.S., & Mohanty, S. (2022). Plant-Derived Natural Products for Drug Discovery: Current Approaches and Prospects. *Nucleus*, *65*, 399–411. https://doi.org/10.1007/s13237-022-00405-3

[58] Ekor, M. (2014). The Growing Use of Herbal Medicines: Issues Relating to Adverse Reactions and Challenges in Monitoring Safety. *Frontiers in Pharmacology*, *4*, 177. https://doi.org/10.3389/fphar.2013.00177

[59] Chaudhary, A., & Singh, N. (2012). Intellectual Property Rights and Patents in Perspective of Ayurveda. *Ayu*, *33*(1), 20–26. https://doi.org/10.4103/0974-8520.100298

[60] Karimi, A., Majlesi, M., & Rafieian-Kopaei, M. (2015). Herbal versus Synthetic Drugs; Beliefs and Facts. *Journal of Nephropharmacology*, *4*(1), 27–30.

[61] Willcox, M., Diallo, D., Sanogo, R., Giani, S., Graz, B., Falquet, J., & Bodeker, G. (2015). Intellectual Property Rights, Benefit-Sharing and Development of "Improved Traditional Medicines": A New Approach. *Journal of Ethnopharmacology*, *176*, 281–285. https://doi.org/10.1016/j.jep.2015.10.041

[62] Gangal, A., Sethiya, N.K., Duseja, M., Shukla, R.K., Bisht, D., Rana, V.S., Lalhlenmawia, H., Azizov, S., & Kumar, D. (2025). Green Nanotechnology: Nanoparticle Synthesis Using *Curcuma amada*, *Curcuma caesia*, *Curcuma longa*, and *Curcuma zedoaria*. *Green Chemistry Letters and Reviews*, *18*(1), 2449122. https://doi.org/10.1080/17518253.2024.2449122

[63] Heinrich, M., Jalil, B., Abdel-Tawab, M., Echeverria, J., Kulić, Z., McGaw, L.J., Pezzuto, J.M., Potterat, O., & Wang, J. (2022). Best Practice in the Chemical Characterization of Extracts Used in Pharmacological and Toxicological Research: The ConPhyMP-Guidelines. *Frontiers in Pharmacology*, *13*, 953205. https://doi.org/10.3389/fphar.2022.953205

[64] Busia, K. (2024). Herbal Medicine Dosage Standardisation. *Journal of Herbal Medicine*, *46*, 100889. https://doi.org/10.1016/j.hermed.2024.100889

[65] Hu, Z., Yang, X., Ho, P.C.L., Chan, S.Y., Heng, P.W.S., Chan, E., Duan, W., Koh, H.L., & Zhou, S. (2005). Herb–Drug Interactions: A Literature Review. *Drugs*, *65*(9): 1239–1282. https://doi.org/10.2165/00003495-200565090-00005

[66] Chen, X., Sneed, K.B., Pan, S., Cao, C., Kanwar, J.R., Chew, H., & Zhou, S. (2012). Herb–Drug Interactions and Mechanistic and Clinical Considerations. *Current Drug Metabolism*, *13*(5), 640–651. https://doi.org/10.2174/138920021209050640

[67] Prieto-Garcia, J.M., Graham, L., Alkhabbaz, O., & Mazzari, A.L.D.A. (2023). Potential Pharmacokinetic Interactions of Common Cardiovascular Drugs and Selected European and Latin American Herbal Medicines: A Scoping Review. *Plants*, *12*(3), 623. https://doi.org/10.3390/plants12030623

[68] Kaptchuk, T.J. (2002). The Placebo Effect in Alternative Medicine: Can the Performance of a Healing Ritual Have Clinical Significance? *Annals of Internal Medicine*, *136*(11), 817–825. https://doi.org/10.7326/0003-4819-136-11-200206040-00011

[69] Li, Y., Du, P., Zhang, X., Ren, C., Shi, X., Dong, X., & Zhang, C. (2023). Qualified Placebo for Trials of Herbal Medicine Treatment in Rare Diseases? A Cross-Sectional Analysis. *Orphanet Journal of Rare Diseases*, *18*, 373. https://doi.org/10.1186/s13023-023-02987-w

[70] Aqil, F., Munagala, R., Jeyabalan, J., & Vadhanam, M.V. (2013). Bioavailability of Phytochemicals and Its Enhancement by Drug Delivery Systems. *Cancer Letters*, *334*(1), 133–141. https://doi.org/10.1016/j.canlet.2013.02.032

[71] Nicolescu, A., Babotă, M., Barros, L., Rocchetti, G., Lucini, L., Tanase, C., Mocan, A., Bunea, C.I., & Crişan, G. (2023). Bioaccessibility and Bioactive Potential of Different Phytochemical Classes from Nutraceuticals and Functional Foods. *Frontiers in Nutrition*, *10*, 1184535. https://doi.org/10.3389/fnut.2023.1184535

[72] Singh, J., Luqman, S., & Meena, A. (2020). Emerging Role of Phytochemicals in Targeting Predictive, Prognostic, and Diagnostic Biomarkers of Lung Cancer. *Food and Chemical Toxicology*, *144*, 111592. https://doi.org/10.1016/j.fct.2020.111592

[73] Cote, B., Elbarbry, F., Bui, F., Su, J.W., Seo, K., Nguyen, A., Lee, M., & Rao, D.A. (2022). Mechanistic Basis for the Role of Phytochemicals in Inflammation-Associated Chronic Diseases. *Molecules*, *27*(3), 781. https://doi.org/10.3390/molecules27030781

[74] Borgert, C.J., Fuentes, C., & Burgoon, L.D. (2021). Principles of Dose-Setting in Toxicology Studies: The Importance of Kinetics for Ensuring Human Safety. *Archives of Toxicology*, *95*(12), 3651–3664. https://doi.org/10.1007/s00204-021-03155-4

[75] Gross, A.S., Harry, A.C., Clifton, C.S., & Pasqua, O.D. (2022). Clinical Trial Diversity: An Opportunity for Improved Insight into the Determinants of Variability in Drug Response. *British Journal of Clinical Pharmacology*, *88*(6), 2700–2717. https://doi.org/10.1111/bcp.15242

[76] Fogel, D.B. (2018). Factors Associated with Clinical Trials That Fail and Opportunities for Improving the Likelihood of Success: A Review. *Contemporary Clinical Trials Communications*, *11*, 156–164. https://doi.org/10.1016/j.conctc.2018.08.001

[77] Klatte, K., Pauli-Magnus, C., Love, S.B., Sydes, M.R., Benkert, P., Bruni, N., Ewald, H., Jimenez, P.A., Bonde, M.M., & Briel, M. (2021). Monitoring Strategies for Clinical Intervention Studies. *Cochrane Database of Systematic Reviews*, *2021*(12), MR000051. https://doi.org/10.1002/14651858.MR000051.pub2

[78] Negi, A., Pare, A., & Meenatchi, R. (2021). Emerging Techniques for Adulterant Authentication in Spices and Spice Products. *Food Control*, *127*, 108113. https://doi.org/10.1016/j.foodcont.2021.108113

[79] Kumar, B.R. (2017). Application of HPLC and ESI-MS Techniques in the Analysis of Phenolic Acids and Flavonoids from Green Leafy Vegetables (GLVs). *Journal of Pharmaceutical Analysis*, *7*(6), 349–364. https://doi.org/10.1016/j.jpha.2017.06.005

[80] Pitt, J.J. (2009). Principles and Applications of Liquid Chromatography-Mass Spectrometry in Clinical Biochemistry. *Clinical Biochemist Reviews*, *30*(1), 19–34.

[81] Altemimi, A., Lakhssassi, N., Baharlouei, A., Watson, D.G., & Lightfoot, D.A. (2017). Phytochemicals: Extraction, Isolation, and Identification of Bioactive Compounds from Plant Extracts. *Plants (Basel)*, *6*(4), 42. https://doi.org/10.3390/plants6040042

[82] Sethiya, N.K., Trivedi, A., Patel, M.B., & Mishra, S.H. (2010). Comparative Pharmacognostical Investigation on Four Ethanobotanicals Traditionally Used as Shankhpushpi in India. *Journal of Advanced Pharmaceutical Technology & Research*, *1*(4), 388–395. https://doi.org/10.4103/0110-5558.76437

[83] Chaudhary, S., Walia, V., Singh, V., & Sethiya, N.K. (2020). Thin Layer Chromatographic Analysis of Mangiferin (A Bioactive Antioxidant from Dietary Plant Sources): A Mini Review. *JPC—Journal of Planar Chromatography–Modern TLC*, *33*, 341–352. https://doi.org/10.1007/s00764-020-00044-x

[84] Rana, V.S., Sethiya, N.K., Bisht, D., Chaudhary, S.K., & Singhal, M. (2023). Pharmacognostical, Physicochemical, and Phytochemical Investigation on Allium Wallichii Kunth (An Unexplored and Underutilized Traditional Aromatic Plant from Himalayan Regions). *Indian Journal of Natural Products and Resources*, *14*(2), 285–291. https://doi.org/10.56042/ijnpr.v14i2.4213

[85] Peng, Y., Ao, M., Dong, B., Jiang, Y., Yu, L., Chen, Z., Hu, C., & Xu, R. (2021). Anti-Inflammatory Effects of Curcumin in the Inflammatory Diseases: Status, Limitations and Countermeasures. *Drug Design, Development and Therapy*, *15*, 4503–4525. https://doi.org/10.2147/DDDT.S327378

[86] Zhang, L., Li, C., Kakar, M.U., Khan, M.S., Wu, P., Amir, R.M., Dai, D., Naveed, M., Li, Q., Saeed, M., Shen, J., Rajput, S.A., & Li, J. (2021). Resveratrol (RV): A Pharmacological Review and Call for Further Research. *Biomedicine & Pharmacotherapy*, *143*, 112164. https://doi.org/10.1016/j.biopha.2021.112164

[87] Wanwimolruk, S., & Prachayasittikul, V. (2014). Cytochrome P450 Enzyme Mediated Herbal Drug Interactions (Part 1). *EXCLI Journal*, *13*, 347–391.

[88] Iasonos, A., & O'Quigley, J. (2014). Adaptive Dose-Finding Studies: A Review of Model-Guided Phase I Clinical Trials. *Journal of Clinical Oncology*, *32*(23), 2505–2511. https://doi.org/10.1200/JCO.2013.54.6051

[89] Suroowan, S., Abdallah, H.H., & Mahomoodally, M.F. (2021). Herb–Drug Interactions and Toxicity: Underscoring Potential Mechanisms and Forecasting Clinically Relevant Interactions Induced by Common Phytoconstituents via Data Mining and Computational Approaches. *Food and Chemical Toxicology*, *156*, 112432. https://doi.org/10.1016/j.fct.2021.112432

[90] Gouws, C., & Hamman, J.H. (2020). What Are the Dangers of Drug Interactions with Herbal Medicines? *Expert Opinion on Drug Metabolism & Toxicology*, *16*(3), 165–167. https://doi.org/10.1080/17425255.2020.1733969

[91] Choi, Y.H., & Chin, Y.W. (2021). Multifaceted Factors Causing Conflicting Outcomes in Herb–Drug Interactions. *Pharmaceutics*, *13*(1), 43. https://doi.org/10.3390/pharmaceutics13010043

[92] Alomar, M., Tawfiq, A.M., Hassan, N., & Palaian, S. (2020). Post Marketing Surveillance of Suspected Adverse Drug Reactions through Spontaneous Reporting: Current Status, Challenges and the Future. *Therapeutic Advances in Drug Safety*, *11*, 2042098620938595. https://doi.org/10.1177/2042098620938595

[93] Surana, A.R., Agrawal, S.P., Kumbhare, M.R., & Gaikwad, S.B. (2021). Current Perspectives in Herbal and Conventional Drug Interactions Based on Clinical Manifestations. *Future Journal of Pharmaceutical Sciences*, *7*, 103. https://doi.org/10.1186/s43094-021-00256-w

[94] Dere, W.H., & Suto, T.S. (2009). The Role of Pharmacogenetics and Pharmacogenomics in Improving Translational Medicine. *Clinical Cases in Mineral & Bone Metabolism*, *6*(1), 13–16.

[95] Balkrishna, A., Sharma, N., Srivastava, D., Kukreti, A., Srivastava, S., & Arya, V. (2024). Exploring the Safety, Efficacy, and Bioactivity of Herbal Medicines: Bridging Traditional Wisdom and Modern Science in Healthcare. *Future Integrative Medicine*, *3*(1), 35–49. https://doi.org/10.14218/FIM.2023.00086

[96] Sardana, S., Gupta, R., Madan, K., Bisht, D., Rana, V.S., Bhargava, S., & Sethiya, N.K. (2023). Advance Drug Delivery and Combinational Drug Approaches for Hepatoprotective Action of Berberine: A Progressive Overview with Underlying Mechanism. *RPS Pharmacy and Pharmacology Reports*, *2*(1), rqad002. https://doi.org/10.1093/rpsppr/rqad002

[97] Zheng, S., Prell, R., Sheng, J., Wang, Y., & Hamuro, L. (2022). Changing the Drug Development and Therapeutic Paradigm with Biologic Drug Combinations and Bispecifics: How to Choose Between These Two Approaches? *Clinical and Translational Science*, *15*(9), 2096–2104. https://doi.org/10.1111/cts.13345

[98] Hassen, G., Belete, G., Carrera, K.G., Iriowen, R.O., Araya, H., Alemu, T., Solomon, N., Bam, D.S., Nicola, S.M., Araya, M.E., Debele, T., Zouetr, M., & Jain, N. (2022). Clinical Implications of Herbal Supplements in Conventional Medical Practice: A US Perspective. *Cureus*, *14*(7), e26893. https://doi.org/10.7759/cureus.26893

[99] Dewi, M.K., Chaerunisaa, A.Y., Muhaimin, M., & Joni, I.M. (2022). Improved Activity of Herbal Medicines through Nanotechnology. *Nanomaterials* (Basel). *12*(22), 4073. https://doi.org/10.3390/nano12224073

[100] Awlqadr, F.H., Majeed, K.R., Altemimi, A.B., Hassan, A.M., Qadir, S.A., Saeed, M.N., Faraj, A.M., Salih, T.H., Al-Manhel, A.J.A., Najm, M.A.A., Tsakali, E., Impe, J.F.M.V., El-Maksoud, A.A.A., & Abedelmaksoud, T.G. (2025). Nanotechnology-Based Herbal Medicine: Preparation, Synthesis, and Applications in Food and Medicine. *Journal of Agriculture and Food Research*, *19*, 101661. https://doi.org/10.1016/j.jafr.2025.101661

3 Competitive Landscape Analysis and Market Dynamics of Herbal Medicines

Kartikeya Joshi, Ayush Tyagi, Naresh Suyunari, Dhruv Ghoshal, and Dhruv Bhardwaj
Department of Economics, Atma Ram Sanatan Dharma College, South Campus, University of Delhi, New Delhi, India.

3.1 INTRODUCTION

Herbal medicines or supplements are natural compounds from plants' leaves, bark, roots, seeds, or flowers that people can use for medicinal purposes. Herbal medicines contain active ingredients from natural plants. They may offer therapeutic benefits when people use them as complementary medicine [1].

Nature was used by humans in the early years to implement suitable conditions for survival, thus improving the quality of life over the years. Since those ancient years, knowledge of herbal medicine, based on tradition, has continued to serve with a view to curing the ailments of humankind. The World Health Organization has estimated that 60% of the world's population depends on herbal medicine, and about 80% of the population in developing countries depends almost totally on it for their primary healthcare needs. Phytocompounds and their chemical analogues have provided abundant clinically useful drugs in the treatment of chronic and acute diseases. Research is still continuing in the search for newer therapeutic agents from plants of medicinal value. The herbal industry accounts for about USD 100 billion, with decent growth potential worldwide. The WHO has stated that medicinal plants, herbal raw material, and herbal drugs are growing at an annual growth rate of about 15% [2].

The use of herbal medicine has dramatically increased worldwide over the last 30 years because of its supposed efficiency, availability, and general acceptability. It is estimated that approximately 80% of the general population, especially in developing countries, uses medicinal herbs for primary healthcare. The use of herbal medicines has also received a warm welcome in most developed countries; as complementary and alternative medicine, also known as CAM, it is already mainstream in the United Kingdom and in the rest of Europe, similar to North America and Australia. One of the many reasons to seek herbal therapy, the most important one in these developed countries, is that it also promotes healthier living. Herbal drugs are, therefore, considered an equilibrated and moderate approach to healing, and people using them as home remedies and over-the-counter drugs spend huge amounts of money (in excess of billions of dollars) on herbal products. This may, in part, explain why sales of herbal medicines are booming and represent a significant proportion of the world drug market [3].

Herbal medicine is used for a lot of conditions, from allergies to asthma, eczema, premenstrual syndrome, rheumatoid arthritis, fibromyalgia, migraine, menopausal symptoms, chronic fatigue, irritable bowel syndrome, and even cancer. It is always advisable that herbal supplements be taken under the guidance of a trained provider. Indeed, one multicenter study found that 90% of persons with arthritis use alternative therapies, such as herbal medicine [4].

Herbal medicines have been used as a mainstay of healthcare for centuries and continue to play an important role in contemporary medicine. The most frequently put forward reasons that are cited for the use of traditional herbal medicine include cost, being more aligned with the ideology of the patient, allaying concerns about the negative effects of chemical or synthetic medicines, satisfying a desire for more personalized healthcare, and allowing greater public access to health information. The major use of herbal medicines is for health promotion and therapy for chronic, as opposed to life-threatening, conditions. Again, traditional medicines are largely perceived as natural and safe—meaning not toxic [5].

3.2 MARKET OVERVIEW AND GROWTH PROJECTIONS

The herbal medicine industry experienced significant growth over the past few years, largely driven by increasing global demand for natural and alternative healthcare solutions. Herbal medicine is rooted in history through various traditional practices; over the years, it has evolved to become an integral part of global healthcare, with a growing consumer preference for plant-based remedies. The rapid growth in health awareness and interest in holistic treatments increased the demand for herbal medicines and widened its market across both established and emerging economies. The global herbal medicine market stood at USD 216.40 billion in 2023. The market will grow at a decent pace, from USD 233.08 billion in 2024 to a staggering USD 437 billion by 2032, at a compound annual growth

DOI: 10.1201/9781003522065-3

rate (CAGR) of 8.17% for the period [6]. In 2023, Europe occupied the dominating market share of 44.82%, thereby showcasing that the region contributed significantly to the global herbal medicine landscape. (Please note that these future projections differ from website to website because of different methodologies; the most reliable data have been used.)

Herbal medicines have been applied over generations in the form of roots, stems, leaves, flowers, or seeds of plants. They were adopted by local communities to relieve various ailments. They include common herbs such as ginkgo, ginseng, ginger, turmeric, and chamomile. They have become very popular among people due to their health benefits. Due to changes in lifestyle and the resulting stress, obesity, and cardiovascular problems, people are increasingly reaching out to herbal remedies as opposed to conventional medicines. According to the WHO, 80% of the world's population uses traditional health practitioners, and there is a belief among people that natural products are comparatively safer and have fewer side effects [7].

On the basis of category, the market segmentation of herbal medicine includes herbal pharmaceuticals, herbal functional foods, herbal beauty products, and herbal dietary supplements. Among them, the herbal pharmaceuticals segment contributed the dominant share. During the projection period, the segment size is expected to increase at an accelerated pace because of an expanding aging population and rising awareness among consumers. Moreover, the herbal medicine industry will be of a larger size due to some other factors, such as herbal suppliers making great improvements, minimal or no side effects, and the U.S. Food and Drug Administration (FDA) adopting the Current Good Manufacturing Practices for dietary supplements. With increased demand for herbal remedies and informed perceptions of their safety and efficacy, come a close look at how such trends are taking shape in the global market. To that end, a clearer understanding of the regional distribution of herbal medicine markets can shed light on which cultural, economic, and regulatory factors help drive growth and acceptance in different regions [8].

3.2.1 Regional Market Distribution

The herbal medicine market displays significant regional variations influenced by cultural preferences, regulatory environments, and consumer awareness. Each region exhibits distinctive demand drivers and market dynamics that shape the growth of herbal products.

- European Market Dynamics—Europe holds the largest share of the herbal medicine market, primarily driven by the increasing demand for natural ingredients in personal care products and heightened health consciousness among consumers. The trend toward natural and organic products, especially turmeric, is expected to bolster market growth further, as it is recognized for its health benefits [9].
- Asia-Pacific Growth Potential—The Asia-Pacific region continues to show robust growth, fueled by a cultural affinity for herbal remedies, increasing disposable incomes, and rising consumer awareness regarding natural health options. Countries like India and China are at the forefront, where traditional systems such as Ayurveda and traditional Chinese medicine play a pivotal role [9].
- North American Market Trends—In North America, particularly in the United States, the herbal medicine market is witnessing a shift toward natural products due to increasing health awareness and consumer preference for alternative medicine as alternatives to synthetic drugs. This growing trend, termed the "green wave," has facilitated the successful integration of herbal products into the mainstream market [9].
- Latin America and Middle East and Africa Market Outlook—The Latin American market is expanding due to growing consumer interest in herbal products, driven by higher disposable incomes and a shift toward healthier lifestyles. In the Middle East and Africa, while market awareness regarding herbal medicine is still developing, rising financial stability is anticipated to support an increase in herbal product consumption [9].

As these regional markets gain momentum, it becomes crucial to examine the market dynamics and regulatory environment of herbal medicine in each region.

3.3 MARKET DYNAMICS

The herbal medicines market is experiencing significant growth, driven by increasing consumer demand for natural and alternative healthcare options. Factors such as rising health consciousness, expanding product offerings, and supportive government policies are contributing to this momentum. As the market evolves, understanding the dynamics that shape it—such as consumer preferences, competitive landscapes, and regulatory frameworks—becomes essential. With these factors in mind, let's take a closer look at the supply chain that supports the herbal medicines industry.

3.3.1 Demand Drivers

Over the past decade, herbal medicine has emerged more as an alternative to antibiotic medicine. It is becoming much more popular among patients as well as health professionals. According to the WHO, almost 10–50% of the population in developed countries uses herbal medicine regularly in some form. The major reason for this is that they offer better immunity than synthetic drugs. However, herbal medicines can be the only inexpensive and accessible treatment alternative in developing countries such as China, Japan, India, Vietnam, South Africa, and Bangladesh. The medicines are majorly adopted by consumers for the treatment of cough, cold, gastrointestinal disorders, and painful conditions such as joint pain, rheumatic diseases, and stiffness. This continuing trend to consume herbal supplements is likely to provide growth opportunities to the market. A number of factors create substantial growth opportunities in the herbal products market [10]. Some of them are:

- Consumer Shift to Natural and Holistic Health Solutions: The rising awareness among consumers about health and wellness triggers an incremental shift toward natural and holistic health solutions. Consumers increasingly look for herbal products as alternatives to synthetic medications, driven by rising concerns over side effects and a desire for more sustainable, eco-friendly options [11].
- Rising Demand for Immuno-Boosters: The increased demand for immune stimulants found new business opportunities during the COVID-19 pandemic, especially in herbal supplements and remedies. People want to immunize themselves and gain general health and well-being, which increases the sales of herbal products made up of known immunomodulatory ingredients [11].
- Distribution Channels Expansion: The expansion of distribution channels, mainly with the rise of electronic commerce, has led to increased accessibility of herbal products. The ease of buying through online retail channels directly from the convenience of home has propelled market growth, with exposure to a wider market and entry into new markets [11].
- Government Support and Regulatory Frameworks: Substantial regulatory support and frameworks for herbal products have given strong impetus to market growth. With a view to instilling consumer confidence, thereby fostering market expansion, governments and respective regulatory agencies are implementing standards and guidelines that ensure the safety, quality, and efficacy of herbal products [11].
- Product Innovation and Diversification: The herbal products market is ripe for product innovation and diversification. New formulations will hit the market, aiming to serve emerging consumer trends and preferences by targeting particular health needs and specific demographics [11, 12].

However, alongside these opportunities for growth, the herbal products market also faces key threats that could challenge its continued expansion and success.

3.3.2 Herbal Products Market—Key Threats

Key threats to the herbal products market include a number of concerns, some of which will have an adverse effect on the growth and profitability of the market in the future. Some of them include:

- Regulatory Challenges and Compliance: The herbal products market suffers from a number of regulatory challenges in terms of safety, quality, and efficacy standards. Compliance with rigorous regulations and requirements that vary between regions can prove to be very complicated and costly for manufacturers. Noncompliance or issues with regulations can lead to unwanted product recalls, litigation, fines, and loss of brand reputation [11].
- Contamination and Adulteration: Herbal products could be purposely contaminated or adulterated with unauthorized or even dangerous substances, increasing the risk to consumer safety and trust. Poor quality control, inferior sourcing of raw materials, and insufficient testing procedures can very quickly contaminate the product, thus affecting its integrity and the effectiveness of herbal products [11].
- Misleading Marketing Claims and Misinformation: Such advertising claims that may be misleading or even contain wrong information regarding the health benefits and efficacy of herbal products are likely to dent consumer trust and confidence. Exaggeration of facts or unproven claims might lead to consumer dissatisfaction, as well as compliance and legal fallouts, which can mar market credibility and growth [11].
- Synthetic Alternatives: Herbal products face competition from synthetic alternatives, including pharmaceutical drugs and OTC medications. The availability of pharmaceutical alternatives, with their advantages of standardized dosages and guaranteed efficacy, can therefore threaten the market share of herbal products when consumers show a propensity for quick relief and medical solutions over results [11, 13].

3.3.3 Supply Chain

The herbal medicine supply chain incorporates cultivation, harvesting, processing, and, finally, distribution of herbal products to the retail stage. This is necessary to ensure that herbs are available while conserving quality and safety. This supply chain mostly involves farmers, manufacturers, distributors, and retailers. In the chain, there are a number of different players involved in different segments. Between them, there needs to be effective coordination and management in order to ensure that, with growing consumer demands, safety and product integrity are maintained. Here is a basic overview of various stages of the supply chain of herbal medicine production:

- Supplier Stage—The supply chain starts with the suppliers who provide the raw materials needed for production. This stage includes supplier selection, procurement procedures, and supplier relationship

FIGURE 3.1 Supply chain network in herbal medicine.

management to ensure quality and timely availability from the suppliers [14].

- Manufacturing Stage—It is at this stage that the raw materials are value-added into finished products through processes such as assembly, quality control, and packaging. Ineffective manufacturing processes bear direct consequences on product quality and the overall performance of the supply chain [14, 15].
- Distribution Stage—After the production of the products, they are then sent to various places, such as warehouses and retail stores. This is the stage where logistics management comes into play: an integration of transportation, warehousing, and inventory control in moving products to the end consumer efficiently and effectively at the least cost [16, 17].
- Retail Stage—The retail stage is where products are sold directly to consumers, either through physical stores or online platforms. Effective retail strategies must balance the processes involved in sales, customer engagement, and inventory turnover to optimize profitability [18].
- Consumer Stage—Finally, the products are ready for consumption by the consumer, who is the final judge of the success of the whole supply chain. It is very important to know the consumer's preferences, feedback, and purchasing patterns for continuous improvement in the supply chain and market trends [18].
- Return Stage—This is also known as reverse logistics. This stage works on dealing with returns and recycling the products or disposing of them. This step has a big role in the management of waste and the sustainability of the whole chain; it greatly reduces the chances of environmental problems and raises the general efficiency of the chain. Now, let's delve into the regulatory environment that governs the herbal medicine industry, as it directly influences how these products are managed, marketed, and distributed across different regions [17]. The supply chain network in herbal medicine is shown in Figure 3.1.

3.3.4 Regulatory Environment

Regulation of herbal medicines is defined as a principle, rule, or law designed to control manufacturers and producers of herbal medicines. For example, a regulation would state that herbal medicines must have been proven to be

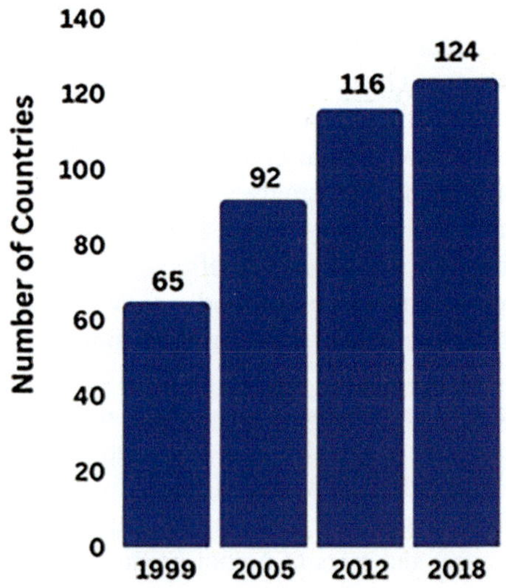

FIGURE 3.2 Global overview: herbal medicine laws and regulations in 124 countries (2018).

safe, effective, and of good quality before reaching the public. The WHO conducted several surveys and published reports, thereby capturing the progress obtained by the member states within three distinct phases: before and after the first WHO Traditional Medicine Strategy, from the first global survey to the second global survey (2005–2012), and from the second to the most recent timeline survey (2012–2018). Due to these reports and the data shown in Figure 3.2, it clearly depicts the increase in the number of countries that have regulations for herbal medicines over the years [19].

Development of natural product medication has its technical and monetary hurdles. These range from formulation to quality assurance, safety, therapeutic efficacy, promotion, and administrative issues, etc., which have to be faced by international regulatory agencies like EMEA, ICH, AYUSH, DCGI, WHO, and the U.S. FDA. They are trying to bring this herbal drug into the regulatory pipeline with relevant knowledge and information pertaining to putting it under the new drug application (NDA) approval process [20]. Regulatory environments for herbal medicines vary greatly from country to country and region to region, reflecting traditional usage, public health priorities, and the integration of herbal products within healthcare systems. The concern over herbal medicines is now pervasive, and the increasing usage rates, most likely associated with meaningful health outcomes, compel most countries to take an active interest in regulating them in one form or another. Herbal products, in the United States, are classified as drugs or dietary supplements, depending on the intended use of the product, and hence have to comply with the regulations followed for this specific category. Herbal products that claim to affect human physiology require an NDA for such claims. On the other hand, the European Medicines Agency has by and large clarified the requirement of registering an herbal medicine [21, 22].

These include a full marketing authorization process under Directive 2001/83/EC Article 8 Inotropic drugs that deal with issues of safety, quality, and efficacy [23, 24]. Countries are also putting their efforts together in order to develop uniform standards that can improve the regulatory framework for herbal products worldwide. An evaluation of existing regulations shows no resemblance across regions, and this adversely affects international trade and market growth for herbal products. The regulations are also put in place to help facilitate research activities on herbal medicines. Most developing countries do not have well-developed regulatory frameworks for herbal medicines. Improved regulations are considered in order to protect the health of the public, particularly in view of concerns over unregulated products being sold that may contain harmful substances. Some countries like Nigeria have recognized the need for comprehensive laws addressing both traditional and modern practice in herbalism [25, 26].

Regulatory changes in herbal medicine mostly result in both positive and negative effects on market dynamics. Greater scrutiny or a much tighter regulatory framework in place ensures standards of safety and efficacy, therefore fostering consumer trust and possible expansion of the market. On the other hand, the clearances and concomitant regulatory requirements are expensive, discouraging new entrants and thus limiting product availability in the market—a custodial ramification for growth. If consumers become aware of the safety and efficacy of regulated products, a shift could occur from unregulated to regulated herbal medicines. Thus, market sales can also see a remarkable boost with such a transition; this fact is exemplified through the growing popularity of herbal products meeting the predefined safety standards. In this regard, regulatory requirements pose great challenges for herbal medicine manufacturers. This may place a high barrier on the small and emerging companies that would wish to invest in clinical trials, for which expensive tests have to prove efficacy and safety. Further, in certain regions or with lax enforcement, there are disparities in the quality and safety of products, especially where informal markets hold sway [27, 28].

Winds of change in the regulatory scenario are progressively becoming global movements toward the harmonization of regulations to facilitate international trade in herbal medicines. Countries are beginning to adopt the same standards, and therefore, the process of entering markets where products are in compliance becomes easier [29]. The region-wise overview of countries with regulatory policies on herbal medicines is shown in Figure 3.3.

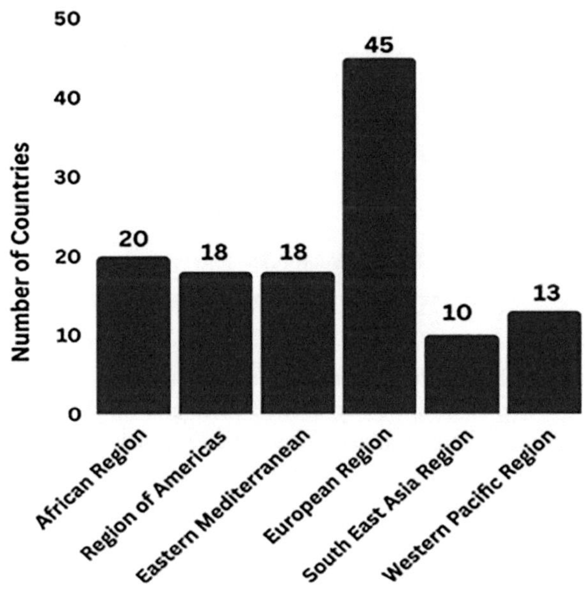

FIGURE 3.3 Region-wise overview of countries with regulatory policies on herbal medicines.

3.4 COMPETITIVE LANDSCAPE

It is incidental that, within this competitive environment, a number of leading players have developed a strong presence in the global herbal medicine market. Firms like Dabur India Limited, Saniju Medical & Pharma, and Blackmores Ltd. have secured leading positions through their innovative product lines and strategic marketing efforts. These companies take advantage of their immense brand presence and extensive product portfolios to gain relevant market share, thereby serving the diverse needs of consumers across several regions. Their success is proof enough of their ability to feel the pulse of market trends and to adapt accordingly.

3.4.1 Dabur India Limited [Annual Sales: USD 115.2 Billion]

Established in the year 1884, Dabur India Limited is considered one of the largest consumer goods firms, with a wide range of products in Ayurvedic and natural segments. It is one of the most diversified firms, with its product offerings ranging from Ayurvedic medicines to personal care and food products, including Dabur Chyawanprash and Dabur Honey. Dabur further exhibits the company's strong emphasis on quality, innovation, and sustainability and is emerging as a trusted name in the herbal medicine market, with constant expansion in product lines and market presence through strategic partnerships and research-driven initiatives [30].

3.4.2 Sanjiu Medical & Pharmaceutical Co. Ltd. [Annual Revenue: USD 17.87 Billion]

Sanjiu Medical & Pharmaceutical Co., Ltd., founded in the year 1993, is among the top manufacturers of herbal medicines. The company carries in its kitty a large basket of quality traditional Chinese medicine, health supplements, and pharmaceuticals. It has gained various awards and certifications for safety and efficacy by strictly paying attention to quality, innovation, and functionality. The company emphasizes research and development in its ever-expanding global presence and, in doing so, upholds the tradition of medicine to the glory of being a trustworthy herbal solutions provider for various health needs [30].

3.4.3 Guangzhou Pharmaceutical Co. Ltd. [Annual Turnover: USD 10.5 Billion]

Guangzhou Pharmaceutical is one of the biggest producers of herbal medicines, manufacturing an extensive range of traditional Chinese herbal medicines and modern pharmaceuticals and healthcare products. Established in 1997, GPC has been the premier standard for quality, innovation, and sustainability culture. Rooted in continuous research and development, GPC is expanding its product line and gaining wider acceptance across the world through carefully chosen strategic partnerships and market initiatives. By marrying age-old wisdom with the might of modern science, GPC shares safe, effective medicine that is easily accessible—while upholding the truth of traditional Chinese medicine [30].

3.4.4 Potter's Herbals [Annual Revenue: USD 7.63 Billion]

Potter's Herbals is a well-established UK firm started in 1812 by Henry Potter. It has the pride of being among the most extensive and most popular herbal remedy producers across the world, with its main blend being traditional wisdom and modern science. Potter's Herbals has a very large portfolio of high-quality products that have earned its tinctures, capsules, and creams awards for their efficacy, purity, and responsible sourcing. With a history of over two centuries, Potter's Herbals is a leading name in the herbal medicine market, now part of the record number of over-the-counter medicinal drugs being developed and marketed in compliance with suitable regulatory standards [30].

3.4.5 Blackmores [Annual Revenue: USD 290 Million]

Established in 1932, Blackmores is one of the oldest prominent herbal medicine companies in Australia. The company offers a broad portfolio of natural health products such as vitamins, minerals, herbal products, and nutritional resources. Blackmores is known for its attitude toward quality and innovation, maintaining the purity, transparency, and efficacy of its products through strict testing and the selection of high-quality ingredients. Recent developments include global expansion through strategic partnerships and investments in research and development that further cement its place as a trusted brand for consumers needing natural health solutions [30].

3.5 CONSUMER BEHAVIOR

Consumer behavior, also called purchase behavior, is defined as "all those acquisition, consumption, and disposition activities of individuals or households that add value to their lives, including the processes that precede and follow these activities" [31]. Such consumer behavior may be influenced by many factors, for example, the psychological factors related to perception, motivation, beliefs, and attitudes that determine how the consumers view and approach brands. Social influences further impinge on purchase behavior when the consumers look up to their families, peer pressure, and culture for direction. For instance, economic factors, including levels of income and general economic conditions, have a larger bearing on what exactly consumers would be willing to buy. Other factors are personal characteristics such as age, gender, lifestyle, and personality differences, all of which result in a variety of individual preferences and buying habits. Consumers' choice is further influenced by other situational factors, such as the shopping environment and specific circumstances surrounding a purchase [32].

A good knowledge of the behavior of consumers is certainly what assists a business in developing appropriate marketing policies, altering product ranges, and creating consumer experiences that can be related to its target audiences. This translates into better interaction and provides more satisfied consumers in a continuously competitive market. Monitoring trends and patterns in consumer behavior, forecasting demand, and making prudent decisions about product design, price, promotion, and distribution require detailed data analysis and understanding on the part of the marketers [31].

When we talk about herbal medicines and the consumer behavior toward them, the stats are fascinating. The use of herbal medicines has risen at an increasing rate, thus capturing a lot of markets globally. COVID-19 has been a sort of "cherry on top" for herbal medicines and drugs in the pharmaceutical industry. So the consumer behavior has been positive and increasing over these past few years as shown in Figure 3.4, which shows the growth in sales of plant-based herbal medicines (2010–2022).

The report mentioned earlier is a Sweden-based report showcasing the rise in sales of plant-based herbal medicines from 2010 to 2022. It shows that the Swedish people have been adopting herbal medicines. There has been a shift in the usage of herbal medicines over a decade. Starting from 2010, the herbal medicine sales were at 202.21 million Swedish Krona. 2017 was the year when the sales value crossed the 300 million mark, demonstrating increased herbal medicine consumption. This then dropped to 294.4 million SEK in 2020, which was the COVID year. The pandemic led to shifts in consumer priorities and spending. Many people focused on essential goods and healthcare, potentially reducing their spending on herbal remedies. The economic impact of the pandemic caused financial strain for many individuals, leading to reduced discretionary

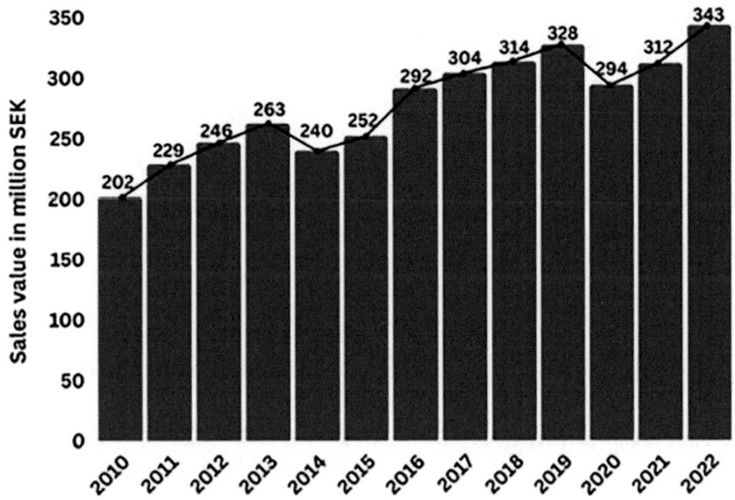

FIGURE 3.4 Growth in sales of plant-based herbal medicines (2010–2022).

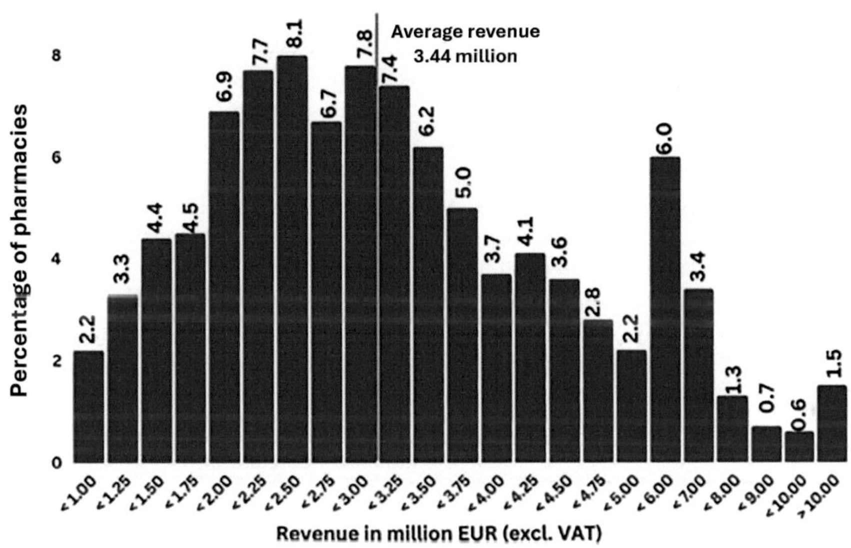

FIGURE 3.5 Average net revenue of pharmacies in Germany.

spending on herbal medicines. Moreover, the pandemic disrupted global supply chains, affecting the availability of herbal medicines and leading to reduced sales, but this reduction was reversed the very next year, in 2021. This report points out the positive consumer behavior toward herbal medicines over the years in Sweden [33].

3.5.1 Pharmacies by Revenue Category

The average net revenue of a pharmacy in Germany is 3.44 million euros per year (excluding VAT) as shown in Figure 3.5, but this figure varies greatly between individual pharmacies. Around 60% of pharmacies fall short of the average revenue, while some large pharmacies greatly exceed it [34].

Now, this is a report showing the average net revenue of pharmacies in Germany. Germany has a strong tradition of scientific research into herbal medicines. Many universities and research institutions conduct studies to investigate the efficacy and safety of various herbal products. Herbal medicines are regulated under the German Medicines Act (Arzneimittelgesetz) and the European Union's herbal medicinal products directive. This ensures that herbal products are subject to stringent quality control, safety, and efficacy standards. Overall, herbal medicines play a significant role in the German healthcare system, combining traditional knowledge with modern regulatory practices. Their integration into conventional medicine, widespread acceptance, and ongoing research reflect the importance of herbal remedies in promoting health and wellness in Germany. The average revenue of a German pharmacy is 3.44 million euros. Different pharmacies fall under different categories; some are below this average, and some are doing exceptionally well, so there is variation [34].

TABLE 3.1
Fiscal Operating Results of German Pharmacies Over Three Years (2021–2023)

	2021	2022	2023
Net revenue in thousand EUR	2,020	2,050	2,375
Cost of sales	1,537	1,600	1,869
Gross profit	483	450	506
Personal costs	204	211	248
Other tax-deductible costs	150	148	164
Fiscal operating result (pretax)	138	100	104

3.5.2 Optimal Results of a Typical Pharmacy

Table 3.1 shows the fiscal operating result of German pharmacies over three years. Fiscal operating result refers to the financial outcome of a company's operations during a specific fiscal period, usually reported in terms of profit or loss. It provides insight into how well the business is performing in terms of its core activities, excluding non-operating income and expenses such as interest, taxes, and extraordinary items. There has been a decline in fiscal operating results because of the increasing trend of herbal drugs and medicines. Can you guess why? There have been many German pharmaceuticals that are actively engaged in the production of herbal medicines. For that purpose, there is a requirement for raw materials, including different natural plant species, which may or may not be available within the geographical boundary. They may also seek help from neighboring nations for raw materials, which adds to their costs. This is the reason there has been a decline in FOR from 138 thousand euros in 2021 to 100 thousand euros in 2022 [34].

If we talk about the United States, it has also been showing a positive trend in consumer behavior. Americans are also inclined toward herbal medicines post-COVID-19. While the American pharmaceutical industry is predominantly focused on synthetic pharmaceuticals and biopharmaceuticals, herbal medicine has gained increasing recognition and popularity among consumers. The herbal supplement market in the United States is substantial, with sales reaching billions of dollars annually. Popular products include herbs like echinacea, ginkgo biloba, turmeric, and ashwagandha. Many consumers view herbal products as complementary to conventional medicine, often using them alongside prescription medications or over-the-counter drugs.

- The revenue generated in the over-the-counter (OTC) pharmaceuticals market in 2024 is estimated to be approximately USD 41.24 bn. The majority is expected to be generated from homeopathy and herbal medicines.
- It is projected that the market will experience an annual growth rate of 3.66% from 2024 to 2029 (CAGR 2024–2029).
- When seen globally, the United States stands out as the country generating the highest revenue in the OTC pharmaceuticals market, with approximately USD 41.24 bn in 2024. The United States is experiencing a growing trend in the OTC pharmaceuticals market toward natural and herbal remedies [35].

3.6 INNOVATION AND DEVELOPMENT

The growth in interest in herbal medicine is invariably accompanied by increasing challenges facing herbal medicine research. A key point of contention here is that a large proportion of herbal medicinal products have very scanty pharmacokinetic, pharmacological, and clinical data to ensure their safety and efficacy. Indeed, performing a clinical trial for herbal medicines poses its peculiar challenges with regard to financial factors, ethical issues, standardization of the product, design of the study, and other matters related to this field. Although the World Health Organization did issue operational guidelines in 2005 to assist regulatory requirements for clinical trials involving herbal products, a host of challenges remain in the definition of inclusion and exclusion criteria that would correspond with both modern and traditional medicine diagnoses, the maintenance of double-blind studies, the choice of appropriate controls, sample size calculations, and standardization of treatments [36].

It is this complexity and the challenges imposed that the creation of appropriate standards must address for effective advancement in the field of herbal medicines. This will be achieved through the design of methodologies capable of attacking the barrier between traditional knowledge and modern medical approaches.

In the area of extraction techniques, impressive steps have been made to improve the inefficiencies of the conventional methods. Traditional techniques, such as maceration, Soxhlet extraction, and hydro-distillation, always carry the burden of applying tedious treatment times, high temperatures, and high volumes of solvents. Most of the very novel extraction techniques have been developed in the past decades and rely on ultrasound-assisted extraction (UAE), microwave-assisted extraction (MWAE), ultrasound-microwave-assisted extraction (UMWAE), supercritical fluid extraction (SFE), and pressurized liquid extraction (PLE). In this respect, these techniques revolutionize the course of herbal product development: They increase the rate of extraction and selectivity for given bio active compounds, thereby augmenting the efficiency and quality of the herbal medicine [37].

The development of herbal products remains unstandardized due to variations in content arising from growth conditions, time of harvesting, and processing methods employed. Standardization would involve some constant parameters and definitive qualitative and quantitative values that guarantee the quality, safety, and reproducibility of herbal medicine. There are four major keys to standardization: marker-based standardization, process control standards, storage standards, and cheminformatics approaches. However, standardization is not easy, and it faces problems such as identification of markers, nonavailability of defined rules and regulations, and a need for universally adopted technical standards [38].

Nanotechnology is developing as one of the important areas in herbal medicine due to its new concepts that can increase the efficacy and bioavailability of herbal formulations. These nano-sized drug delivery systems may therefore overcome the limitations of plant medicines by increasing therapeutic activity, promoting the sustained release of active constituents, and reducing required doses, thereby minimizing side effects. Nanotechnological development, in the form of nanocarriers, polymeric nanoparticles, and solid lipid nanoparticles, significantly influences herbal product development, potentiation of plant extracts, and improvement of treatment outcomes [39].

In designing clinical trials for herbal medicines, there exist challenges that must be addressed in order for results to be valid and reliable. One of the complex issues associated with herbal formulations is that, very often, there is more than one active ingredient whose effects can differ due to processing, source, and storage. To address such challenges, to take one example, strict adherence by researchers to context-specific elaborations of the CONSORT statement in the conduct and reporting of clinical studies relating to the efficacy and safety of herbal medicines is necessary [40].

In herbal medicine trials, it is still very challenging to maintain double-blind conditions due to the typical organoleptic properties observed with many herbal products. Researchers are expected to provide detailed descriptions of controls for potential biases so that studies can be replicated. Another problem is the enormous difference between modern and traditional concepts of disease, so it may be necessary to have a sort of "double classification." Moreover, the choice of comparison groups is also a problem because of the difficulty for the placebo or inactive treatment to mimic whatever properties the herbal medicine might possess. Such influences should be borne in mind by researchers when setting up their studies in order to provide results that are valid and generalizable [40].

Systematic reviews and meta-analyses are important in the validation of herbal medicines by providing a better understanding of their efficacy and safety by synthesizing evidence from several clinical trials. However, it is challenging to conduct such reviews and analyses. Studies on herbal medicine exhibit variations with respect to herbal formulations used, populations involved, and outcomes measured. This challenge can be solved by research scientists by making a very careful selection of studies with strict inclusion criteria and applying the right statistics to analyze the data. In spite of these challenges, systematic reviews and meta-analyses are vital in the establishment of a firm evidence base for the integration of herbal treatment into modern medicine and help point out issues where further research is needed [41].

The other important challenge facing the herbal medicine industry is quality control, which gets complicated because of the formulation of herbal products and variability in the raw materials used. There are a number of factors that affect quality and uniformity—geographical origin, harvest time, and processing—and control of quality must be very stringent and followed scrupulously all along the value chain so as to have consistency and safety in herbal medicines. Furthermore, there is every probability of adulteration or contamination at any link in the supply chain; therefore, herbal products need rigorous testing and inspection to ensure that products manufactured are as per the required quality [42].

This is blurring the distinction in modern scientific research and opening a new horizon in healthcare for the integration of traditional herbal knowledge. Innovations in extraction techniques, standardization processes, and nanotechnology are paving the pathway toward the development of herbal formulations for their use as constituents of modern medicine. The herbal medicine field is in the process of further change and progress that may bring considerable impacts on the delivery of global health through the avenue of such recommendations, which are safe, effective, and reliable.

3.7 CONCLUSION

The future of the herbal medicine sector depends on how effectively it organizes the complex interplay of traditional approaches and modern achievements. So the sector can keep on building and extending its platform in global healthcare through the provision of natural, effective, and easily procurable treatments that can meet the fast-changing needs of consumers at a global scale, with continuous emphasis on research, quality, and consumer confidence. This will mean cooperation between the industry, independent researchers, and regulatory authorities to ensure that the best standards of safety and efficacy are met by herbal medicines if they are going to secure their place within a future model of healthcare.

REFERENCES

[1] Ewumi, O. (2022, October 6). *What is herbal medicine, and what are the benefits?* https://www.medicalnewstoday.com/articles/herbal-medicine

[2] Mirzaeian, R., Sadoughi, F., Tahmasebian, S., & Mojahedi, M. (2021). The role of herbal medicines in health care quality and the related challenges. *Journal of Herbmed Pharmacology*, *10*(2), 156–165. https://www.sciencedirect.com/topics/agricultural-and-biological-sciences/herbal-medicines

[3] Islam, Md Rezaul, et al. (2013, December 26). Herbal medicine: Current trends and future prospects. *National Center for Biotechnology Information (NCBI).* https://www.ncbi.nlm.nih.gov/pmc/articles/PMC3887317/

[4] *Herbal Medicine.* (n.d.). *Mount Sinai health system.* https://www.mountsinai.org/health-library/treatment/

herbal-medicine#:~:text=What%20is%20herbal%20medicine%20good,%2C%20and%20cancer%2C%20among%20others

[5] World Health Organization. (2019). Herbal medicine: An overview. In *WHO Global Report on Traditional and Complementary Medicine 2019*, Chapter 3 (pp. 85–102). Geneva: WHO. https://www.ncbi.nlm.nih.gov/books/NBK92773/

[6] *Herbal Medicine Market Size, Share | Growth Analysis [2032].* (n.d.-b). https://www.fortunebusinessinsights.com/herbal-medicine-market-106320#:~:text=The%20global%20herbal,44.82%25%20in%202023

[7] Abdel-Aziz, S. M., Aeron, A., & Kahil, T. A. (2016). Health benefits and possible risks of herbal medicine. In *Springer eBooks* (pp. 97–116). https://doi.org/10.1007/978-3-319-25277-3_6

[8] Pandey, P., Pal, R., Koli, M., Malakar, R. K., Verma, S., Kumar, N., & Kumar, P. (2024). A traditional review: The utilization of nutraceutical as a traditional cure for the modern world at current prospectus for multiple health conditions. *Journal of Drug Delivery and Therapeutics*, *14*(3), 154–163. https://doi.org/10.22270/jddt.v14i3.6445

[9] *Herbal Medicine Market Size, Share | Growth Analysis [2032].* (n.d.). https://www.fortunebusinessinsights.com/herbal-medicine-market-106320

[10] Liu, Y., Wang, Y., Kong, J., Jiang, X., Han, Y., Feng, L., Sun, Y., Chen, L., & Zhou, T. (2023). *An effective antimicrobial strategy of colistin combined with the Chinese herbal medicine shikonin against colistin-resistant Escherichia coli.* https://www.semanticscholar.org/paper/An-effective-antimicrobial-strategy-of-colistin-the-Liu-Wang/86160a018b50d2fbb03f08102a65e7ec693a9844

[11] *Global Herbal Products Market Size, Trends, Share 2033—CMI.* (2025, January 21). *Custom market insights.* https://www.custommarketinsights.com/report/herbal-products-market/#:~:text=Product%20Innovation%20and,and%20target%20demographics

[12] *Discover Thousands of Collaborative Articles on 2500+ Skills.* (n.d.). https://www.linkedin.com/pulse/herbal-supplements-market-growth-future-demand-regional-b3bme/

[13] Ltd, M. D. F. (n.d.). Global herbal medicine market size, share, trends, growth forecast report by type (Herbal functional foods, herbal dietary supplements, herbal beauty products and herbal pharmaceuticals), source, form and region (North America, Europe, Asia Pacific, Latin America, Middle East, and Africa), industry analysis (2024 to 2032). *Market Data Forecast.* https://www.marketdataforecast.com/market-reports/herbal-medicine-market

[14] Adewumi, M., Akinsola, G., & Olawoye, O. O. (2020). *Structure and conduct of medicinal plants supply chain in Kwara State.* https://www.semanticscholar.org/paper/b928f60f3bd13142d666c65dd014d1404f237a56

[15] Ye, S., & Business School of Yulin Normal University. (2017). Analysis on the operating cost of Chinese herbal medicine enterprises from the perspective of supply chain. In *International Conference on Humanities Science, Management and Education Technology (HSMET 2017): Vol. 96* (pp. 1029–1030). https://download.atlantis-press.com/article/25872905.pdf

[16] Haiju, H., Zhu, W., Yakun, L., & Jia, Z. (2023). *A four-dimensional evolutionary game analysis of quality cheating behavior in traditional Chinese medicine supply chain.* https://www.semanticscholar.org/paper/A-Four-dimensional-Evolutionary-Game-Analysis-of-in-Haiju-Zhu/e2d54d3c91859771f2f89a0559374823122ca681

[17] Booker, A., Fitzgerald, M., Heinrich, P., & Scheid, V. (2018). *Complexity in Chinese herbal medicine supply chains.* https://www.semanticscholar.org/paper/Complexity-in-Chinese-herbal-medicine-supply-chains-Booker-Fitzgerald/54260c0bf54e3b5894efddb2c4a85db3df056bf0

[18] Octavia, A., Sriayudha, Y., & Ali, H. (2020). *Innovation capability and supply chain management: Empirical study of Indonesian traditional herbal medicine products.* https://www.semanticscholar.org/paper/Innovation-Capability-and-Supply-Chain-Management%3A-Octavia-Sriayudha/c00b85d5baf86878d56a6b5cee614bcb6d58d51f

[19] World Health Organization. (2019). *WHO Global Report on Traditional and Complementary Medicine 2019.* Geneva: World Health Organization. https://www.who.int/publications/i/item/9789241515436

[20] Hossain, C. M., Gera, M., & Ali, K. A. (2022, December 7). Current status and challenges of herbal drug development and regulatory aspect: A global perspective. *Asian Journal of Pharmaceutical and Clinical Research.* https://innovareacademics.in/journals/index.php/ajpcr/article/download/46134/27360.

[21] Ahmad, N. H., Ahmad, N. I., Abdullah, S. N., Jamaludin, N. A., & Fauzi, N. A. M. (2015). Issues and challenges of public sector accounting in Malaysia: Implementation of accrual accounting. *Proceedings of the 10th Malaysian National Economic Conference (PERKEM) 2015*, Johor Bahru: PERKEM. https://www.ukm.my/fep/perkem/pdf/perkem2015/PERKEM_2015_3A1.pdf

[22] Lara-Sanchez, J. (2024). Cancer stem cells in glioblastoma: Therapeutic target and drug resistance. *Current Topics in Medicinal Chemistry*, *10*(6), 487–498. https://www.ingentaconnect.com/content/ben/ctm/2024/00000010/00000006/art00014

[23] Chang, J., & Lin, M. (2024). Nanomaterials in neurodegenerative disease treatment: Current status and future prospects. In *Nanomedicine for the Treatment of Disease* (pp. 139–155). Amsterdam: Elsevier. https://www.sciencedirect.com/science/article/abs/pii/B9780128243855000091

[24] Yan, W. (2011). *Analysis on regulations of traditional herbal medicinal products for human use in United Kingdom.* https://www.semanticscholar.org/paper/3154d649df8534133d083fabaa3364c5362a3898

[25] Demeke, H., Hasen, G., Sosengo, T., Siraj, J., Tatiparthi, R., & Suleman, S. (2022). Evaluation of policy governing herbal medicines regulation and its implementation in Ethiopia. *Journal of Multidisciplinary Healthcare*, *15*, 1383–1394. https://doi.org/10.2147/jmdh.s366166

[26] Krishna, P. D., Gowrav, M., Bhaskaran, M., & Mr, K. (2024b). Current regulations of Herbal medicines in the US and EU. *Current Traditional Medicine*, *10*(6). https://www.ingentaconnect.com/content/ben/ctm/2024/00000010/00000006/art00014

[27] Zhang, Y., & Liu, X. (2024). AI-driven innovations in healthcare: Current trends and future perspectives. In *Advances in Artificial Intelligence and Data Science*, edited by S. Wang and J. Li (pp. 455–470). Springer. https://link.springer.com/chapter/10.1007/978-981-99-7703-1_30

[28] Li, X., Wang, Q., & Chen, D. (2014). Treatment of coronary heart disease with traditional Chinese medicine. *Chinese*

Journal of Integrative Medicine, *20*(4), 243–248. https://doi.org/10.1007/s11655-014-2024-y

[29] Bhattacharjee, B., Sandhanam, K., Ghose, S., Barman, D., & Sahu, R. K. (2023). Market overview of herbal medicines for lifestyle diseases. In *Role of Herbal Medicines* (pp. 597–614). https://doi.org/10.1007/978-981-99-7703-1_30

[30] Emergen Research. (n.d.-c). *Top 10 Companies in Herbal Medicine Market in 2024*. Emergen Research. https://www.emergenresearch.com/. https://www.emergenresearch.com/blog/top-10-companies-in-herbal-medicine-market-in-2024

[31] QuestionPro. *Consumer behavior: Definition, importance, and types*. https://www.questionpro.com/blog/consumer-behavior-definition/

[32] Statista. (2023, November 16). *Sales value of plant-based medicine in Sweden 2010–2022*. https://www.statista.com/statistics/789343/sales-value-of-plant-based-medicine-in-sweden/

[33] ABDA—Federal Union of German Associations of Pharmacists. (2024). *Zahlen, Daten, Fakten 2024: The German Pharmaceutical Market*. Berlin: ABDA, 2024. https://www.abda.de/fileadmin/user_upload/assets/ZDF/Zahlen-Daten-Fakten-24/ABDA_ZDF_2024_Brosch_english.pdf

[34] Statista. (n.d.). *OTC pharmaceuticals—US | Statista market forecast*. https://www.statista.com/outlook/hmo/otc-pharmaceuticals/united-states#:~:text=OTC%20Pharmaceuticals%20%2D%20United%20States&text=The%20revenue%20generated%20in%20the,(CAGR%202024%2D2029)

[35] Singh, H., Sharma, S., Chauhan, S., Kaur, I., Nagar, H., & Vihar, P. (2014). *Clinical trials of traditional herbal medicines in India: Current status and challenges*. https://www.semanticscholar.org/paper/016056194dc0e7d6006bf5cbf307428425b02858

[36] Kaur, J., & Kaur, S. (2020). Phytochemical screening and antioxidant activity of medicinal plants. *Asian Journal of Pharmaceutical Sciences*, *10*(4), 243–250. https://ajpsonline.com/AbstractView.aspx?PID=2020-10-4-8

[37] Ahire, T., Gosavi, S., Sonawane, S., Chaudhari, R., & Bornare, P. (2023). Standardization of herbal medicines and the guidelines for the standardization. In *Current Nutraceuticals*. https://www.semanticscholar.org/paper/445291f5fa44a81c06fe5ec1bb756d1c2b55dbce

[38] Lekshmi, & Chandran, S. (2018). *Role of nanotechnology in herbal medicine*. https://www.semanticscholar.org/paper/e39c036485c9d4baaa1f03f1c9f653b922705982

[39] Hassan, H. A., & Abdel-Aziz, S. M. (2016). Protective effects of green tea extract against complications of diabetes in rats. *Journal of Physiology and Biochemistry*, *72*(1), 93–101. https://pubmed.ncbi.nlm.nih.gov/26681895/

[40] Smith, J. A., & Doe, P. (2014). Antioxidant properties of medicinal plants: A review. *Journal of Medicinal Plants Research*, *8*(5), 925069. https://doi.org/10.1155/2014/925069

[41] Alamgir, A. N. M. (2020). Standardization of medicinal plants: A review. *Future Journal of Pharmaceutical Sciences*, *6*(1), 91. https://fjps.springeropen.com/articles/10.1186/s43094-020-00091-5

[42] Muyumba, N. W., Mutombo, S. C., Sheridan, H., Nachtergael, A., & Duez, P. (2021). Quality control of herbal drugs and preparations: The methods of analysis, their relevance and applications. *Talanta Open*, *4*, 100070. https://www.sciencedirect.com/science/article/pii/S2666831921000400

4 Integrating Traditional Wisdom and Modern Science

Evidence-Based Approaches and Reverse Pharmacology in Herbal Medicine

Anjana Sathyan, Anu Ruhila, and Galib Ruknuddin
All India Institute of Ayurveda, New Delhi

4.1 BACKGROUND

Traditional medicines (TMs) have been a vital component of communities for ages in addressing the health and well-being needs of people. TM has significantly contributed to modern science and medicine, where traditional products and practices have been translated into effective treatment entities for various health conditions for many centuries. The practice of TM is native to particular cultures and geographic areas; however, the system of TM is still considered scientifically invalidated or understudied. According to WHO report, 124 member states (accounting for 64% of the total) had laws or regulations on herbal medicine, with the WHO European Region having the highest number of countries with a national regulatory framework. Furthermore, the special regulatory requirement, i.e., traditional use without demonstrated harmful effects and reference to safety data in documented scientific research on similar products, was reported by the highest number of member states [1]. Thus, the promotion of TM faces several obstacles worldwide, mostly in industrialized countries, despite global restructuring and a long history of traditional use. The rising concern over quality and quality control, pharmaceuticals, scientific research and validation, intellectual property rights, and biopiracy has been impeding the global development of TM.

The significant worldwide rising prevalence of the usage and acceptance of TMs as a preferred healthcare option necessitates the development and execution of rigorous scientific research and regulatory processes to provide an evidence base for the system. Researchers and medical professionals are working toward developing TM as evidence-based medicine (EBM), further supporting the effectiveness and safety of the system. Validating and evaluating clinical expertise with the advanced research tools forms the foundation of EBM, which has been considered the mandate to establish the safety and efficacy of TM. Drug discovery strategies based on natural products and TMs are reemerging as attractive options to address the increasing demand for therapeutics for various newly emerging health challenges. Herbal medicinal products with robust scientific evidence that are rationally formulated and standardized are being considered promising alternatives to conventional medicines. The reverse pharmacology (RP) approach is an innovative approach in drug discovery done from "bedsides to benches." Understanding the mechanisms of action at various levels of biological structure and optimizing the safety, effectiveness, and acceptability of natural product leads based on pertinent knowledge are the goals of RP (Figure 4.1). With the RP approach, traditional drugs with a long history of effective usage are validated for their clinical efficacy using modern or conventional methods of clinical research, further cutting the time and expense spent on conventional drug discovery approaches. The chapter discusses and elaborates on the concepts of EBM for TM, further deliberating on the available approaches and future aspects in this segment.

4.2 CONCEPT OF EVIDENCE-BASED MEDICINE IN HERBAL THERAPIES

The foundations of TM are deeply rooted in centuries of empirical knowledge derived from extensive practice and careful documentation by ancient seers. Historically, this knowledge was passed down through generations largely based on experiential learning, rather than the systematic approach characteristic of modern EBM [2]. Evidence is crucial in two key aspects of traditional herbal medicine, i.e., safety and efficacy. There's a common misconception that herbal medicines are inherently less toxic and can be used under any condition [3, 4]. However, this overlooks critical challenges such as variations in dose, dosage forms, the individual constitution (~*Prakriti*) of the patient, and potential drug-to-drug interactions, all of which warrant further investigation. Safety assessments are essential, as they help identify and mitigate potential risks associated

DOI: 10.1201/9781003522065-4

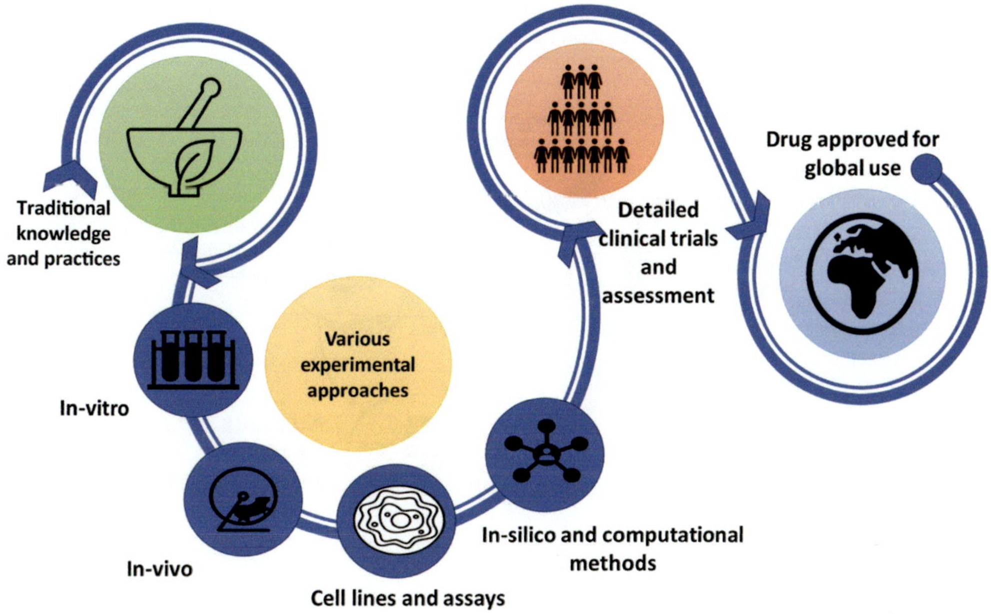

FIGURE 4.1 Reverse pharmacology approach in TM.

with the use of herbal medicines, whereas evidence is essential for establishing the efficacy of herbal medicines. Additionally, robust quality control methods are crucial for the standardization of various herbal formulations [5].

Ayurveda takes a holistic approach to treatment, emphasizing the need to consider both the disease and the patient's unique condition before prescribing any medication. This philosophy necessitates specific, tailored research questions in scientific studies due to the complex interplay of variables involved. By compartmentalizing each aspect of patient care, researchers can adopt a more focused approach, enhancing the clarity and relevance of their findings [6]. A significant distinction between modern medicine and herbal formulations lies in the characterization of their active components. Modern pharmaceuticals are typically defined by a clear biochemical profile of a single molecule or a well-defined compound mix, facilitating a detailed understanding of their pharmacological activities [7]. On the contrary, TM makes extensive use of therapeutic herbs. The formulations are not based on a thorough understanding of the active chemicals in medicinal herbal formulations but rather on empirical knowledge. These herbal medicines consist of complex mixtures of potentially hundreds to thousands of compounds. Currently, there are no comprehensive methods available that can fully elucidate the complete biochemical profiles and the collective pharmacological activities of all compounds present in herbal medicine. This gap underscores the challenges faced in integrating traditional herbal formulations into the defined framework of modern medical research. Furthermore, the batch-to-batch variation of the active chemicals/compounds in the herbal formulation emerges as a fundamental challenge to carrying out validated randomized clinical trials in TM. The issues also encompass the product-to-product variation arising from different manufacturers and brands, further hindering the fair comparison of the findings among similar studies. These discrepancies lead to a lack of reproducibility among clinical trials and even challenge the reliability of extrapolating clinical trial results.

Earlier, people used crude plants in different forms to treat various diseases. However, with the advancement and development in technology, the curiosity toward identifying the details of plant extracts increased, which led to the introduction of a new phytochemical era. At the beginning of the 18th century, researchers were successful in the isolation of several phyto-molecules of pharmacological importance with the help of various fundamental chemical techniques of separation and chromatography. It was in the 20th and 21st centuries that the research and development of phytochemistry received immense thrust and yielded several phyto-molecules of immense importance to medical science [8]. The introduction of phytochemistry revolutionized TM by allowing for the molecular breakdown of herbs, thereby providing a deeper understanding of the bioactive constituents and their mechanisms of action [9–11]. Phytochemical analysis (PCA) plays a crucial role in enhancing the understanding of herbs and herbal formulations, offering a scientific basis for several key aspects of TM (Figure 4.2).

PCA explains the rationale behind the preprocessing of potent single herbal drugs, such as *Kupeelu* [12], which is necessary to reduce their inherent toxicity. Second, it provides insights into the reasons for selecting specific herbs in compound formulations, exemplified by studies on polyherbal formulations such as *Amruthotharam kashayam*, which reveal the synergistic interactions between its components [13]. Third, it aids in the phytochemical screening of different modified dosage forms of various polyherbal

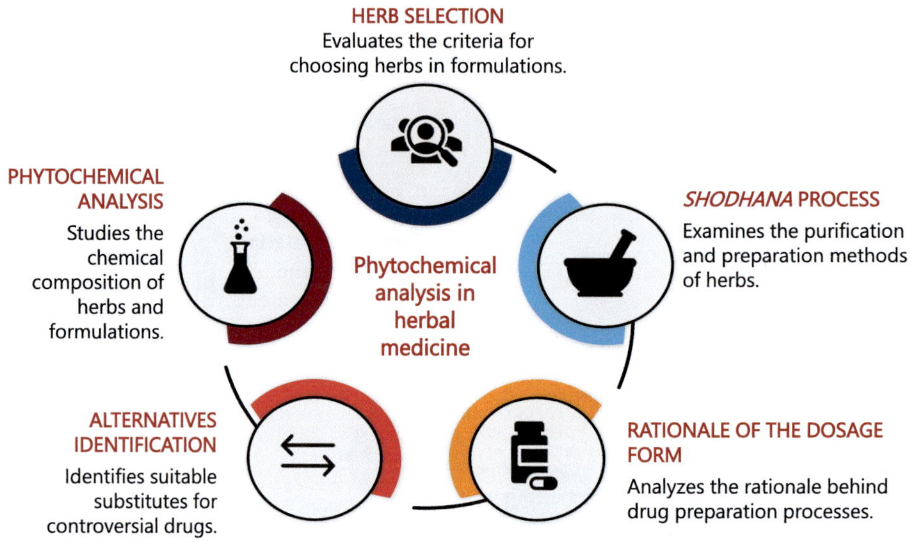

FIGURE 4.2 Phytochemical analysis in herbal medicine.

formulations mentioned in ancient texts of TM, further establishing their potent efficacy [14]. Furthermore, this analytical method helps identify possible substitutes for controversial drugs or moderately available drugs, such as *Pashanabheda*, thereby addressing issues of scarcity and sustainability [15]. Finally, the comprehensive phytochemical profiling of compound formulations not only elucidates the bioactive components involved but also enhances the reproducibility and standardization of herbal medicines [16]. Some studies on various aspects of phytochemical analysis are listed in Table 4.1.

4.3 ASSESSING THE EFFICACY OF HERBAL MEDICINES VIA PRECLINICAL TESTING

The efficacy of herbal medicines is rigorously assessed through preclinical experimental methods, including both *in vitro* and *in vivo* analyses [25]. These experimental studies offer insights into the pharmacological activity of herbal medicines. The development and assessment of herbal medicines necessitate a thorough research strategy that includes clinical trials and experimental procedures to verify the effectiveness and mode of action, as well as chemometric testing of herbs for quality, consistency, and identification of bioactive components [26]. Experimental studies are conducted in a controlled environment, allowing for the precise isolation and examination of specific components of herbal medicines at cellular or molecular levels. This controlled setting facilitates the manipulation of experimental conditions to explore the impact of various variables on the biological activity of these compounds.

In vivo studies are indispensable for evaluating the efficacy of herbal formulations, as they provide comprehensive insights that are unattainable with *in vitro* methods alone. Well-documented reports of pharmacological activity in animals may be viewed as having scientific rationale [27]. Over the years, Ayurvedic formulations have been studied through experimental studies using various animal models such as Wistar strain albino rats [28, 29], Swiss male mice [30, 31], New Zealand White rabbits [32], and Drosophila melanogaster [33, 34] to simulate human disease conditions. This approach offers a more dynamic biological environment where the interactions and systemic effects of herbal formulations can be observed.

However, *in vitro* and *in vivo* experiments do not clearly explain the specific mechanisms behind pharmacological activity. To overcome this limitation, recent research has increasingly utilized *in silico* studies. These computational analyses help predict and analyze the modes of action of herbal medicines, providing deeper insights into the interactions between the phytochemicals (ligands) present in herbal medicines and the disease targets. By combining *in vitro*, *in vivo*, and *in silico* methods, researchers can achieve a more comprehensive understanding of how herbal medicines function, which can enhance the development and application of these treatments. Several studies that utilize *in vitro*, *in vivo*, and *in silico* experimental methods for assessing the efficacy of various herbs and herbal formulations are summarized in Table 4.2.

4.4 INTEGRATED APPROACH AND FUTURE RESEARCH ASPECTS OF RESEARCH IN HERBAL MEDICINE

Future research in herbal medicine is poised to advance significantly across multiple domains, including the pharmaceutical, preclinical, and clinical areas (Figure 4.3). In the pharmaceutical realm, new drug dosage forms of traditional formulations are increasingly leveraged. In Ayurveda, the primary method of preparing herbal

TABLE 4.1
Practices in Evidence Synthesis Based on Phytochemical Analysis

	Context of the Study	Brief of the Study	Findings and Results	Ref.
1	Rationale behind the preprocessing of single herbal drugs to reduce their inherent toxicity	*Semecarpus anacardium* L.f. (*Bhallataka*) has to be processed before it is used in a formulation or administered to a patient. Using the drug in its unprocessed state can lead to irritation, blisters, toxicity, and contact dermatitis. The study involved a chemical analysis of both raw and purified samples of *Bhallataka* using LC/MS analysis.	There were three derivatives of anacardic acid found in both raw and purified samples of *Bhallataka*. The derivatives are anacardic acid monoene, anacardic acid diene, and anacardic acid triene. However, the abundance of the compounds was less in purified samples. Additionally, a compound, cardanol (m/z 297.154), was present in the purified samples. This cardanol might have been formed from the decarboxylation of anacardic acid during the purification process.	[17]
2		*Vatsanabha* (*Aconitum heterophyllum* Wall. ex Royle) is purified by soaking cut pieces in cow urine for seven days, with daily removal of old urine and the addition of fresh urine. A study compared this traditional method with a new one using lab-made urine instead of cow urine. Aconite root samples underwent toxicity evaluation and alteration assessment through animal studies and thin layer chromatography (TLC). This approach aimed to establish a chemical lab-scale method and to contrast it with the conventional Ayurvedic Shodhana process, while also examining the advantages and disadvantages of the traditional Shodhana method.	As compared to the traditional method, the new method was found to be less effective in terms of the change in the conversion of the compounds pseudo-aconitine and aconitine to much less toxic substances such as veratroyl pseudo-aconine and benzoyl-aconine, respectively. The new chemical purification could not convert toxic alkaloids into less toxic alkaloids.	[18]
3	The rationale behind selecting various herbs in polyherbal formulations	*Amruthotharam kashayam* is a formulation composed of plant parts of *Tinospora cordifolia* (Willd.) Hook.f. & Thomson, *Terminalia chebula* Retz., and *Zingiber officinale* Roscoe in specific proportions. Different spectroscopic techniques were used to assess the formulation, which included comparative chemical profiling of the ingredient plants through HPTLC analysis. Additionally, the major chemical constituents were characterized using Fourier transform infrared spectroscopy (FTIR) and tandem mass spectrometry.	The chemical profile of an individual plant extract is different from that of the compound formulation. FTIR analysis results suggested the synergistic effect of the ingredients in the compound formulation. Chromatographic and mass spectroscopic studies showed that this change in chemical profile has happened during the preparation of the formulation.	[13]
4		*Triphala churna* is a powder formulation composed of *Phyllanthus emblica* L. (*Amalaki*), *Terminalia chebula* Retz (*Haritaki*), and *Terminalia bellirica* (Gaertn.) Roxb (*Vibhitaki*). The formulation is prepared using both equivalent and nonequivalent proportions of ingredients. The antioxidative and cytotoxic effects of these two variants are examined. Tests, including antioxidant DPPH activity, free radical scavenging, and total phenolic content (TPC), were performed, alongside the chemical characterization of the individual components of *Triphala* to identify and quantify each compound.	A significantly higher synergistic antioxidant effect and cytotoxicity were observed in the nonequivalent proportion of *Triphala*. Spectroscopic analysis revealed a wide range of compounds, including fatty acid esters, triterpenes, and aminoglycosides, in the individual ingredients.	[19]

(Continued)

TABLE 4.1 ***(Continued)***
Practices in Evidence Synthesis Based on Phytochemical Analysis

	Context of the Study	Brief of the Study	Findings and Results	Ref.
5	To identify and evaluate the substitutes for controversial drugs or moderately available drugs for their safety and efficacy	In a study, the Ayurvedic substitution pair *Musta* (*Cyperus rotundus* L.) and *Ativisha* (*Aconitum heterophyllum* Wall. ex Royle), which is relatively unavailable were studied for biochemical and metabolic similarities and differences.	It was found that the superimposed high-performance liquid chromatography (HPLC) profiles of both drugs bore a striking resemblance in terms of peak profiles and fingerprints, except that *Aconitum heterophyllum* Wall. ex Royle had two additional peaks at retention times of 11.8 and 29.6 minutes. Moreover, the common peaks were significantly larger in *Ativisha* than in *Musta*, indicating higher concentrations of phytoconstituents.	[20]
6		Ayurveda classics advocate the substitution of *Pippali* (*Piper longum* L.) with *Maricha* (*Piper nigrum* L.), both belonging to the same genus, which may be due to the presence of common phytoconstituents. An attempt was made to perform phytochemical screening of both drugs to develop and validate methods for quantitatively estimating piperine from various geographical varieties of *P. nigrum* and *P. longum*.	Alkaloids were identified in the preliminary phytochemical experiments in both *P. nigrum* and *P. longum* fruits. Piperine was found in higher amounts in *P. nigrum* than in *P. longum*.	[21]
7	Phytochemical profiling of herbal formulations	A comparative analysis of aqueous and methanolic extracts of the polyherbal formulation, *Triphala*, is conducted. The antioxidant activity was assessed using the following methods: DPPH, ferric-reducing antioxidant power (FRAP), superoxide dismutase (SOD), and catalase assay. The antibacterial potential was also determined by broth dilution and agar well diffusion assays.	Both extracts displayed adequate radical-scavenging activity such as ascorbic acid. Methanolic extracts showed superior antioxidant activity compared to aqueous extracts and also exhibited antibacterial effects against two gram-positive bacteria (*Bacillus subtilis* and *Staphylococcus aureus*) and two gram-negative bacteria (*Escherichia coli* and *Pseudomonas aeruginosa*), similar to the performance of ampicillin.	[22]
8		In the study, the *Ashwagandhadi lehyam* was assessed for its quality and efficacy through phytochemical screening and physicochemical evaluation. The phytochemical screening was done using FTIR analysis, HPTLC fingerprinting, and heavy metal determination using AAS, followed by the determination of alcohol content. The formulation was tested against *Escherichia coli, Staphylococcus aureus*, aerobic bacteria, yeasts, and mold; oral toxicity, was evaluated and antiepileptic activity was measured using the MES method.	HPTLC analysis of *Ashwagandhadi lehyam* revealed the presence of withanolide functional groups in the formulation. In comparison with normal phenytoin, the formulation exhibited fewer epileptic seizures across different phases and demonstrated a stronger anti-inflammatory effect.	[23]
9	Identification of adulterants/ variations in commercially available polyherbal formulations	The study was carried out to comprehensively analyze and compare different samples of *Sahacharadi kwatha* formulation for their organoleptic and physicochemical properties, phytochemical content, microbial load, and sodium benzoate levels using both titrimetric and HPLC methods.	The research uncovered notable differences in organoleptic and physicochemical characteristics, HPTLC profiles, and microbial counts among the samples tested, finding sodium benzoate levels surpassing FDA- and API-approved limits. It underscores quality parameter disparities and preservative levels across various market versions of *Sahacharadi kwatha* while highlighting the absence of regulated standards for the preparation of Ayurvedic medicines on the market.	[24]

HPTLC: high-performance thin-layer chromatography; AAS: atomic absorption spectroscopy (also called atomic absorption spectrometry).

TABLE 4.2
Preclinical Experimental Studies on Herbal Medicines

	Animal/ Experimental Model/s	Study Objective	Brief of the Study and Findings	Ref.
			In vivo experimental studies	
1.	**Wistar strain albino rats**	Antipyretic activity of *Guduchi ghrita* formulations in albino rats against yeast-induced pyrexia	• The pyrexia induced by yeast was evaluated by measuring the rectal temperature at consistent intervals. • The *Guduchi ghrita* samples, along with the vehicle, notably reduced the increase in temperature three hours post yeast injection. At the 6- and 9-hour marks, these samples showed a highly significant reduction in temperature rise compared to both the yeast control and vehicle control groups.	[35]
2.		Hepatoprotective activity of *Punarnavashtak kwatha* (PNK) against CCl (4)-induced hepatotoxicity in rats	• The serum liver marker enzymes, such as aspartate transaminase, serum alanine transaminase, serum alkaline phosphatase, and serum bilirubin, were significantly decreased after the administration of PNK, and an increase in protein level was also observed. • Thiopentone-induced sleeping time was also decreased in the PNK-treated animals compared with the CCl (4)-treated group. • A notable antioxidant effect was observed, marked by heightened activities of glutathione, superoxide dismutase, and catalase, along with a reduction in thiobarbituric acid reactive substance levels when compared with the CCl (4)-treated group.	[36]
3.		Wound healing activity of *Ashwagandha* (*Withania somnifera* (L.) Dunal) leaves in the form of paste and gel in excision and incision wound models on Wistar strain albino rats	• The effectiveness of wound healing was evaluated by measuring the percentage of wound closure and the duration of epithelialization in the excision wound model. Additionally, the tensile strength of the healed wounds was determined using the incision wound model. • Both forms of *Ashwagandha* demonstrated a notable enhancement in wound contraction percentage, a decrease in epithelialization duration, and an increase in skin tensile strength compared to the normal control group.	[28]
4.	**Swiss male mice**	To assess the effects of *Chyawanprash* (Chy) at the doses of 1% and 2% w/w of the diet for 15 successive days in mice with memory deficits	• The formulation was studied using exteroceptive memory models (Morris water maze, Hebb-Williams maze, and elevated plus maze) and interoceptive memory models (scopolamine amnesia and alprazolam amnesia). Brain acetylcholinesterase activity, brain thiobarbituric acid reactive substances (TBARS), and reduced glutathione levels (GSH) were also estimated. • Administration of the formulation for 15 consecutive days significantly protected the animals from developing memory impairment. • There was a significant decrease in brain TBARS and an increase in GSH levels after administration of the formulation, indicating decreased free radical generation and increased scavenging of free radicals, respectively.	[30]
5.		To explore the curative effects of *Saraswatharishtam* (SWRT) on scopolamine-induced memory impairment in adult Wistar Swiss albino mice and to compare the effects with piracetam	The effect of scopolamine on neurobehavioral dysfunction was assessed by observing the duration of transfer latency, working memory errors, and reference memory errors using the elevated plus maze, rectangular maze, and radial arm maze. • The elevated plus maze, radial maze, and rectangular maze experiments revealed that SWRT significantly reversed the memory loss in a mouse model. • SWRT could scavenge ROS through its antioxidant properties, thereby regulating the body's natural antioxidant defense mechanisms. • Piracetam significantly decreases the latency period compared to animals fed normal saline, indicating heightened cortical excitability in brain function, which results in pathological fatigue. On the other hand, SWRT restores normal function after scopolamine-induced temporary impairment.	[37]

(Continued)

TABLE 4.2 *(Continued)*
Preclinical Experimental Studies on Herbal Medicines

	Animal/ Experimental Model/s	Study Objective	Brief of the Study and Findings	Ref.
6.		To assess the effect of *Amalaki rasayana* (AR) and *Rasa sindoora* (RS) on the dendritic processes of neurons, glial activation, and the formation of corpora amylacea, along with the cognitive and behavioral traits of mice	• Animals that were fed AR or RS showed marked enhancement in episodic, working, and reference-spatiotemporal memory. • Microscopic analysis showed a notable rise in dendritic spine density within the apical dendrites of hippocampal pyramidal cells neurons.	[38]
7.	**Rabbit**	Effect of *Ashwagandha* on lipid peroxidation in stress-induced animals	• The aqueous suspension of root extract was tested for its impact on lipid peroxidation (LPO) caused by LPS from gram-negative bacteria and PGN from gram-positive bacteria in two species: rabbits and mice. • Research indicated that the oral administration of *Ashwagandha* root extract (100 mg/kg) inhibited the increase in LPO in rabbits and mice.	[39]
8.		To evaluate the anti-inflammatory effect of *Triphala* (TA) aqueous extract on experimental uveitis in the rabbit	• The control group rabbits exhibited a significantly higher degree of ocular inflammation than those treated with Prednisolone and *Triphala* groups. • Total proteins, as estimated 24 hours after intravitreal injection of endotoxin, showed significantly lower values in the *Triphala*-treated group as a compared to the control.	[32]
9.		To assess the efficacy of *Trikatu* on the bioavailability and pharmacokinetics of rifampicin in rabbits	• In a crossover study, the animals were administered a single dose of rifampicin (24 mg/kg, p.o.) alone or in combination with a single dose of *Trikatu* (500 mg/kg, p.o.), followed by a washout period of 7 days. • Co-administration of a single dose of *Trikatu* significantly decreased rifampicin's plasma levels during the absorption phase and its C_{max}. In contrast, multiple doses of *Trikatu* lowered C_{max} but delayed T_{max} for rifampicin. This indicates that while *Trikatu* does not alter the extent of rifampicin's bioavailability, it does reduce its rate of bioavailability.	[40]
10.	**Equine Model (horse)**	The clinical efficacy of *Ashwagandha* root extract as an adaptogen was studied against various types of stress in horses	• The animals were subjected to different types of stress, including exercise-induced stress, separation, and noise stress, on three different days and were evaluated for various hematological, biochemical, hormonal, and immunological parameters for 21 days. • The total erythrocyte count, total leucocyte count, hemoglobin content, lymphocyte percentage, reduced glutathione, and superoxide dismutase activities were significantly increased, along with a decrease in cortisol, epinephrine, glucose, triglycerides, creatinine, IL-6, alanine aminotransferase, and aspartate aminotransferase, in the *Ashwagandha*-treated groups when compared to the control group.	[41]
11.	***Drosophila melanogaster***	*In vivo* effects of AR and RS on longevity, development, fecundity, stress tolerance, and heterogeneous nuclear ribonucleoprotein (hnRNP) levels	• Research indicated that the size of salivary glands, hnRNP levels in larval tissues, and thermotolerance in larvae and adult flies notably increased with both formulations. However, improvements in median lifespan and starvation resistance were observed only with AR. • Providing AR to both larvae and adults enhanced fecundity, whereas RS decreased it. When larvae were fed regular food and adults received AR supplements, there was no impact on fecundity; however, a similar feeding strategy with RS supplements resulted in improved fecundity.	[42]
12.		The efficacy of *Guduchi* and *Madhuyashti* in providing tolerance to different stresses and underlying mechanisms was studied	• Drosophila larvae and flies given *Guduchi* or *Madhuyashti* displayed improved thermo-protection, which was associated with elevated heat shock gene expression, even in the absence of heat stress. • The feeding of *Guduchi* or *Madhuyashti* also enhanced the expression of antimicrobial peptides, offering improved tolerance to pathogen assaults.	[34]
13.		To evaluate the efficacy of *Dhanwantaram kashaya* (DK) and *Saraswatarishta* (SA) in different neurodegenerative conditions in Alzheimer's (AD) and Huntington's disease (HD) models of *Drosophila melanogaster*	• A 0.25% concentration of both formulations enhances longevity in wild-type flies. • The larval development and lifespan of adults in Eye-GAL4>Aβ42 (AD) and GMR-GAL4>127Q (HD) larvae and flies raised on 0.25% & 0.50% DK and 0.25% & 1.00% SA showed significant improvement. • The decrease in Aβ plaques and polyQ aggregates points to disease suppression, highlighting their potential in treating AD and HD.	[43]

(Continued)

TABLE 4.2 *(Continued)*
Preclinical Experimental Studies on Herbal Medicines

	Animal/ Experimental Model/s	Study Objective	Brief of the Study and Findings	Ref.
14.		To study the effects of five Ayurveda herbs—Zandopa (containing *Mucuna pruriens* [L.] DC., *Withania somnifera* [L.] Dunal, *Centella asiatica* [L.] Urb., *Sida cordifolia* L., and *Bacopa monnieri* [L.] Wettst.)—in a Parkinson's disease (PD) model of fruit flies	• The potential of the herbs to enhance climbing ability was examined in the PD model of flies, focusing on the loss of function of phosphatase and tensin-induced putative kinase 1 (PINK1). • Flies treated with *B. monnieri* showed a notable increase in climbing ability compared to untreated PINK1 mutant flies. In contrast, wild-type flies exhibited a considerable decline in climbing ability when treated with the same herb, as well as with *Centella asiatica*.	[33]
15.	**Guinea pig**	Antitussive Activity of the water-extracted carbohydrate polymer (CP) from *Terminalia chebula* Retz	• Research revealed a significant reduction in cough efforts induced by citric acid following the oral administration of a polysaccharide fraction at a dose of 50 mg/kg body weight. The overall antitussive activity of CP was quantitatively greater than that of codeine, indicating that it is the most potent antitussive agent in clinical use.	[44]
16.		Efficacy of *Haridra khanda* (HK) and *Manjisthadi kwatha* (MK) in the management of allergic rhinitis in histamine-challenged guinea pigs	• Both formulations significantly reduced eosinophil counts and sneezing rates in guinea pigs. • In the membrane stability experiment, both HK and MK lowered RBC lysis rates.	[45]
17.	**Syrian Hamster model**	Evaluation of Ayush-64 (a polyherbal formulation) and its ingredients for SARS-CoV-2 in a Syrian hamster infection model	• The formulation was found to lack significant antiviral activity; yet it demonstrated considerable potential to improve pulmonary conditions and enhance the pro-inflammatory cytokine response in the hamster model. Assessment of its four main components revealed that *A. scholaris* exhibited strong antiviral and immunomodulatory effects, significantly influencing T cell differentiation *in vitro*.	[46]
18.		To assess the prophylactic effect of two intranasal formulations, *Anu taila* and *Tila taila*, in the hamster model of SARS-CoV-2 infection	• It was found that the prophylactic intranasal instillation of oil formulations exhibited reduced viral load in the lungs and resulted in reduced body weight loss and lung pneumonitis. • Histopathological examination showed less lung pathology in the *Anu taila* group compared to the control infected group, along with diminished expression of pro-inflammatory cytokine genes, including Th1 and Th17 cytokines, in both intranasal formulations.	[47]
19.	**Zebrafish**	To evaluate the anti-nociceptive potential of a standardized methanolic extract of *Bacopa monnieri* (BME) in an experimental adult zebrafish (*Danio rerio*) model of pain	• BME exhibits an antioxidant effect through radical quenching activity in *in vitro* studies. • The nociceptive effect was significantly reduced by BME at 1.25 mg/ml, whereas BME at 2.5 mg/ml exhibited significant antinociceptive activity against glutamate, formalin, capsaicin, and cinnamaldehyde when compared to control and sham-group animals.	[48]
20.		Phytochemical evaluation, embryotoxicity, and teratogenic effects of *Curcuma longa* extract	• The toxicity effects were discovered to be dose-dependent at dosages above 62.50 µg/mL, while at 125.0 µg/mL, mortality of embryos was observed, and physical body deformities of larvae were recorded among the hatched embryos at higher concentrations. • At elevated concentrations, the extract exhibited a severe teratogenic effect, leading to physical deformities, including kinked tails, bent trunks, and enlarged yolk sac edema. • The calculated therapeutic index (TI) values were roughly equivalent across different concentration points tested.	[49]
			***In vitro* experimental studies**	
1.	*In vitro* enucleated rat lenses	Evaluation of the anticataract potential of *Triphala* in a selenite-induced experimental model of cataract	• *Triphala* was found to significantly restore glutathione levels and reduce malondialdehyde levels, along with restoring the activities of antioxidant enzymes compared to the controls. • *In vivo*, TF at 25 mg/kg led to only 20% of subjects developing nuclear cataracts, in contrast to 100% in the control group.	[50]

(Continued)

TABLE 4.2 *(Continued)*
Preclinical Experimental Studies on Herbal Medicines

	Animal/ Experimental Model/s	Study Objective	Brief of the Study and Findings	Ref.
2.	MDAMB-231 cells (breast cancer cell line), HeLa (cervical cancer cell line), DLD1, HT29, HCT-116 (colon cancer cell lines), MG-63 (osteosarcoma cell line), and HEK-293 (human embryonic kidney)	To investigate the anticancer potential of the aqueous extract of *Haritaki churna* (HC) on several cancer cell lines	• The aqueous extract of HC exhibits a dose-dependent loss of cellular viability and alterations in cellular morphology against MDAMB-231, HeLa, and MG-63 cells, with an IC50 of 53.1 ± 4.96 µg/ml, 79.35 ± 4.95 µg/ml, and 97.04 ± 4.09 µg/ml, respectively. • Additionally, the phytochemicals in the mixture show that ellagic acid plays a major role in the anticancer properties of the aqueous extract of HC.	[51]
3.	Acetylcholinesterase inhibitory capacity, α-amylase inhibition assay, α-glucosidase inhibition assay	*In vitro* enzyme inhibitory and cytotoxic studies with *Evolvulus alsinoides* (Linn.) Linn. leaf extract for the management of Alzheimer's disease and diabetes mellitus	• The methanolic extract inhibits α-amylase with an IC50 of 1.33 ± 0.05 µg/mL. The water extract inhibits α-glucosidase and acetylcholinesterase with IC50 values of 3.58 ± 0.02 µg/mL and 4.46 ± 0.03 µg/mL, respectively. • Cytotoxicity studies with the SH-SY5Y cell line substantiate the inhibition potential of the water extract with an IC50 of 103.0035 µg/mL.	[52]
		***In silico* studies on single herb/polyherbal formulations**		
1.	Beta-site amyloid precursor protein cleaving enzyme 1 (BACE1)	To analyze the pharmacological activity of bioactive compounds in *Medhya rasayana* plants against BACE1, employing a structure-based docking approach	• Out of 876 compounds, 11 were identified as potential hits based on docking scores, binding energies, and interactions with the essential residues of BACE1. • Among the 11 compounds evaluated, two stood out: "Convolidine" from the plant *Convolvulus pleuricaulis* Choisy and "N-(4-hydroxybutyl) phthalimide" from *Glycyrrhiza glabra*. Both compounds met the pharmacological criteria outlined by Lipinski's rule of five and ADMET property predictions.	[53]
2.	ACE2, TMRSS2	*In silico* study of phytoconstituents from *Piper longum* L. and *Ocimum sanctum* on ACE2 and TRMPSS2 proteins	• Among the screened phytochemicals, vicenin 2, rosmarinic acid, and orientin demonstrated the strongest efficacy in molecular interaction and drug-likeness properties related to the ACE2 and TMPRSS2 host receptor proteins.	[54]
3.	Network pharmacology approach	To explore the metabolites of *Trigonella foenum-graecum* L. using LC-MS/MS-based metabolite screening, followed by ADME screening and network pharmacology exploration of the mechanism of action in hyperlipidemia and hypolipidemia, with a neighborhood-based combination synergy approach	• The pathway analysis showed enrichment for different pathways such as the MAPK pathway, JAK-STAT pathway, adipocytokine pathway, type 2 diabetes mellitus, insulin signaling pathway, and mTOR signaling pathway, all of which are connected to hyperlipidemia and hyperglycemia. • The synergy network revealed 23 targets interacting with 13 compounds, based on a network neighborhood approach.	[55]

ADME: absorption, distribution, metabolism, and excretion; ADMET: absorption, distribution, metabolism, excretion, and toxicity; DPPH: 2,2-diphenyl-1-picrylhydrazyl; HPTLC: high-performance thin-layer chromatography; LPS: lipopolysaccharide; MAPK: mitogen-activated protein kinase; MES: maximal electroshock seizure (test/model); PGN: peptidoglycan; RBC: red blood cell(s); ROS: reactive oxygen species medicines involves aqueous extraction, although other forms such as oils, ghee, and fermented preparations are also utilized. Despite the diversity in these formulations, water-based extracts remain the foundational element across all types [56]. These various formulations, each suited to specific therapeutic needs, are instrumental in the management of diseases within Ayurvedic practice. The array of formulations employed in Ayurveda is detailed in Figure 4.3. Over the years, advancements in pharmaceutical science have significantly influenced Ayurveda pharmaceutics. Innovations such as reduced bottled *kashayam*, *kashayam* tablets, *kashaya* sachets, antitussive granules, and oil-containing soft gelatin capsules have emerged, replacing traditional *ghrita* formulations or oil preparations for internal administration. The government is committed to promoting Ayurvedic medicine and enhancing access to it, while also encouraging innovation in the medicinal plants recognized in Ayurvedic literature. The number of patents granted in the past five years reflects a trend of ongoing research and development in this field. The growth of research and development (R&D) in Indian herbal research organizations and the establishment of manufacturing units outside India indicate a strong capability to develop new products and processes [57]. For instance, the Central Council for Research in Ayurvedic Sciences (CCRAS) has been awarded 18 patents, showcasing the continuing advancements in Ayurveda.

Batch-to-batch consistency is a critical aspect of standardization in Ayurvedic pharmaceutics, although achieving it can be challenging due to the complex nature of the formulations. Typically, the quality of each Ayurvedic formulation is assured through a quality by testing (QbT) method, in which each batch is tested post-preparation using quality-control parameters outlined in the Ayurvedic Pharmacopoeia of India (API) and the Ayurvedic Formulary of India (AFI), and the General Guidelines for Drug Development of Ayurvedic Formulations (CCRAS guidelines) [58, 59]. While effective, this approach primarily focuses on end-product testing. In contrast, the quality by design (QbD) approach offers a more proactive strategy. QbD involves designing and optimizing manufacturing processes to ensure quality at every stage of formulation. This method is not only applicable to complex preparations but can also enhance the production of simpler products such as decoctions. By establishing predefined design parameters, the QbD approach uses design of experiments (DoE) techniques to systematically study the impact of each variable on the final product. This ensures that quality is built into the product from the start rather than merely inspected at the end, leading to more consistent and reliable formulations. In preclinical research, the integration of *in silico* and *in vitro* methods continues to expand. Computational modeling and simulation (*in silico*) are becoming essential for predicting pharmacokinetics and dynamics, identifying potential drug targets, and minimizing the risk of failure in later stages. Concurrently, *in vitro* studies, particularly those involving diverse animal models, are crucial for assessing toxicity and pharmacological activity. On the clinical front, the "omics" approaches like genomics, proteomics, metabolomics, and transcriptomics, will be necessary to understand how herbal medicines interact at molecular and cellular levels within the human body [60]. These modalities have proven effective in establishing the fundamental principles of Ayurveda. A genomics study on body constitution yielded promising results, suggesting that the phenotypic classification used in India's TM has a genetic basis. Furthermore, the practice of Prakriti, which has been in use for many centuries, aligns with the concept of personalized medicine [61]. These methods provide insights into the mechanisms through which

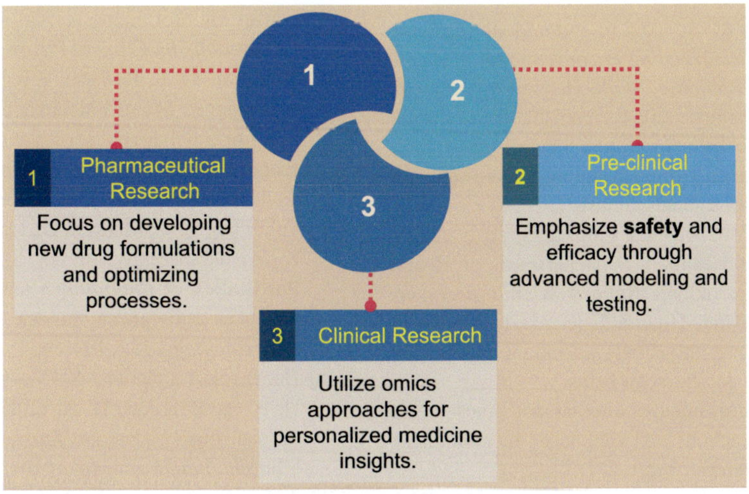

FIGURE 4.3 Future aspects of research in herbal medicine.

these treatments exert their effects, including identifying molecular biomarkers that can predict therapeutic outcomes.

4.5 CONCLUSION

The EBM and RP approach to traditional wisdom have brought a therapeutic revolution through the thoughtful integration of high-throughput screening, combinational chemistry, and understanding the effects of the novel herb-induced compounds through human gene expression and proteomics studies. Over the centuries, a vast number of therapeutic agents have been developed by evaluating new chemical entities originating from herbs or having substantial therapeutic use in TM, using the principles of RP, further providing an evidence base to TM at a global scale. The emergence of newer approaches to integrating clinical and experimental knowledge into leads through transdisciplinary exploratory studies, followed by experimental and clinical research to further develop these leads into therapeutic candidates, has created a paradigm shift in the sector of drug research and discovery. India, with its pluralistic healthcare system, provides a virtual gold mine for such novel approaches and discoveries. The current unmet healthcare needs in communicable and noncommunicable diseases, as well as the emergent new challenges and health crises, can only be met through the discovery of safe, effective therapeutics from the validated, clinically and experientially established herbals and medicines from the diverse traditional systems of medicine using RP approaches.

REFERENCES

[1] World Health Organization. (2019). *WHO Global Report on Traditional and Complementary Medicine 2019*. World Health Organization. [Internet]. Available from: https://iris.who.int/bitstream/handle/10665/312342/9789241515436-eng.pdf?sequence=1

[2] World Health Organization. (2023). *Traditional medicine has a long history of contributing to conventional medicine and continues to hold promise*. World Health Organization. [Internet]. [cited 25 February 2024]. Available from: https://www.who.int/news-room/feature-stories/detail/traditional-medicine-has-a-long-history-of-contributing-to-conventional-medicine-and-continues-to-hold-promise

[3] National Center for Complementary and Integrative Health, National Institutes of Health, U.S. Department of Health and Human Services. (2024). *Natural doesn't necessarily mean safer, or better*. [Internet]. [cited 26 November 2024]. Available from: https://www.nccih.nih.gov/health/know-science/natural-doesnt-mean-better

[4] Ekor, M. (2014). The growing use of herbal medicines: Issues relating to adverse reactions and challenges in monitoring safety. *Frontiers in Pharmacology*, 4, 177.

[5] Jyothi Raga, P. M., Vivek, P., & Harinarayanan, C. M. (2023). Need for standardization of Ayurveda formulations. *International Journal of Ayurveda and Pharma Research*, 11(10), 40–45.

[6] Dornala, S. N., & Dornala, S. S. (2024). Criteria to select a research topic for postgraduate and doctoral studies in Ayurveda. *Journal of Ayurveda and Integrative Medicine*, 15(4), 100978.

[7] Southey, M. W., & Brunavs, M. (2023). Introduction to small molecule drug discovery and preclinical development. *Frontiers in Drug Discovery*, 3, 1314077.

[8] Sinha, D., Odoh, U. E., Ganguly, S., Muhammad, M., Chatterjee, M., Chikeokwu, I., & Egbuna, C. (2023). Phytochemistry, history, and progress in drug discovery. In *Phytochemistry, Computational Tools and Databases in Drug Discovery* (pp. 1–26). Elsevier.

[9] Vitale, S., Colanero, S., Placidi, M., Di Emidio, G., Tatone, C., Amicarelli, F., & D'Alessandro, A. M. (2022). Phytochemistry and biological activity of medicinal plants in wound healing: An overview of current research. *Molecules*, 27(11), 3566.

[10] Bommakanti, V., Puthenparambil Ajikumar, A., Sivi, C. M., Prakash, G., Mundanat, A. S., Ahmad, F., Haque, S., Prieto, M. A., & Rana, S. S. (2023). An overview of herbal nutraceuticals, their extraction, formulation, therapeutic effects, and potential toxicity. *Separations*, 10(3), 177.

[11] Camilleri, E., & Blundell, R. (2024). A comprehensive review of the phytochemicals, health benefits, pharmacological safety, and medicinal prospects of *Moringa oleifera*. *Heliyon*, 10(6), e27807.

[12] Mitra, S., Shukla, V. J., & Acharya, R. (2011). Effect of *Shodhana* (processing) on *Kupeelu* (*Strychnos nux-vomica* Linn.) with special reference to strychnine and brucine content. *Ayu*, 32(3), 402–407.

[13] Sulaiman, C. T., Anju, K., Anandan, E. M., & Balachandran, I. (2021). Synergistic interactions of phytochemicals in polyherbal formulation enhance the chemical transformations of active constituents. *Journal of Applied Spectroscopy*, 88(1), 181–186.

[14] Dahanayake, J. M., Perera, P. K., Galappatty, P., Perera, H. D. S. M., & Arawwawala, L. D. A. M. (2019). Comparative phytochemical analysis and antioxidant activities of Tamalakyadi decoction with its modified dosage forms. *Evidence-Based Complementary and Alternative Medicine*, 2019, 6037137.

[15] Sulaiman, C. T., Jisha, C. K., Pradeep, K. M., & Balachandran, I. (2022). Phytochemical characterization and evaluation of antiurolithiatic activity of selected source plants of *Pashanabheda*. *Clinical Phytoscience*, 8(1), 12.

[16] Sulaiman, C. T., Ramesh, P. R., Manoj, K., Mathew, K. M., Anandan, E. M., & Balachandran, I. (2020). Chemical profiling of a polyherbal formulation by tandem mass spectroscopic analysis with multiple ionization techniques. *Future Journal of Pharmaceutical Sciences*, 6, 1–8.

[17] Sulaiman, C. T., Mathew, D., Pradeep, K. T., Raveendran, L. K., Anandan, E. M., Manoj, S., & Balachandran, I. (2021). Purification of *Bhallathaka* (*Semecarpus anacardium* L.f.) enhances anti-cancer activity. *Regulatory Toxicology and Pharmacology*, 122, 104898.

[18] Deore, S. L., Moon, K. V., Khadabadi, S. S., Deokate, U. A., & Baviskar, B. A. (2013). Evaluation of toxicity of 'Vatsanabha' (*Aconitum ferox*, Ranunculaceae) before and after *Shodhana*. *Journal of Young Pharmacists*, 5(1), 3–6.

[18] Akter, R., Khan, S. S., Kabir, M. T., & Halder, S. (2022). GC–MS-employed phytochemical characterization, synergistic antioxidant, and cytotoxic potential of Triphala methanol extract at non-equivalent ratios of its constituents. *Saudi Journal of Biological Sciences*, 29(6), 103287.

[19] Venkatasubramanian, P., Kumar, S. K., & Nair, V. S. (2010). *Cyperus rotundus*, a substitute for *Aconitum heterophyllum*: Studies on the Ayurvedic concept of *Abhava Pratinidhi Dravya* (drug substitution). *Journal of Ayurveda and Integrative Medicine*, 1(1), 33–39.

[20] Hamrapurkar, P. D., Jadhav, K., & Zine, S. (2011). Quantitative estimation of piperine in *Piper nigrum* and *Piper longum* using high-performance thin-layer chromatography. *Journal of Applied Pharmaceutical Science*, 1, 117–120.

[21] Parveen, R., Shamsi, T. N., Singh, G., Athar, T., & Fatima, S. (2018). Phytochemical analysis and in vitro biochemical characterization of aqueous and methanolic extract of Triphala, a conventional herbal remedy. *Biotechnology Reports*, 17, 126–136.

[22] Parveen, R., Shamsi, T. N., Singh, G., Athar, T., & Fatima, S. (2018). *[Duplicate of Reference 22—should be removed]*

[23] Gopal, S., & Ittiachen, L. (2024). Analysis of quality parameters and preservative concentrations in *Sahacharadi Kwatha*: A comparative study of three commercial brands. *Journal of Ayurveda and Integrative Medicine*, 15(5), 101029.

[24] Park, K. S., Park, K. I., Hwang, D. S., Lee, J. M., Jang, J. B., & Lee, C. H. (2014). A review of in vitro and in vivo studies on the efficacy of herbal medicines for primary dysmenorrhea. *Evidence-Based Complementary and Alternative Medicine*, 2014, 296860.

[25] Chau, F., Fung, K., Koon, C., et al. (2011). Bioactive components in herbal medicine: Experimental approaches. In I. F. F. Benzie & S. Wachtel-Galor (Eds.), *Herbal Medicine: Biomolecular and Clinical Aspects* (2nd ed., Chapter 20). CRC Press/Taylor & Francis. Available from: https://www.ncbi.nlm.nih.gov/books/NBK92759/

[26] World Health Organization. (2000). *General guidelines for methodologies on research and evaluation of traditional medicine.* [Internet]. Available from: https://iris.who.int/bitstream/handle/10665/66783/W?sequence=1

[27] Bhosale, S. D., Mayya, J. A., Gowda, H. K., Kumar, V., Yadav, P., Galib, R., & Prajapati, P. K. (2023). Wound healing activity of *Withania* leaf paste and its gel in Wistar albino rats: An experimental evaluation. *Wound Practice and Research*, 31(4), 182–189.

[28] Ghatpande, N. S., Misar, A. V., Waghole, R. J., Jadhav, S. H., & Kulkarni, P. P. (2019). *Tinospora cordifolia* protects against inflammation-associated anemia by modulating inflammatory cytokines and hepcidin expression in male Wistar rats. *Scientific Reports*, 9(1), 10969.

[29] Parle, M., & Bansal, N. (2011). Antiamnesic activity of an Ayurvedic formulation, *Chyawanprash*, in mice. *Evidence-Based Complementary and Alternative Medicine*, 2011, 898593.

[30] Deole, Y. S., Chavan, S. S., Ashok, B. K., Ravishankar, B., Thakar, A. B., & Chandola, H. M. (2011). Evaluation of antidepressant and anxiolytic activity of *Rasayana Ghana Tablet* in albino mice. *Ayu*, 32(3), 375–379.

[31] Gupta, S. K., Kalaiselvan, V., Agrawal, S. S., Srivastava, S., & Saxena, R. (2011). Prevention of endotoxin-induced uveitis in rabbits by *Triphala*, an Ayurvedic formulation. *International Journal of Current Biomedical and Pharmaceutical Research*, 1(2), 20–23.

[32] Jansen, R. L., Brogan, B., Whitworth, A. J., & Okello, E. J. (2014). Effects of five Ayurvedic herbs on locomotor behaviour in a *Drosophila melanogaster* Parkinson's disease model. *Phytotherapy Research*, 28(12), 1789–1795.

[33] Singh, S., & Tapadia, M. G. (2019). Molecular basis for efficacy of *Guduchi* and *Madhuyashti* feeding under different environmental stressors in *Drosophila. Cell Stress and Chaperones*, 24(3), 549–565.

[34] Ashok, B. K., Ravishankar, B., Prajapati, P. K., & Bhat, S. D. (2010). Antipyretic activity of *Guduchi Ghrita* formulations in albino rats. *Ayu*, 31(3), 367–370.

[35] Shah, V. N., Shah, M. B., & Bhatt, P. A. (2011). Hepatoprotective activity of *Punarnavashtak Kwath*, an Ayurvedic formulation, against CCl_4-induced hepatotoxicity in rats and on the HepG2 cell line. *Pharmaceutical Biology*, 49(4), 408–415.

[36] Prabhu, J., Prabhu, K., Chaudhuri, A., Rao, M. R. K., Selvi, V. K., Balaji, T. K., & Dinakar, S. (2020). Neuroprotective effect of Ayurvedic formulation *Saraswatharishtam* on scopolamine-induced memory impairment in an animal model. *Pharmacognosy Journal*, 12(3).

[37] Verma, B., Sinha, P., & Ganesh, S. (2022). Ayurvedic formulations *Amalaki Rasayana* and *Rasa Sindoor* improve age-associated memory deficits in mice by modulating dendritic spine densities. *Journal of Ayurveda and Integrative Medicine*, 13(4), 100636.

[38] Dhuley, J. N. (1998). Effect of *Ashwagandha* on lipid peroxidation in stress-induced animals. *Journal of Ethnopharmacology*, 60(2), 173–178.

[39] Karan, R. S., Bhargava, V. K., & Garg, S. K. (1999). Effect of *Trikatu*, an Ayurvedic prescription, on the pharmacokinetic profile of rifampicin in rabbits. *Journal of Ethnopharmacology*, 64(3), 259–264.

[40] Priyanka, G., Anil Kumar, B., Lakshman, M., Manvitha, V., & Kala Kumar, B. (2020). Adaptogenic and immunomodulatory activity of *Ashwagandha* root extract: An experimental study in an equine model. *Frontiers in Veterinary Science*, 7, 541112.

[41] Dwivedi, V., Anandan, E. M., Mony, R. S., Muraleedharan, T. S., Valiathan, M. S., Mutsuddi, M., & Lakhotia, S. C. (2012). In vivo effects of traditional Ayurvedic formulations in a *Drosophila melanogaster* model relate to therapeutic applications. *PLoS One*, 7(5), e37113.

[42] Prabhu, J., Prabhu, K., Chaudhuri, A., Rao, M. R. K., Selvi, V. K., Balaji, T. K., & Dinakar, S. (2020). *[Duplicate of Reference 37—should be removed]*

[43] Nosalova, G., Jurecek, L., Chatterjee, U. R., Majee, S. K., Nosal, S., & Ray, B. (2013). Antitussive activity of the water-extracted carbohydrate polymer from *Terminalia chebula* on citric acid-induced cough. *Evidence-Based Complementary and Alternative Medicine*, 2013, 650134.

[44] Bhowmik, R., Shaharyar, M. A., Kanakal, M. M., Sarkar, A., Farhana, S. A., Hussain, S. M., Khan, A., Mandal, P., Roshan, S., Mitra, A., & Karmakar, S. (2024). Ayurvedic herbal formulations *Haridra Khanda* and *Manjisthadi Kwath (Brihat)* in the management of allergic rhinitis: A pharmacological study. *Heliyon*, 10(11), e31937.

[45] Rizvi, Z. A., Madan, U., Tripathy, M. R., Goswami, S., Mani, S., Awasthi, A., & Dikshit, M. (2023). Evaluation of *Ayush-64* (a polyherbal formulation) and its ingredients in the Syrian hamster model for SARS-CoV-2 infection reveals the preventative potential of *Alstonia scholaris. Pharmaceuticals*, 16(9), 1333.

[46] Rizvi, Z. A., Tripathy, M. R., Sharma, N., Goswami, S., Srikanth, N., Sastry, J. L. N., Mani, S., Surjit, M., Awasthi, A., & Dikshit, M. (2021). Effect of prophylactic use of intranasal oil formulations in the hamster model of COVID-19. *Frontiers in Pharmacology*, 12, 746729.

[47] Sharma, M., Gupta, P. K., Gupta, P., & Garabadu, D. (2022). Antinociceptive activity of standardized extract of *Bacopa monnieri* in different pain models of zebrafish. *Journal of Ethnopharmacology*, 282, 114546.

[48] Alafiatayo, A. A., Lai, K. S., Syahida, A., Mahmood, M., & Shaharuddin, N. A. (2019). Phytochemical evaluation, embryotoxicity, and teratogenic effects of *Curcuma longa* extract on zebrafish (*Danio rerio*). *Evidence-Based Complementary and Alternative Medicine*, 2019, 3807207.

[49] Gupta, S. K., Kalaiselvan, V., Srivastava, S., Agrawal, S. S., & Saxena, R. (2010). Evaluation of anticataract potential of *Triphala* in selenite-induced cataract: In vitro and in vivo studies. *Journal of Ayurveda and Integrative Medicine*, 1(4), 280–286.

[50] Khan, M. R. U. Z., Yanase, E., & Trivedi, V. (2023). Extraction, phytochemical characterization, and anticancer mechanism of *Haritaki churna*: An Ayurvedic formulation. *PLoS One*, 18(5), e0286274.

[51] Sundaramoorthy, P. M. K., & Packiam, K. (2020). In vitro enzyme inhibitory and cytotoxic studies of *Evolvulus alsinoides* leaf extract: A plant recognized as *Dasapushpam* in Ayurveda. *BMC Complementary Medicine and Therapies*, 20, 1–12.

[52] Nadh, A. G., Revikumar, A., Sudhakaran, P. R., & Nair, A. S. (2022). Identification of potential lead compounds against BACE1 through in silico screening of phytochemicals of *Medhya Rasayana* plants for Alzheimer's disease management. *Computers in Biology and Medicine*, 145, 105422.

[53] Jindal, D., & Rani, V. (2023). In silico studies of phytoconstituents from *Piper longum* and *Ocimum sanctum* as ACE2 and TMPRSS2 inhibitors: Strategies to combat COVID-19. *Applied Biochemistry and Biotechnology*, 195(4), 2618–2635.

[54] Banerjee, S., Bhattacharjee, P., Kar, A., & Mukherjee, P. K. (2019). LC–MS/MS analysis and network pharmacology of *Trigonella foenum-graecum* against hyperlipidemia and hyperglycemia with combination synergy. *Phytomedicine*, 60, 152944.

[55] Savrikar, S. S., & Ravishankar, B. (2010). *Bhaishajya Kalpana*: The Ayurvedic pharmaceutics—an overview. *African Journal of Traditional, Complementary and Alternative Medicines*, 7(3), 174–184.

[56] Sahoo, N., Manchikanti, P., & Dey, S. H. (2011). Herbal drug patenting in India: Intellectual property potential. *Journal of Ethnopharmacology*, 137(1), 289–297.

[57] Joshi, V. K., Joshi, A., & Dhiman, K. S. (2017). *The Ayurvedic Pharmacopoeia of India*.

[58] Central Council for Research in Ayurvedic Sciences. (2018). *General guidelines for drug development of Ayurvedic formulations*. Ministry of AYUSH, Government of India. [Internet]. Available from: https://ccras.nic.in/wp-content/uploads/2024/07/CCRAS_Guideline-of-Drug-Development.pdf

[59] Prasher, B., Gibson, G., & Mukerji, M. (2016). Genomic insights into Ayurvedic and Western approaches to personalized medicine. *Journal of Genetics*, 95(1), 209–228.

[60] Govindaraj, P., Nizamuddin, S., Sharath, A., Jyothi, V., Rotti, H., Raval, R., Nayak, J., Bhat, B. K., Prasanna, B. V., Shintre, P., Sule, M., Joshi, K. S., Dedge, A. P., Bharadwaj, R., Gangadharan, G. G., Nair, S., Gopinath, P. M., Patwardhan, B., Kondaiah, P., Satyamoorthy, K., & Thangaraj, K. (2015). Genome-wide analysis correlates with Ayurveda *Prakriti*. *Scientific Reports*, 5, 15786.

5 Stakeholders and Their Roles in the Clinical Trials of Herbal Medicines

Navigating Collaboration and Compliance

Ramneek Khaira[1], *Paranjeet Kaur*[1], *and Roqia Bashery*[2]
[1]Chitkara College of Pharmacy, Chitkara University, Punjab, India
[2]Faculty of Pharmacy, Kabul University, Kabul, Afghanistan

5.1 INTRODUCTION TO HERBAL MEDICINES

Herbal medicines are organic compounds produced from plants that are used in traditional local or regional medicine to treat various ailments. The organic chemicals used to make these goods might originate from any portion of the plant, whether it is raw or processed. Every culture on the planet has used herbal remedies. There are numerous distinct traditional medical systems, and each one's philosophy and methods are shaped by local social and environmental factors as well as geographic location. All of these systems do, however, concur on a holistic outlook on life. Prominent herbal medicine systems such as traditional Chinese medicine and Ayurveda uphold the fundamental principle that prioritizing health over illness is important. Herbal medicine, sometimes referred to as herbalism or botanical medicine, is a branch of medicine that relies on the topical or dietary application of plants or plant extracts [1].

Because traditional herbal therapy comes from a natural source and has fewer negative effects, it has been used extensively for thousands of years in both developed and developing nations. Originally, these treatments were made up of unrefined herbal formulations, including tinctures, teas, poultices, powders, and other mixtures. Much of modern medicine has its roots in the ancient practice of using plants for medicinal purposes, which predates human history. The majority of early drugs, including aspirin (willow bark), digitoxin (from foxglove), morphine (from opium poppy), and quinine (from cinchona bark), were developed based on clinical, pharmacological, and chemical examinations of these traditional remedies, which were primarily derived from plants. For primary healthcare, between 75% and 80% of the world's population still relies on herbal medicine, particularly in underdeveloped nations [2].

Over 80% of people on the planet today rely on conventional medicines made from plants. About 121 medicinal products have been developed in the past century using traditional knowledge gathered from a variety of sources [2]. Due to the widespread and increasing use of naturally derived substances worldwide, herbal remedies require a thorough evaluation of their pharmacological qualities and safety concerns. This evaluation cannot be based solely on custom or pretended millennial beliefs; explanatory and pragmatic studies are helpful and complementary in the acquisition of trustworthy data for patients and healthcare providers alike [3].

5.1.1 Safety Considerations

There are many safety issues related to herbal use worldwide such as herbal medication registration; herbal medicine regulation; herbal-related problems such as adverse effects, interactions, nephrotoxicity, hepatotoxicity; and counterfeit and adulterated herbal medicines.

In addition to the notable rise in use across the globe, the safety of herbal medicine has been brought to light. There are currently misconceptions and biases regarding the efficacy of herbal treatment. Publicity, impartial interpretation, and objective understanding are therefore required [4]. Certain medicinal plants include elements that make them inherently hazardous, and when taken improperly, they can have negative effects. Adverse occurrences related to their use may also result from other reasons such as herb–drug interactions, noncompliance with good manufacturing practice (GMP), inadequate regulatory measures, and adulterations. While most *in vitro* investigations on single isolated cells, primarily using extracts other than aqueous ones, produce contradictory results and hence prolong the debate on herbal medication safety, many *in vivo* tests on aqueous extracts largely support the safety of herbal medicines. In order to facilitate reasoned conversations about the safety of herbal medicines for their intended purposes, it is planned that toxicity studies will mirror the traditional applications of these products. While various attempts continue to establish the safety of herbal medicines in man, their cautious and responsible use is required [5]. To ensure the safety, effectiveness, and reproducibility of herbal products, it is essential to maintain their uniform quality and standardization. Strong quality control procedures are necessary to safeguard customers' health and safety. Strict testing and quality assurance procedures can

DOI: 10.1201/9781003522065-5

reduce hazards like adulteration, contamination, and dangerous compounds, guaranteeing the safety of herbal medicine products for consumption. Inadequate quality control can put consumers' safety at serious risk. Herbal medicine items that include contaminants, adulterants, or improper formulations run the risk of causing toxicity, unpleasant reactions, or other health issues. The levels of active substances may be tracked and managed using standardized manufacturing procedures and testing techniques, guaranteeing that the products consistently provide the intended therapeutic benefits. This commitment helps build trust among consumers, healthcare professionals, and regulatory authorities [6].

5.1.2 Clinical Trials in Herbal Medicine Research

The WHO provides guidelines to support clinical research projects to assess the safety and effectiveness of traditional medicines. The general principle for the clinical trial studies of herbal medicine is the same as that of synthetic medicine. Still, these studies have primary objectives such as validation of safety and efficacy claims, development of new herbal medicines, finding new indications for already existing herbal medicines, or changing the route of administration or dose. Another goal is to test the clinical activity of some purified or semi-purified chemicals from herbal medicines. All regulatory requirements needed to support clinical trials of herbal medicine for their safety and effectiveness are issued in operational guidelines by the WHO [7].

The general guidelines for studying and assessing methodologies for traditional medicines were released in 2002. They include information on how to develop protocols, design studies, include populations in studies, obtain informed consent, define endpoints, use statistics, and document clinical trials involving traditional medicines. These guidelines also address formulation issues, standardizing dosages and quality control for herbal products, and identifying acceptable placebos. 2013 saw the release of the WHO traditional medicine strategy "2014–2023," which aims to assist member nations in maximizing their contributions to advancing complementary and alternative medicine for practical use in healthcare. One of the main goals of implementing these strategic objectives was to use legislation to improve the safety, efficacy, and quality of complementary and alternative medicine [7].

5.2 KEY STAKEHOLDERS AND THEIR ROLES

5.2.1 Regulators and Government Bodies

Government regulatory bodies play a vital role in herbal medicine. They regulate and provide licenses, certifications, and funding. The WHO established international regulatory cooperation for herbal medicine [IRCH]in 2006 [8]. It is governed by regulatory authorities responsible for the development and promotion of herbal medicine. In the United States, the Food and Drug Administration (FDA) has control over dietary supplements formulated with herbal ingredients. In Canada, the Natural and Non-prescription Health Directorate labels an eight-figure Natural Product Number or a homeopathic medicine label on licensed herbal medicines. In Europe, the role of European Medicines Agency (EMA) lies in monitoring and advancing research in herbal medicine across Europe [9]. In 2005, approximately USD 33 million was spent on herbal medicine by the National Center for Complementary and Integrative Health (NCCIH) in the United States [10].

5.2.2 Manufacturers and Industry Partners

The manufacturing industry has a key role in establishing the quality of products and the safety of humans during the production. They follow various guidelines given by the WHO: GMP for herbal medicine and Guidelines on Good Agriculture and Collection Practices (GACP) for medicinal plants [11]. Post-harvesting activities are covered by GACP, and manufacturing processes are covered by GMP. Different GMPs are established: WHO-GMP, cGMP in the United States, and EU-GMP. There are various factors that are looked upon during manufacturing such as quality assurance, adulteration, sanitation and hygiene, quality risk misidentification and mislabeling, and interference of pesticides and fertilizers in formulation. The storage areas should be organized and tidy, with prevention of microbial contamination, and raw materials should be stored between 2 °C and 8 °C, while frozen materials should be stored below −18 °C. Aeration and ventilation should be adequate, and protection against animals, insects, and rodents should be ensured. The principle of first in, first out (FIFO) should be followed. In production areas designed under GMP guidelines, proper air exhaust systems should be set up for the heating and boiling of materials in order to avoid fumes and vapors. Proper sampling, weighing, quality assurance, processing, etc., fall under GMP guidelines [12]. The top 10 pharmaceutical companies spend more than USD 0.54 billion for the study of traditional medicine and anti-AIDS drugs derived from 300 plant species, according to the American AIDS Prevention Center, in the year 2003. India exports non-timber forest products, processed or unprocessed, with 70% of export revenue from forests. Researchers and clinical scientists play role in the development of drugs that are beneficial to the society (Table 5.1) (Figure 5.1) [13].

5.2.3 Researchers and Scientists

Researchers should aim at the feasibility and implementation of drugs with the help of scientific training and well-generalized knowledge. Drugs should be beneficial and convenient to society and improve the health index of the country. The development of traditional herbal medicine (THM) has played a significant role. In China, THM has helped to cure severe acute respiratory syndrome (SARS). African flower has been predominantly used for the treatment of HIV; government officials ran randomized controlled trials (RCT), and it progressed down AIDS. During

TABLE 5.1
Role of Manufacturers and Industry Partners in Herbal Medicine

Role	Description
Manufacturing	Factories produce herbal products from raw materials, focusing on maintaining the natural properties of herbs.
Quality control	Laboratories conduct tests to ensure products meet safety, purity, and potency standards.
Quality assurance	Industry partners and manufacturers follow certification standards to maintain consistent product quality.
Distribution	Transport vehicles and logistics networks distribute herbal products to markets and retail outlets.
Industry partnerships	Collaboration between various stakeholders ensures the development and marketing of high-quality herbal medicines.
Certifications	Certifications validate the herbal products' adherence to safety and quality regulations.
Supply chain management	Overseeing the flow of products from raw material sourcing to finished product distribution.

TABLE 5.2
Role of Researchers and Scientists in Herbal Medicine

Role	Description
Plant identification	Identifying medicinal plants and their active compounds for therapeutic uses.
Phytochemical analysis	Analyzing the chemical composition of plants to understand their bioactivity.
Clinical trials	Testing the efficacy and safety of herbal medicines through controlled studies.
Traditional knowledge integration	Collaborating with indigenous communities to integrate traditional herbal knowledge.
Quality control	Ensuring that herbal products meet safety, purity, and potency standards.
Toxicity and safety evaluation	Assessing the potential risks or side effects of herbal compounds.
Formulation and standardization	Developing standardized herbal formulations for consistency in treatment.
Regulatory compliance	Ensuring adherence to local and international regulations for herbal medicines.
Education and training	Teaching and mentoring future herbal medicine practitioners and researchers.
Sustainability and conservation	Promoting the sustainable use and conservation of medicinal plants.

FIGURE 5.1 Role of industry in herbal medicine.

2005, the National Center for Complementary and Alternative Medicine at the National Institutes of Health in the United States spent USD 33 million on herbal medicines. In 2006, Novartis promised to spend over USD 100 million on THM in Shanghai. Research is divided into two validities: internal and external. Internal deals with clinical research, the methods that are used, and issues related to the research (Table 5.2). External focuses on how research can affect a targeted population. For *Echinacea angustifolia* extract for the prevention of parainfluenza virus infection, the researchers ran various clinical trials but could not justify why it should be implemented for mass use [10].

5.2.4 Healthcare and Professionals

Among healthcare professionals, the knowledge of herbal medicine is inadequate. According to the study performed on dietary supplements (DS), herbal medicine (HM), and drug-dietary supplement (DDSIs) and drug-herbal product (DHPIs) interactions, HCPs have better knowledge than general doctors but still show a lack of knowledge in DDSIs and DHIPs. A lack of patient information and medical records on DS and HM (Herbal medicine) was observed. Similar results were present among pharmacists; awareness was there, but the knowledge was low. The requirement for HM courses and workshops on herbal medicines is needed (Table 5.3) (Figure 5.2) [14].

TABLE 5.3
Role of Healthcare Professionals in Herbal Medicine

Role	Description
Prescribing herbal medicines	Healthcare professionals, such as doctors, naturopaths, and herbalists, evaluate patients' conditions and prescribe specific herbal remedies tailored to individual needs.
Consultation and patient education	They provide consultations to educate patients about the benefits, risks, and proper usage of herbal treatments, including possible interactions with conventional medicines.
Diagnosis and assessment	Professionals assess the patient's health, including conducting physical exams and understanding medical history, to determine the appropriate herbal treatments for conditions.
Dosage recommendations	Healthcare providers calculate and recommend the correct dosages for herbal medicines based on factors such as age, weight, health status, and specific conditions.
Monitoring and follow-up	They monitor the patient's progress after starting herbal treatments, adjusting doses or treatments based on response, side effects, or new developments in the patient's condition.
Integrating herbal with conventional care	Healthcare professionals integrate herbal treatments with conventional medicine, ensuring a complementary approach that enhances treatment effectiveness without causing interactions.
Regulation and ethical standards	Professionals adhere to ethical guidelines, including ensuring that herbal products prescribed are from regulated, safe sources and meet medical standards.
Research and clinical trials	Some healthcare professionals are involved in clinical trials and research to expand the scientific understanding of herbal medicine, validating its effectiveness for medical use.
Collaboration with herbalists and pharmacies	They work closely with specialized herbalists and pharmacists to ensure patients receive the highest quality and properly compounded herbal medicines.
Professional development	Continuous learning and staying updated with the latest research in herbal medicine to improve patient outcomes and ensure safety.

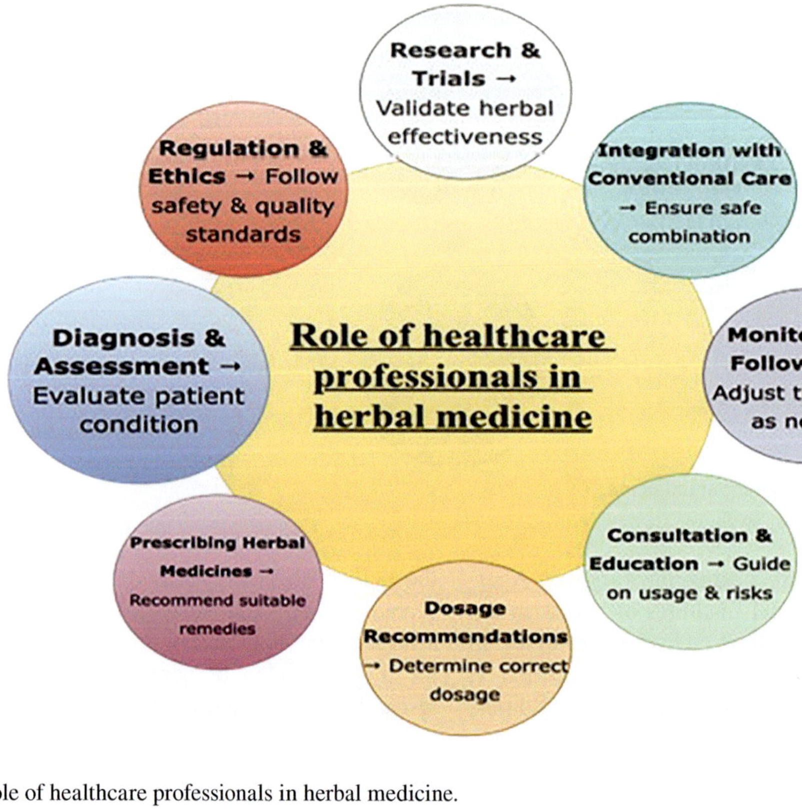

FIGURE 5.2 Role of healthcare professionals in herbal medicine.

5.2.5 Collaboration and Compliance

5.2.5.1 Navigating Ethical and Legal Aspects

Ethical aspects of herbal medicine should follow the four principles stated in the principles of biomedicine. The four principles are beneficence (the norms stating relieving, lessening, or preventing harm to give benefits and balancing benefits against risks and costs), respect for autonomy (to respect norms and to support various autonomous decisions), nonmaleficence (to avoid any cause of harm), justice (the norms to provide justice for the equal distribution of benefits, risks, and costs). The rules are categorized as substantial rules, authority rules, and procedural rules. To regulate ethical and legal aspects along with other aspects, principles should be followed: care (the formal practitioner should have adequate knowledge in order to serve the people who are consuming medicine and the sources from which it has been manufactured and distributed), respect (to respect different social and societal changes, whether they follow different spiritual methods to deal; respect for different communities), honesty (the gaps in knowledge should be stated and revealed; the efficacy of the drug, the duration of treatment, etc.; all small and big details must be shared with patients), and fairness (to deliver the same treatment despite different economic and social backgrounds) [15].

In the United States, the government funded research activities in 2015, allocating a total of $124 million to the NCCIH. The total United States funding for health research and development activities increased to $158,716 million. In India, in 2014, the Ministry of AYUSH (Ayurveda, Yoga and Naturopathy, Unani, Siddha, and Homeopathy) was established. It deals with the improvement of development and research in traditional medicines in India [16]. In the annual report of AYUSH, comprehensive development of conservation, cultivation, collection, processing, marketing, research, and extension support systems for medicinal plants is described. AYURGYAN has been approved for the period from FY 2021–2022 to FY 2025–2026 for proper education and research in herbal medicine. AYURSWASTHYA has been established for financial requirements. The State Council of the People's Republic of China published its first white paper in 2016, which states the laws and regulations in herbal medicine and the enhancement in research and development in order to bring sustainable development (Table 5.4) [17].

5.2.5.2 Ensuring Transparency and Data Sharing

The introduction of open science in traditional, complementary, and integrated medicine (TCIM) can lead to

TABLE 5.4
Navigating Ethical and Legal Aspects in Herbal Medicine

Aspect	Description	Example
Identifying herbal sources	Ensuring herbs are ethically sourced, respecting biodiversity, local customs, and sustainable practices.	Example: Sourcing echinacea from certified organic farms that follow sustainable harvesting practices to avoid overexploitation.
Quality control and certification	Products must undergo quality checks to confirm safety, efficacy, and consistency, with certification validating adherence to standards.	Example: Herbal teas being certified by USDA Organic or ISO certifications for herbal supplements to ensure quality and authenticity.
Patient safety and informed consent	Ensuring patients are informed of potential risks, side effects, and drug interactions associated with herbal treatments before use.	Example: A doctor advising patients about the blood-thinning effects of ginkgo biloba, especially if the patient is on anticoagulant medications.
Adhering to legal regulations	Manufacturers and professionals must comply with laws regulating the sale, production, and claims related to herbal products.	Example: The FDA regulating herbal supplements in the United States under the Dietary Supplement Health and Education Act (DSHEA), ensuring product safety.
Marketing and labeling compliance	Products must be labeled with accurate information about ingredients, dosages, and health claims. Advertising must not make unsupported claims.	Example: An herbal sleep remedy clearly labeled with ingredients such as valerian and chamomile, without making unverified claims of "curing insomnia."
Clinical trials and research	Conducting rigorous clinical studies to gather scientific evidence on the efficacy and safety of herbal medicines.	Example: Clinical trials investigating the use of St. John's Wort for treating mild depression, ensuring ethical standards in patient treatment.
Continuous monitoring and reporting	Ongoing surveillance for product safety post-market, with required reporting of adverse reactions to authorities for evaluation.	Example: Reporting cases of liver toxicity potentially linked to certain kava supplements, prompting regulatory review and possible product recall.

TABLE 5.5
Ensuring Transparency and Data Sharing in Herbal Medicine

Aspect	Description	Example
Access to research data	Ensuring that research findings, including clinical trials and studies on herbal products, are publicly accessible to enhance trust and the validation of efficacy and safety.	Example: Publishing results of clinical trials on turmeric's anti-inflammatory effects in public databases like ClinicalTrials.gov for wider access.
Ingredient disclosure	Providing comprehensive information on all ingredients used in herbal products to allow consumers and healthcare providers to make informed decisions.	Example: Herbal supplement manufacturers listing all active and inactive ingredients on product labels, including potential allergens and fillers.
Regulatory compliance reporting	Sharing data with regulatory bodies to ensure compliance with safety and quality standards in herbal product manufacturing.	Example: Manufacturers providing batch testing results and certifications to the FDA or other regulatory agencies to verify product safety and quality.
Adverse event reporting	Establishing systems for consumers and healthcare providers to report adverse effects from herbal products, facilitating data collection for safety monitoring.	Example: Implementing a system in which patients can report side effects of herbal supplements through platforms like MedWatch in the United States.
Transparency in marketing claims	Ensuring all marketing and health claims made about herbal products are substantiated by scientific evidence and clearly communicated to consumers.	Example: Herbal product advertisements include disclaimers about the lack of FDA evaluation for claims of efficacy, ensuring consumers are aware of limitations.
Collaboration in research	Encouraging partnerships between universities, healthcare institutions, and herbal product manufacturers to promote shared research and knowledge.	Example: Collaborative studies between universities and herbal manufacturers investigate the efficacy of adaptogens, with findings shared in peer-reviewed journals.
Consumer education and resources	Providing accessible educational materials about herbal products, including potential benefits, risks, and responsible usage guidelines.	Example: Websites offering comprehensive guides on using herbal remedies, including dosage information and contraindications.

transparency and data sharing. According to UNESCO, open science can be defined as a set of principles and rules through which scientific research is made accessible to all fields for the benefit of scientists and society as well. It is not limited to scientific research knowledge; it can be extended to production knowledge so that it is inclusive, equitable, and sustainable. Various methods can be used. Study registration involves the registration of the research plan and includes hypotheses, methodologies, etc., which helps TCIM enhance the quality and credibility of the research by providing crucial information about the existence and primary outcomes of the research (Table 5.5). Reporting guidelines can be used in TCIM; they provide a standard framework for reporting the methods and results of TCIM research. This is useful for maintaining transparency and allows researchers to read, evaluate, and replicate the research effectively. Open science in TCIM follows reporting guidelines without meta-analysis (SWIM), with meta-analysis (PRISMA), and for RCT (CONSORT). Open science states that data should be used in the scientific community and by the public. In TCIM, data sharing includes sharing of clinical trials, raw data from studies on herbal medicines, and patient-reported outcomes. For data sharing, the FAIR (findability, accessibility, interoperability, reuse) principle for digital assets is used, which is as follows: Findability states that metadata and data should be accessible to humans as well as computers; accessibility states how a user can access the data, which includes authentication and authorization; interoperability refers to how the data are integrated with other data, especially metadata that can be used to interoperate for analysis, storage, and processing; reuse states that metadata or data should be well defined so that they can be replicated and combined easily [18–20].

5.3 CONCLUSION, FUTURE OUTCOMES, AND CHALLENGES OF HERBAL MEDICINE

The acceptance of herbal medicine within conventional healthcare is on the rise. This may foster collaborative treatment plans that combine the strengths of both pharmacological and herbal therapies, enhancing patient care. Continued scientific exploration will deepen our understanding of the pharmacodynamics and pharmacokinetics of herbal compounds. This enhances the research and development of herbal products and prioritizes safety and efficacy. Advances in genomics and pharmacogenomics hold the potential for personalized herbal treatments tailored to individual genetic profiles, optimizing therapeutic outcomes while minimizing adverse effects. As the demand for herbal products grows, the need for sustainable sourcing and cultivation practices becomes critical. Efforts to conserve medicinal plants and promote biodiversity will help ensure these resources are available for future generations. It also enhances collaboration between countries and cultures, leading to a shared understanding of herbal medicine. This could promote cross-border research initiatives that enrich the field as a whole.

A significant challenge in herbal medicine is the lack of standardized dosages and formulations. Variability in active ingredient concentrations can lead to inconsistent therapeutic effects and raise safety concerns. Ensuring quality control and standardization is needed. The regulatory framework for herbal medicines is inconsistent across different regions. Many areas lack comprehensive guidelines regarding the safety, efficacy, and quality of herbal products, making it difficult for consumers to make informed decisions. While many herbal remedies are generally safe, adverse effects can occur, particularly when herbs interact with conventional medications. Raising awareness about potential side effects and contraindications is essential. The commercialization of traditional herbal practices raises ethical concerns, particularly regarding the potential exploitation of indigenous knowledge and resources. It is vital to respect traditional practices and ensure fair compensation for indigenous communities. Misconceptions and misinformation about herbal medicine can breed skepticism among healthcare professionals and patients. Educational initiatives and awareness campaigns are crucial for fostering a balanced understanding of herbal therapies.

REFERENCES

[1] Sam, S. (2019). Importance and effectiveness of herbal medicines. *Journal of Pharmacognosy and Phytochemistry, 8*(2), 354–357.
[2] Aejazuddin, D. (2016). Herbal medicine: A comprehensive review. *International Journal of Pharmaceutical Research, 8*(1).
[3] Firenzuoli, F., & Gori, L. (2007). Herbal medicine today: Clinical and research issues. *Evidence-Based Complementary and Alternative Medicine, 4*, 37–40.
[4] Zhang, J., Onakpoya, I. J., Posadzki, P., & Eddouks, M. (2015). The safety of herbal medicine: From prejudice to evidence. *Evidence-Based Complementary and Alternative Medicine: eCAM, 2015*, 316706.
[5] Mensah, M. L., Komlaga, G., Forkuo, A. D., Firempong, C., Anning, A. K., & Dickson, R. A. (2019). Toxicity and safety implications of herbal medicines used in Africa. *Herbal Medicine, 63*(5), 1–13.
[6] Wang, H., Chen, Y., Wang, L., Liu, Q., Yang, S., & Wang, C. (2023). Advancing herbal medicine: Enhancing product quality and safety through robust quality control practices. *Frontiers in Pharmacology, 14*, 1265178.
[7] Parveen, A., Parveen, B., Parveen, R., & Ahmad, S. (2015). Challenges and guidelines for clinical trial of herbal drugs. *Journal of Pharmacy and Bioallied Sciences, 7*(4), 329–333.
[8] World Health Organization. (2006). International Regulatory Cooperation for Herbal Medicines (IRCH). Available at: WHO IRCH official page
[9] Sharma, S. (2015). Current status of herbal product: Regulatory overview. *Journal of Pharmacy and Bioallied Sciences, 7*(4), 293–296.
[10] Tilburt, J. C., & Kaptchuk, T. J. (2008). Herbal medicine research and global health: An ethical analysis. *Bulletin of the World Health Organization, 86*, 594–599.
[11] Liu, F. X., & Salmon, J. W. (2010). Herbal medicine regulation in China, Germany, and the United States. *Integrative Medicine, 9*(6), 123–140.
[12] World Health Organization. (2007). WHO guidelines on good manufacturing practices (GMP) for herbal medicines. In *WHO guidelines on good manufacturing practices (GMP) for herbal medicines.*
[13] Varshney, N., Jain, D., Janmeda, P., & Mitra, D. (2021). Role of medicinal plants in pharmaceutical sector: An overview. *Global Journal of Bio-Science and Biotechnology, 10*(2), 18–24.
[14] Stanojević-Ristić, Z., Mrkić, I., Ćorac, A., Dejanović, M., Mitić, R., Vitković, L., . . . & Valjarević, A. (2022). Healthcare professionals' knowledge and behaviors regarding drug–dietary supplement and drug–herbal product interactions. *International Journal of Environmental Research and Public Health, 19*(7), 4290.
[15] Beauchamp, T. L., & Childress, J. F. (1994). *Principles of biomedical ethics.* Edicoes Loyola.
[16] Research!America (2022). US investments in medical and health research and development 2016–2020.
[17] Director, M. G. B., Director, M. J. B., Director, M. A. K. M., Director, M. J. R. S., Director, M. K. S., Modi, J., & Roads, M. S. (2018). Annual report 2017–2018. Ministry of AYUSH, Government of India, New Delhi.
[18] Ng, J. Y., Wieland, L. S., Lee, M. S., Liu, J. P., Witt, C. M., Moher, D., & Cramer, H. (2024). Open science practices in traditional, complementary, and integrative medicine research: A path to enhanced transparency and collaboration. *Integrative Medicine Research, 13*(2), 101047.
[19] Pal, S. K., & Shukla, Y. (2003). Herbal medicine: Current status and the future. *Asian Pacific Journal of Cancer Prevention, 4*(4), 281–288.
[20] Saggar, S., Mir, P. A., Kumar, N., Chawla, A., Uppal, J., & Kaur, A. (2022). Traditional and herbal medicines: Opportunities and challenges. *Pharmacognosy Research, 14*(2).

6 Quality Control, Standardization, and Safety Evaluation Strategies of Herbal Medicines for Clinical Trials

Dhanashree Sharma[1], *Navratan Soni*[1], *Mukul Yadav*[1], *Meena Chandran*[2], *and David Paul*[1]

[1]Department of Pharmaceutical Analysis, National Institute of Pharmaceutical Education and Research (NIPER)–Kolkata, West Bengal, India

[2]Department of Pharmaceutical Chemistry, St. James College of Pharmaceutical Sciences, Riverbank, Chalakudy, Kerala, India

ABBREVIATIONS

ABCD1	ATP binding cassette subfamily D member 1
ADRs	Adverse drug reactions
ATR	Average true range
AUC	Area under the curve
AYUSH	Ayurveda, Yoga and Naturopathy, Unani, Siddha and Homeopathy
CAGR	Compound annual growth rate
Cmax	Maximum concentration
CMs	Complementary medicines
DDT	Dichlorodiphenyltrichloroethane
DILI	Drug-induced liver toxicity
DNA	Deoxyribonucleic acid
E.C.	European Commission
EFSA	European Food Safety Authority
EMA	European Medicines Agency
EU	European Union
FAO	Food and Agriculture Organization
FDA	Food and Drug Administration
FTIR	Fourier transform infrared spectroscopy
GC	Gas chromatography
GC-MS	Gas chromatography–mass spectrometry
GC-TOF	Gas chromatography time-of-flight
GMPs	Good Manufacturing Practices
GST	Glutathione S-transferase
HC	Health Canada
HMBC	Heteronuclear multiple bond correlation
HMPC	The Committee on Herbal Medicinal Products
HMPs	Herbal medicinal products
HMs	Herbal medicines
HPFBI	Health Products and Food Branch Inspectorate
HPLC	High-performance liquid chromatography
HPLC-MS	High-performance liquid chromatography–mass spectrometry
HPTLC	High-performance thin-layer chromatography
HR-MS/MS	High-resolution tandem mass spectrometry
HSQC	Heteronuclear single quantum correlation
IAEA	International Atomic Energy Agency
ICH	International Council for Harmonization
IND	Investigational new drug
LC-MS	Liquid chromatography–mass spectrometry
LC-MS/MS	Liquid chromatography–mass spectrometry/mass spectrometry
LC QTOF-MS	Liquid chromatography quadrupole time-of-flight mass spectrometry
LD50	Lethal dose 50
LOAEL	Lowest observed adverse effect level
LOEL	Lowest observed effect level
MHPD	Marketed Health Products Directorate
MS	Mass spectrometry
MTD	Maximum tolerated dose
NCE	New chemical entity
NHPR	Natural Health Product Regulation
NIR	Near-infrared spectroscopy
NMR	Nuclear magnetic resonance
NMT	Not more than
NNHPD	Natural and Non-prescription Health Products Directorate
NOAEL	No observed adverse effect level
NOEL	No observed effect level
OECD	Organization for Economic Cooperation and Development
OPLS-DA	Orthogonal partial least squares discriminant analysis

DOI: 10.1201/9781003522065-6

PCA	Principal component analysis
PCR	Polymerase chain reaction
PgP	P-glycoprotein
PLS-DA	Partial least squares–discriminant analysis
PLSR	Partial least squares regression
PV	Pharmacovigilance
QA	Quality assurance
QC	Quality control
Q-TOF-MS	Quadrupole time-of-flight mass spectrometry
RCTs	Randomized clinical trial
TCM	Traditional Chinese medicine
TGA	Thermogravimetric analysis
TGx-DDI	Toxicogenomic DNA damage–inducing
TLC	Thin-layer chromatography
UHPLC	Ultra-high-performance liquid chromatography
USD	United States dollar
UV	Ultraviolet
WHO	World Health Organization

6.1 INTRODUCTION

Nature has been the source of various medicines that have proved helpful since primitive times. Recent developments in the analytical science and other screening approaches have resulted in policymakers' work on herbal medicines (HMs) [1]. People all over the world have used herbal medications for a variety of purposes, including the treatment and prevention of various illnesses. Over the past decade, there has been a notable increase in interest in pharmaceuticals originating from higher plants, specifically phyto-therapeutics and herbal remedies. It is estimated that around 25% of contemporary medications originate from higher plants, either directly or indirectly. The market for HMs has seen a notable upsurge in interest and acceptability in recent years, both in developed and developing countries. A large fraction of people in the developing parts of the world are thought to receive their treatment primarily from HMs, and in some societies, using herbs in traditional ways is considered an essential component of their culture. Various factors, such as consumer preferences for natural healthcare solutions, the integration of traditional practices with modern medicine, and increasing health and wellness awareness, are responsible for remarkable growth in the global HMs market. The market for HMs was estimated to be worth USD 165.13 billion globally in 2023. It is anticipated to increase at a compound annual growth rate (CAGR) of 11.34%, from USD 189.34 billion in 2024 to USD 386.07 billion by 2032 [2]. Integration of traditional and modern medicine practices fosters innovative herbal formulations. The pharmaceutical industry increasingly focuses on phytochemicals acquired from plants, enhancing the development of HMs to address modern health challenges. The WHO estimates that a similar scenario is experienced in industrialized countries, where 70% to 80% of the population depends on traditional and alternative types of medicine. Therefore, the prevalence of HMs is common in underdeveloped countries.

The HMs are generally considered safe as they are derived from natural sources; it is also important to acknowledge that herbal drugs can cause undesirable effects. These unwanted effects may arise from factors such as complex herbal composition, which can vary depending on geographical origin, cultivation practices, processing methods, lack of standardization, adulteration, etc. The potential for variation between batches of medication is enormous. Hence, a standardization process must be developed for all plant medicines available on the market. Consumers' health is of utmost importance and should be taken care of by assuring the purity and safety of herbal plants and dietary supplements during preparation. Strong foundations for these traits are essential as the herbal sector grows. The future of quality control (QC), standardization, and safety evaluation procedures for HMs in clinical trials is vital for improving their global acceptability and efficacy (Figure 6.1).

The lack of comprehensive pharmacokinetic, pharmacological, and clinical data for many herbal medicinal products (HMPs) significantly contributes to the uncertainty surrounding their safety and efficacy. This difficulty is exacerbated by a failure to satisfy statutory standards for herbal medication research, which complicates regulatory supervision. Critics of HMs worldwide have raised concerns that have been instructive for practitioners. While clinical trials have been initiated, substantial progress is still required to elevate HMs to the standard of evidence-based treatments recognized at the international level. Therefore, this chapter seeks to draw attention to the need for standardization and QC of HMs. Additionally, it discusses the safety concerns associated with herbal medications and the more recent approaches to evaluating their safety.

6.2 THE REGULATORY STATUS OF HERBAL DRUGS RESEARCH

Herbal drugs are widely used to treat various health issues across traditional medicine systems, including Ayurvedic, Homeopathic, Siddha, and Unani practices. This highlights the cultural significance and reliance on herbal remedies in these systems. The classification of herbal drug products varies significantly from one country to another. This variation hinders the efforts by manufacturing companies to sell standardized herbal products worldwide. This variability can lead to confusion regarding these products' regulatory status and safety. The regulatory pathway for bringing herbal drugs to market, as per EU guidelines, is shown in Table 6.1. However, some

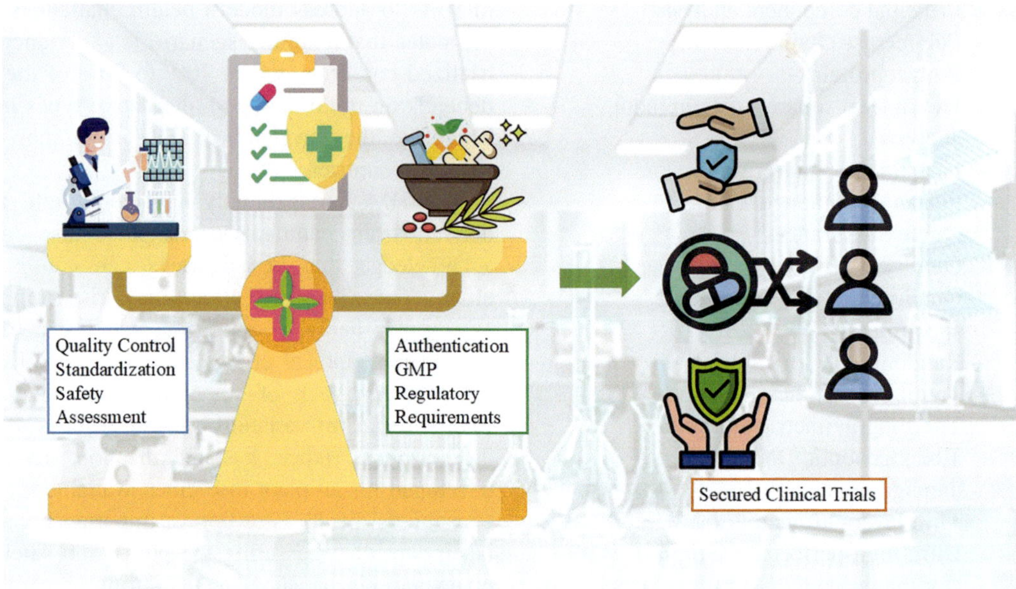

FIGURE 6.1 HMs standardization and clinical trials are vital for improving patient safety.

common categories include functional foods, dietary supplements, and traditional medicines [3].

The regulatory frameworks governing the research and development of herbal drugs vary significantly across different regions, as shown in Table 6.2. In Europe, the EMA is responsible for developing guidelines for HMPs that ensure the quality, safety, and efficacy of HMs, thereby overseeing their research and development. In contrast, the United States categorizes most herbal products as dietary supplements under the DSHEA of 1994, which establishes a regulatory framework that includes post-market notification and safety demonstration requirements. However, the regulatory status of herbal drugs in the North American region is displayed in Table 6.3. In Canada, the Natural Health Products Directorate regulates a large variety of complementary medicines (CMs), including HMPs, emphasizing the need for quality research on their efficacy and safety. Meanwhile, Australia's Therapeutic Products Act of 1989 classifies herbal products and homeopathic remedies as medicines, setting specific standards for their regulation. According to the TGA, differences in the requirements for natural products "classified as supplements" or "classified as medicine" are given in Table 6.4. These diverse regulatory approaches reflect the varying cultural attitudes and scientific standards applied to HMs across different countries, highlighting the need for harmonization in international regulations to facilitate trade and ensure consumer safety.

TABLE 6.1
The Regulatory Pathway for Bringing Herbal Drugs to Market, as per EU Guidelines [4]

Regulatory Authority	Article No.	Primary Requirements of Safety and Efficacy
Traditional use registration	Article 16a (1) of Directive 2001/83/E.C.	a) As long as sufficient safety data and convincing efficacy exist, no clinical studies or tests on safety and efficacy are necessary. b) Evaluate safety and efficacy data primarily from bibliographic sources. c) Their use dates back at least 30 years, including 15 years in the European Union. d) They are not supplied by injection and are intended to be used without a doctor's supervision.
Well-established use marketing authorization	Article 10a of Directive 2001/83/E.C.	a) Demonstrates the scientific research proving that the medication's active ingredients have been used for at least 10 years in the EU for established medical purposes, with proven efficacy and reasonable safety. b) It involves evaluating safety and efficacy data, primarily from bibliographic sources.

6.3 CHARACTERISTICS OF HMS

One of the critical characteristics of HMs is their natural origin, which relies on plant-based ingredients [6]. HMs is derived exclusively from whole plants or their parts, making them vital to traditional medicine practices worldwide. This natural composition is associated with the belief that herbs are inherently safe and can provide therapeutic efficacy across various doses. The extensive variety of secondary metabolites in plant-derived natural products contributes to their distinctive health benefits, allowing for diverse applications in treating different ailments [7]. Consequently,

TABLE 6.2
Overview of Regulations Among Different Countries

Country	India	The United States	Australia	Canada	EU
Regulatory agency	Department of AYUSH	FDA	TGA	HC	EC and national authorities
Regulation	Drugs and Cosmetics Act, 1940	DSHEA	Therapeutic Products Act (TGA), 1989, and the associated Therapeutic Goods Regulations, 1990CDGF	NHPR	Directive 2004/24/E.C.
Summary or current status/ key considerations	Categories in three distinct groups—recognized as classical texts, traditional HMs as NCEs, and formulations that adhere to GMP-compliant standardization practices.	HMs are considered dietary supplements. They do not require rigorous premarket approval. Manufacturers must follow specific labeling requirements.	It ensures product registration, evaluation, and ongoing monitoring of CM. Categorized into two types—registered medicines and listed medicines. Post-market surveillance is conducted through adverse-event reporting and routine inspections.	Overseen by three primary directorates: NNHPD, MHPD, and HPFBI. Licensing of NHPs is mandatory. Safety and efficacy data are mandatory.	Their objectives are to safeguard the free flow of HMPs throughout the EU and to protect public health.

TABLE 6.3
The Regulatory Status of Herbal Products in The North American Region [5]

Classification	The United States	Canada	Mexico
National program on traditional and complementary medicine (TCM)	NA	Yes	Yes
Laws or regulations at the national or state level for TCM	Yes	Yes	Yes
Regulation of HMs	Yes	Yes	Yes
National office for TCM	Yes	Yes	Yes
National committee of experts for TCM	Yes	Yes	Yes
Registration system for HMs	Yes	Yes	Yes
National policies on TCM	No	Yes	Yes
National Research Institute for TCM	Yes	No	No
Population using TCM (including indigenous medicine)	Yes	Yes	Yes

TABLE 6.4
Differences in The Requirements for Natural Products "Classified as Supplements" or "Classified as Medicines" According to TGA

	Classified as Supplements			
Product	**Dietary Supplement**	**NA**	**Supplemented Food**	**Food Supplements**
Allowed route of administration	Oral only	NA	Oral only	Oral only
Premarket approval	No	NA	Yes	Only if considered as "novel foods" safety assessment by EFSA
Therapeutic claims	No	NA	No	No
Recommended dosing	Yes	NA	Yes	Yes
Addition of a new compound	Via NDI notification	NA	Needs to go through the novel food process	Apply for registration
Classified as medicine				
Product	Botanical drugs	CMs	Natural health products	(EMA) HMPs
Premarket approval	Yes	Yes	Yes	Yes
Clinical trial data	Yes	Yes	Yes	Yes
Therapies	Medicine only	Medicines Homeopathy Aromatherapy	Medicine only	Medicine Homeopathy
Reported adverse reactions/poisoning	Yes	Yes	Yes	Yes

herbal, botanical, or phytomedicines have gained popularity as alternatives to synthetic medications. HMs also adopt a holistic approach to healing, which involves the person's entire body, mind, and spirit in the quest for optimal health and wellness [8]. This perspective goes beyond merely alleviating symptoms, focusing on addressing the root causes of health issues. Integrative medicine, which incorporates holistic principles, focuses on improving mental, physical, and spiritual well-being. By recognizing the interconnectedness of these aspects, HMs aim to promote overall wellness rather than just treating specific conditions. This comprehensive approach can lead to more sustainable health outcomes and patient satisfaction.

Another significant characteristic of HMs is their relatively minimal side effects compared to conventional medicines. Although there is a common belief that herbal products are entirely free from adverse reactions, it is essential to acknowledge that some herbs can cause undesirable or severe side effects. Generally, herbal remedies are considered safer than synthetic drugs, as evidenced by the fact that thousands of adverse events result from conventional medications each year. Nevertheless, individuals must be aware that herbal supplements can interact with other pharmaceutical drugs, potentially increasing the risk of adverse side effects. This awareness underscores the importance of consulting with healthcare professionals before incorporating herbal remedies into one's health regimen. The manufacturer should clearly define the raw material, mode of preparation, and product to ensure the quality of the medicinal preparations. The plants are wholly responsible for the phytotherapeutic effect. Most HMPs and their active ingredients are extracted using aqueous or organic solvents such as ethanol or methanol. Phytotherapy is the scientific study of HMs, focused on the therapeutic benefits of plant extracts and their active components. Despite their benefits, the HMs sector faces challenges in standardization, regulatory compliance, and identifying active constituents [9]. Ongoing research is essential to enhance the understanding and integration of HMs into modern healthcare systems [10]. While HMs offer promising therapeutic options, their integration into mainstream healthcare necessitates rigorous research and regulatory frameworks to ensure safety, efficacy, and quality.

6.4 NEED FOR STANDARDIZATION AND QUALITY CONTROL

The use of HMs is commonly and openly accepted in modern medicine. The first known form of medical treatment for humans is the use of herbs, which have been utilized historically in all civilizations. A standardization system must be created for all plant medicines on the market because there is tremendous potential for variance between batches of medication. The time and location of collection affect the therapeutic activity of the phyto-constituents. Other factors like environmental conditions and cultivation techniques also affect the production of particular medicinal plants. This indicates the requirement for appropriate quality-control tests and standardization throughout the preparation to ensure product quality.

Standardization, as defined by the American Herbal Products Association, refers to the information and controls required to ensure product uniformity. This is accomplished by minimizing the inherent diversity of natural product composition using quality assurance procedures applied to agricultural and industrial operations. Herbal medications are typically considered safe due to their long-standing use in diverse cultures. However, some herbal medications can also be highly toxic due to the presence of adulterants and contamination. If herbal medications are not carefully analyzed, they can cause adverse effects and drug-drug or drug-food interactions. Therefore, the primary aim of herbal research is to assess the safety of HMs. Five or more distinct botanical constituents are expected to be seen in a single herbal composition [11]. The absence of any reference material for identification leads to batch-to-batch variation. Standardization provides significant benefits to businesses, including a stable platform on which new technologies may be established and the chance to exchange and improve existing processes. Standardization also plays a crucial role in supporting governments, administrations, regulators, and the legal profession, as standardization supports all legislation, regulation, and policy efforts [12].

The Pharmaceutical Manufacturers Association of the United States defines quality as the elements that directly or indirectly impact the safety, efficacy, and acceptance of a product. Quality refers to satisfying client requirements, both explicitly and implicitly. The term "quality" comes from the Latin word "qualis," which refers to "of what type" and refers to the composition and attributes of objects. When dealing with medications, quality is a crucial factor, but managing herbal medications might be challenging for several reasons. Standardization encompasses all industrial processes, while QC ensures consistent product quality. The WHO acknowledges the growing need for standardization and QC of HMs. Its policies and checklists emphasize the need for national standards, technical guidelines, and methods to evaluate traditional medicine's safety, effectiveness, and quality. The WHO emphasizes the necessity of developing national pharmacopeias and monographs for medicinal plants and cultivating and conserving them for long-term use in botanical medicine. Many challenges are faced during the standardization of botanicals. In contrast, synthetic drugs have comparatively well-defined structures, reference standards for comparison, and established pharmacopoeia assays. QC of herbal drugs poses various challenges, such as plant identification disputes, intentional adulteration, and storage and transportation issues, which should be considered.

6.5 MODERN ADVANCEMENTS IN HERBAL DRUG STANDARDIZATION

The subject of herbal drug standardization has advanced remarkably as a result of the rising demand for secure and efficient natural therapies. HMs, which integrate the therapeutic practices of indigenous systems of medicine passed down through generations, have become increasingly popular worldwide as people seek alternatives to chemical-based pharmaceuticals. This resurgence is partly attributed to the ill effects associated with synthetic medications, the absence of curative modern therapies for various chronic conditions, and the alarming rise of antimicrobial resistance.

Global pharmaceutical companies have also renewed their interest in natural product drug development, recognizing herbs as a potential source of new drug candidates. Consequently, the standardization of herbal medications has advanced significantly because of the fusion of contemporary science and technology with traditional herbal knowledge. One of the critical challenges in ensuring the quality and safety of HMPs has been the accurate identification of genuine plant materials, as the enormous quantity and variety of medicinal plants sold in various markets, coupled with the use of processed or modified forms, have made it difficult to distinguish authentic products from adulterants and counterfeits. Advances in deoxy ribose nucleic acid (DNA) barcoding and metabolomics integrated with chemometrics have emerged as promising solutions, providing reliable and rapid methods for the botanical authentication of herbal ingredients.

6.5.1 The Role of DNA Barcoding and Metabarcoding in Herbal Drug Standardization

Advancements in DNA barcoding and metabarcoding techniques are revolutionizing herbal drug standardization and QC, addressing significant challenges such as adulteration and contamination. The herbal industry has seen a notable increase in demand for natural products, unfortunately leading to an uptick in the use of spurious and unauthorized raw materials [13]. This has led to an increase in the importance of ensuring the authenticity and safety of HMPs. DNA barcoding offers a robust methodology for rapidly and accurately identifying medicinal plants, which is essential in mitigating the risks associated with low-quality herbal products. Moreover, integrating DNA barcoding with advanced technologies such as next-generation sequencing enhances the reliability of quality assurance protocols, allowing for more thorough analyses of herbal samples. Regulatory bodies increasingly recognize the importance of incorporating DNA-based techniques into their standards, streamlining the authentication process and ultimately promoting consumer safety. Combining DNA barcoding with methodologies like metabolomics and proteomics can develop a multifaceted approach to standardization that identifies plant species and assesses the chemical profiles of herbal products, ensuring they meet safety and efficacy standards [13]. This synergy of technologies represents a significant leap forward in safeguarding the integrity of HMPs in the market.

6.5.2 The Role of Chemometrics in Herbal Drug Standardization

Chemometrics plays a critical role in herbal drug standardization by providing robust analytical methods to analyze the complex chemical composition of herbal materials. Through the application of multivariate statistical techniques, chemometrics (Figure 6.2) enables the discovery and assessment of crucial marker compounds, the detection of impurities or contaminants, and the establishment of QC parameters to ensure consistency and reproducibility of HMPs. This analytical approach is essential for developing reliable fingerprints and establishing standardized reference materials, ultimately contributing to the QA and safety of HMs. The major critical challenge in herbal drug standardization is the inherent complexity and variability of the active constituents, which can be influenced by factors such as plant species, geographical origin, harvesting time, and processing methods. Conventional analytical methods, such as HPLC and GC-MS, often cannot fully capture the holistic chemical profile of herbal materials.

Conversely, chemometrics offers a more comprehensive and pattern-oriented approach that considers the entire metabolome rather than focusing solely on individual marker compounds. The benefits of this "pattern-oriented"

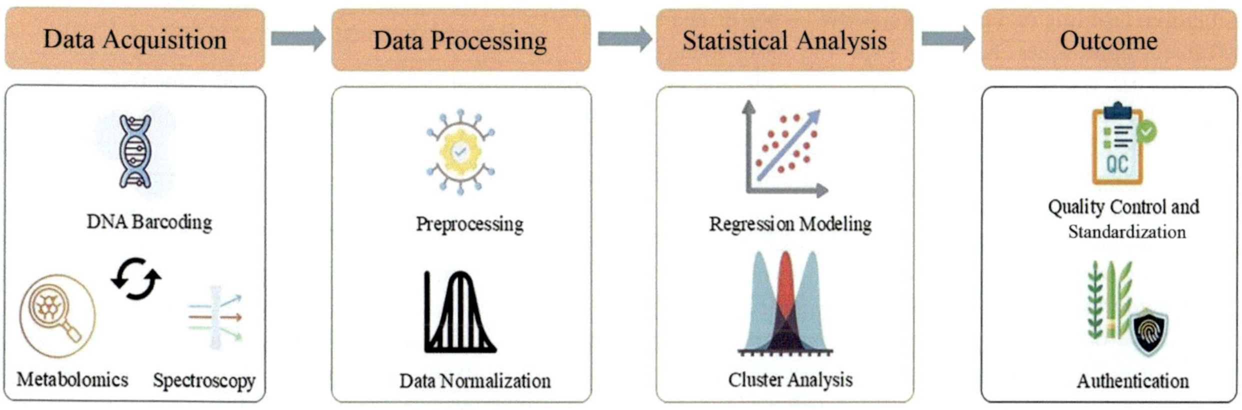

FIGURE 6.2 Role of chemometrics in herbal drug standardization.

approach include species authentication and regional differentiation, both of which are essential for guaranteeing the quality and safety of HMs. Chemometric techniques, such as principal component analysis and partial least squares regression (PLSR), can be applied to analytical data obtained from various spectroscopic methods, including near-infrared spectroscopy (NIR), Raman, and nuclear magnetic resonance (NMR).

Using these multivariate statistical methods, researchers can identify different herbal species' unique chemical signatures or "fingerprints" and detect potential adulterants or substitutions. Furthermore, chemometrics can establish QC parameters, such as batch-to-batch consistency and stability, to ensure the reproducibility and reliability of herbal drug products. Multivariate calibrations are commonly used when developing prediction models for specific analytes. Principal component regression, PLSR, and multiple linear regression are commonly used in the standardization and QC of HMs as popular multivariate calibration methods. By identifying essential marker compounds, detecting contaminants, and establishing QC parameters, chemometrics contributes to HMs' quality assurance and safety, ultimately enhancing their therapeutic efficacy and consumer confidence.

6.5.3 The Role of Metabolomics in Herbal Drug Standardization

Metabolomics comprehensively analyzes a biological system's small molecules or metabolites. This approach allows for the simultaneous analysis and quantification of a wide range of compounds, including primary and secondary metabolites, which can be used as chemical fingerprints for species authentication and geographical origin discrimination. Compared to traditional analytical techniques, metabolomics offers several advantages, such as detecting a broader range of compounds, increased sensitivity, and the potential to identify novel biomarkers for standardization.

Environmental elements such as seasonal dynamics, soil type, precipitation, moisture content, temperature, and stress factors can all impact the metabolome, which reflects the entire collection of metabolites found in a plant. The metabolite profiling of a plant species can be performed by two major approaches: targeted (a specific set of metabolites) and untargeted (all metabolites) metabolite analyses. Metabolite profiling, also known as targeted metabolite analysis, focuses on analyzing a specific group of metabolites in a sample rather than conducting a comprehensive analysis of the entire metabolome. This approach utilizes specific or combined analytical methods, such as GC-MS and LC-MS. Metabolomics has been shown to be helpful in the QC and authentication of HMs in several studies. The targeted approach can be used to compare the metabolic profiles of different batches of cultivars to verify consistency and quality. Environmental factors can lead to differences in metabolite composition, affecting drug potency. It can also assess batches procured during different seasons, revealing the best possible season to cultivate and collect the harvest. It can be beneficial to have the highest amount of metabolite content to prepare a standardized drug [14]. Metabolomic fingerprinting combined with chemometrics enables the creation of a unique metabolite signature, which is valuable for authenticating and identifying counterfeit and substandard products.

6.6 DEVELOPMENT OF QUALITY STANDARDS FOR MEDICINAL HERBS

The rising global demand for natural and herbal products has sparked a growing interest in establishing high-quality standards for Indian medicinal herbs. Traditional Indian medical systems have a rich history of utilizing MS, offering significant information in the selection, preparation, and application of herbal formulations for treating various ailments. Nevertheless, due to various concerns, it is essential to develop quality standards for medicinal herbs to ensure their effectiveness, safety, and authenticity. The process involves several key considerations:

6.6.1 Literature Survey of the Existing Data

The most crucial stage in creating quality standards for herbal medications is that it offers a thorough summary of the current state of the discipline regarding procedures, regulations, and information. The results of this survey show research gaps and variations in the quality and effectiveness of HMs, which can differ significantly depending on several criteria, including formulation, sourcing, and processing. Researchers can create standards for quality assessment, standardization, and safety evaluation by combining the results of multiple investigations. Furthermore, recognizing past and present practices promotes the adoption of verified analytical techniques and helps inform best practices. In the end, a comprehensive literature review offers a strong foundation for evidence-based suggestions that safeguard patient safety and encourage the trustworthy application of herbal remedies in medical settings.

6.6.2 Identification and Authentication of Raw Materials

Although most people believe herbal remedies to be safer than synthetic drugs, there have been numerous adverse effects caused by the absence of controls on herbal remedies, particularly in the past. Various approaches are necessary for accurate identification, including chromatographic procedures, molecular techniques like DNA barcoding, and macroscopic and microscopic examination. These techniques aid in distinguishing authentic herbal products from adulterants or imitations, which may jeopardize their therapeutic effectiveness and safety. Moreover, authentication determines the place of origin, which may impact herbal medications' pharmacological characteristics and chemical makeup. Manufacturers can maintain quality standards that safeguard customers and improve the marketability of HMs by putting in place strong identification and authentication procedures. This systematic approach is crucial to building patient and healthcare provider confidence and encouraging the wise use of herbal treatments.

6.6.3 Standardization of the Extraction Procedure and Chemical Profiling

It is essential for controlling quality and standardizing herbal drugs. Natural and genetic alterations, seasonal fluctuations, soil disparities, and the nutritional condition of the herbal plant can all cause variations in efficacy from batch to batch [15]. A comprehensive quality-control system must be implemented, incorporating strict guidelines and norms, like those used for synthetic pharmaceuticals.

Standardization ensures uniform biological function, chemical composition, and an extensive quality assurance procedure for producing herbal medications. The Indian Pharmacopoeia (IP) encompasses reference standards for macroscopic and microscopic characteristics, fingerprints obtained from thin-layer chromatography (TLC), prescribed limits for heavy metals, and the purity and potency of plant extracts. Researchers encounter challenges in devising analytical methodologies capable of consistently delineating the phytochemical composition, encompassing the quantification of primary constituents and marker/bioactive molecules. Phytochemistry contributes to the standardization of herbal medications by identifying known active components at ideal quantities and preserving their efficacy [16]. Owing to the shortcomings of traditional techniques, metabolomics, in conjunction with systems biology, is being used more frequently to link the biochemical constituents of plant extracts with biological activity. Standardization via metabolite profiling offers advantages over traditional chromatographic techniques, as it enables the simultaneous analysis of numerous molecules during untargeted analysis without requiring prior knowledge of which specific chemical constituents contribute to efficacy. This approach can facilitate the development of robust analytical methods that ensure extracts display uniform biological function and chemical profiles, a prerequisite for therapeutic reliability. Metabolomics, encompassing both targeted and untargeted profiling strategies, is gaining preference across various scientific disciplines for the chemical profiling of herbal drugs.

6.6.4 Selecting the Appropriate Marker Compound

Selecting a marker compound is a critical step in the standardization process, as it serves as a reference point for QC. Marker compounds are typically bioactive constituents or chemical entities unique to herbal material and can be reliably quantified. The selection of a suitable marker compound should be based on various factors, including its stability, abundance, and relevance to the therapeutic activity of the HMs [17].

One of the critical considerations in selecting a marker compound is its stability. The marker compound should withstand the processing steps in producing HMs, such as extraction, purification, and formulation, without undergoing significant degradation [18]. Additionally, the marker compound should be present in sufficient quantities to allow for accurate quantification and facilitate standardization, ensuring the consistency of the final product. The marker must be uncomplicated to quantify using trustworthy analytical methods (such as GC-MS/LC-MS or HPLC) so that amounts can be measured precisely.

6.6.5 Screening for Efficacy and Safety Assessment

The safety assessment of herbal drugs is an essential and multifaceted process in developing quality standards prioritizing consumer health and well-being. The toxicity of medicinal plants and HMs is closely associated with the presence of bioactive compounds in plant material and their potential for toxicity. The matter becomes more intricate when considering complex, heterogeneous herb mixes with unpredictable effects.

Evaluating herbal extracts' pharmacological effects on different biological targets, including cells or enzymes, usually entails various laboratory techniques. Techniques, including enzyme inhibition assays, receptor binding studies, and cell viability assays, are frequently used to evaluate the bioactive substances' effectiveness in the extracts. Researchers can determine the ideal concentrations to induce therapeutic responses by examining characteristics, including dose-response relationships and time-dependent effects. Moreover, the identification of synergistic interactions between various ingredients in complicated herbal mixes is made more accessible by *in vitro* investigations, which might improve overall efficacy [19]. This thorough assessment helps to set consistent quality criteria and standardize the active ingredients in herbal medications.

Acute toxicity assessment requires giving each animal a single dose of the product under investigation. For 14 days, the animals' deaths and toxicity indicators are noted to calculate parameters such as the half-lethal dosage (LD_{50}) and the maximum tolerated dose (MTD). To assess subacute and chronic toxicity, the tested compounds are exposed daily to animals for a variable time. It is possible to estimate toxicity characteristics such as LOAEL, LOEL, NOAEL, and NOEL. Although topical, intraperitoneal, and inhalation routes of administration are also possible, oral administration is the most common method for using medicinal plants and HMs [20].

6.6.6 Stability Assessment

The goal of stability is to guarantee that the medication or product stays within the parameters set forth to ensure its identification, potency, purity, and quality. Environmental influences may impact stability. Stability can also be affected by elements such as pH, particle size, water/solvent characteristics, container type, and the existence of other chemicals because of contamination or purposeful product mixing. Extracts are intricate combinations of several ingredients, each with a unique concentration, consistency, activity, and shelf life. As a result, it becomes challenging to establish the stability of the finished product solely on the activity and the stability of a single component, which causes issues when choosing storage conditions. Every ingredient can greatly impact the stability of the formulation. Degradation of these components could result in unfavorable outcomes and safety hazards. Significant potency loss can be brought about by extensive chemical degradation, which can also alter physical appearance, bioavailability, and solubility. All these factors could contribute to product failure. Therefore, ensuring sufficient stability for long-term storage and safety is essential for patient use. Still, numerous issues need to be addressed, such as polytomous drugs, processing methods, adulteration and contamination, and improper agricultural practices, which can be assessed by Good Manufacturing Practices (GMPs).

6.6.7 Good Manufacturing Practices

The fundamental requirements for GMP applicable to HMs align closely with those established for pharmaceutical products. In 1996, the World Health Organization (WHO) published the *GMP: Supplementary Guidelines for Manufacturing Herbal Medicinal Products.* This document provides essential frameworks to guarantee the quality and safety of HMs. The WHO Expert Committee adopted the revised guidelines in 2006, which were finalized in 2005. This process culminated in their publication as Annex 3 of the WHO Technical Report Series, No. 937 (2006). Implementing these GMP allows manufacturers to establish standardized protocols for sourcing raw materials, processing, and packaging. This approach significantly reduces the risk of contamination, adulteration, and variability in product quality. GMP underscores the necessity of maintaining a controlled environment, implementing rigorous documentation practices, and conducting regular quality-control testing, which ensures that every batch of HMs complies with established specifications. Furthermore, these practices promote transparency and traceability, enabling manufacturers to address any quality concerns that may arise effectively. By integrating GMP into the production of HMs, stakeholders can enhance consumer confidence and ensure that products meet regulatory standards and provide consistent therapeutic benefits, thereby supporting the responsible use of HMs in healthcare.

6.7 WHO GUIDELINES FOR STANDARDIZATION

Standardization of HMs is a necessity of the era due to the variation in safety, efficacy, and stability during manufacturing, storage, packaging, transport, and distribution. The WHO describes the standardization and QC of herbs as a process that involves safety assessment, stability and efficacy assurance, documentation of safety and risk, consumer education, and product promotion. All relevant data and parameters (Figure 6.3) regarding the chemical constituents of herbal drugs must be gathered to achieve phytochemical standardization.

6.7.1 Preliminary Evaluation

6.7.1.1 Sampling

Raw materials and finished goods are essential in sample selection during QC.

6.7.1.2 Foreign Matter Determination

Foreign matter refers to anything other than the biological source of raw pharmaceuticals. Insects, fungi, and

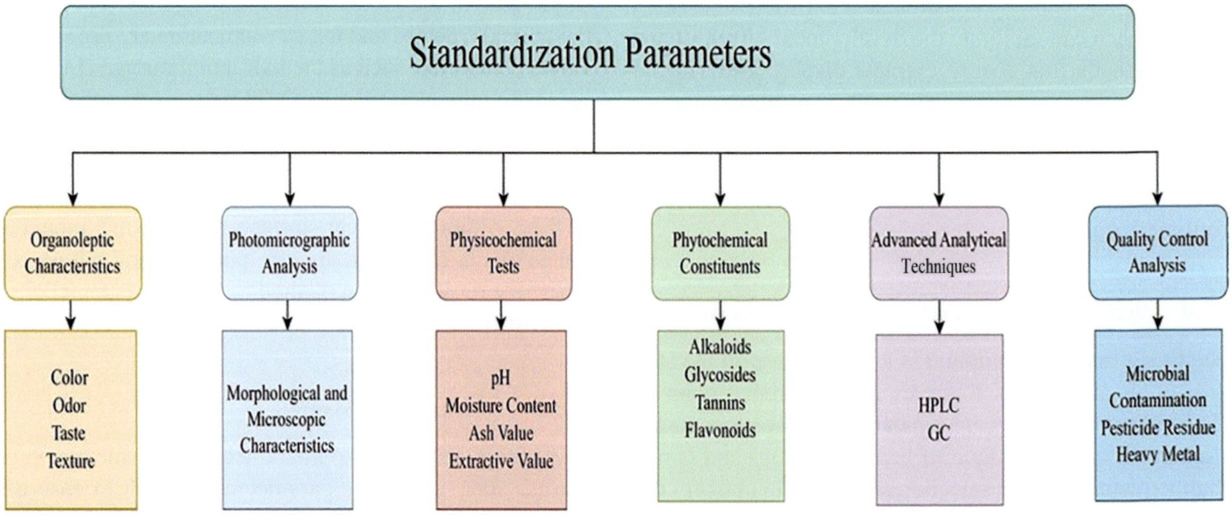

FIGURE 6.3 Standardization parameters as per the WHO.

visible and excluded pollutants, including stones, sand, hazardous foreign bodies, and chemical residues, must be entirely removed. Animal objects include insects, unseen microbiological pollutants that create poisons, and possible contaminants in herbal treatments. Under some circumstances, starch is added purposefully to dilute the sample. This can be detected using microscopy, whereas foreign matter is determined using macroscopic evaluation [21].

6.7.1.3 Microscopical Evaluation

The histological features of a medicinal plant are identified and detailed in microscopic evaluation.

6.7.2 Physical Qualitative Evaluation

6.7.2.1 Ash Value

The ash value determines the quality and purity of a medicine and its identification. The ash value defines a medicine's quality, purity, and identity. Quantitative evaluation of ash value aids in standardization due to its presence in the crude drug. The ash value is utilized to identify foreign inorganic materials as an impurity.

6.7.2.2 Total Ash Value

The determination of the total material that remains after ignition is done by the total ash value and is classified into physiological and nonphysiological values. Physiological ash, which comes from the plant tissue itself, and "nonphysiological ash," which is frequently from environmental contaminants such as sand and soil, are both included in total ash [22].

6.7.2.3 Acid-Insoluble Ash Value

Inorganic factors such as calcium oxalate, silica, and carbonate concentration in the crude medication can impact its total ash value. After acid treatment, these variables are removed, and the acid-insoluble ash value can be determined.

6.7.2.4 Extractive Value

The extractive value of a crude drug determines its chemical constituents and estimates the components that can be extracted using a solvent. Materials without an appropriate chemical or biological assay are subjected to this technique. The methods employed to determine extractive value include cold maceration, hot extraction, and ethanol extraction.

6.7.2.5 Moisture Content and Volatile Matter

Lowering the moisture content will slow down the rate at which microbial infection or chemical changes cause crude medicines to degrade. Medication is heated in an oven at 105°C to assess moisture content until a consistent weight is achieved.

6.7.2.6 Volatile Oil Content

These liquids evaporate quickly and are immiscible with water. They are also known as ethereal or essential oils. They act as a flavoring agent. Herbal medications often employ volatile oil as a flavoring agent. The volatile oil concentration of medications containing volatile components is determined using the toluene distillation/steam distillation technique.

6.7.3 Biological Evaluation

6.7.3.1 Bitterness Value

Intensely bitter medicinal herbs can be used therapeutically as flavoring agents; the amount of bitterness in the extracted material is measured and compared to a standard quinine hydrochloride. The unit of bitterness is represented in units that indicate 1 g of quinine hydrochloride in 2,000 mL.

6.7.3.2 Hemolytic Property

Saponins are usually found in plant families like *Caryophyllaceae*, *Araliaceae*, *Sapindaceae*, *Primulaceae*, and *Dioscoreaceae*. Saponins cause hemolysis by altering erythrocyte membranes, allowing hemoglobin to permeate the surrounding media. They are compared to 1,000 units per gram of saponin R used as a reference material to assess the hemolytic activity of plant materials or saponin preparations.

6.7.3.3 Astringent Property

The quantity of tannins in the herbal drug determines the astringent property. Tannins, also known as tanning chemicals, utilize binding proteins to produce leather from animal hides and form water-insoluble compounds that resist proteolytic enzymes. Application of tannins to living tissue shows a therapeutic action known as an astringent.

6.7.3.4 Swelling Index

The swelling index estimates the milliliters required for 1 g of plant material to swell under specific circumstances. The test process for each plant material requires adding either water or a swelling agent. This index aids in evaluating the medication by providing an estimate of its mucilage content.

6.7.3.5 Foaming Index

Saponins, when shaken with an aqueous decoction, produce foam. The foaming index determines the foaming ability of the herbal drug's aqueous decoction.

6.7.4 Toxicological Evaluation

6.7.4.1 Arsenic and Heavy Metals

Pesticide residue and environmental contamination are two possible sources of heavy metal and arsenic contamination in medicinal plants. Methods used to quantify heavy metals include a limit test for arsenic and a limit test for cadmium and lead. The suggested maximum levels for dry plant products are based on ADI values (lead: 10 mg/kg; cadmium: 0.3 mg/kg). The stain created on $HgBr_2$ paper is compared to a conventional stain.

6.7.4.2 Pesticide Residues

Pesticide levels in herbs are typically limited by the WHO and FAO standards. These are blended with herbs during cultivation. It is well known that pesticides like DDT, benzene hexachloride, toxaphene, and aldrin have severe adverse effects on people. Arsenic and heavy metals were determined. Even tiny levels of arsenic and heavy metals can be hazardous when eliminated from herbal medications. The amount is assessed by comparing the color depth with typical stains [23].

6.7.4.3 Radioactive Contamination

Guidelines have been developed by the WHO in close coordination with other international organizations in the event that a catastrophic nuclear accident results in widespread radioactive contamination. Major nuclear accidents like Chernobyl and Fukushima can offer considerable dangers depending on the exact radionuclide, level of contamination, and quantity consumed. While naturally occurring, radionuclides do not constitute a significant health risk, they can pose one. To ascertain the radioactivity of the plant, one should refer to the guidelines published by the IAEA in Vienna and Australia [23].

6.7.4.4 Microbial Contamination

Chances of contamination occur during processing and at the source. The WHO has set a limit for the number of microorganisms per gram of material, and their values are given in Table 6.5.

6.7.4.5 Aflatoxins Determination

Aspergillus parasiticus and *Aspergillus flavus* produce mycotoxins called aflatoxins. Chromatographic techniques can determine the standard aflatoxin mixtures B1, B2, G1, and G2. The determination limit in the IP method is NMT 2 µg/kg of aflatoxin B1 and total aflatoxins 4 µg/kg, and in the USP method, it is NMT 5 ppb of B1 and total aflatoxins 20ppb.

TABLE 6.5
Number of Microorganisms Per Gram of Material

Type of Microorganism	Finished Product	Raw Material
E.coli	10^1	10^4
Salmonella	—	—
Total aerobic bacteria	10^3	—
Enterobacteria	10^3	—

6.7.5 Analytical Evaluation

TLC can be performed for both qualitative and quantitative evaluation. Allowing simple equipment, having a short development time of 15 min to 1 h, a wide choice of stationary phases, and quick recovery of separated constituents prove TLC to be more advantageous compared to other chromatographic techniques. TLC is a preferred approach for rapid and straightforward examination due to its ability to see isolated components using UV light. The HPLC/UPLC systems with various chemistries are widely used to separate and isolate alkaloids and glycosides. Owing to the existence of naturally volatile chemical components, gas chromatography has established itself as a crucial instrument for HM quality monitoring. It is used to find chemical compounds that are volatile and thermostable [24].

6.8 ANALYTICAL TECHNIQUES FOR Q-MARKER INVESTIGATION

The FDA and the European Medical Association have recommended using chromatographic techniques for the standardization and QC of herbal medications due to their efficient separation capability. The primary techniques used are TLC, HPLC, GC, HPTLC, LC-MS/MS, and capillary electrophoresis. Along with these techniques, other analytical tools such as NMR spectroscopy and vibrational spectroscopy methods, which involve fourier transform infrared spectroscopy (FTIR), NIR, and Raman spectroscopy, have been widely used due to their ability to generate spectra that contain valuable information for standardizing HMs.

6.8.1 NMR

The interaction of the molecules with radio waves gives rise to unique resonance and so-called fingerprints, and proton (^{1}H) and carbon (^{13}C) NMR spectroscopy are often used for QC of HMs. 2D-NMR approaches, including J-resolved, HSQC, and HMBC, can be analytical fingerprints for HM standardization [23]. This method is nondestructive, noninvasive, ecologically friendly, quick, and easy to use regularly, requires minimal sample preparation, and may offer structural information of complicated mixtures without the need for purification. It is also ideal for metabolite fingerprinting of HMs. However, NMR is also limited by low sensitivity, restricted substances for detection, and signal overlapping in complex HMs. Recently, the usefulness of ^{1}HNMR in conjunction with PCA and OPLS-DA has been beneficial in differentiating three Curcuma species: *Curcuma longa, Curcuma xanthorrhiza*, and *Curcuma manga* from various sources in Indonesia. Differentiation between the red bell peppers distributed in Korea and Vietnam was done using ^{1}H

NMR, which aided in determining the different metabolites present in each [25].

6.8.2 Capillary Electrophoresis

Capillary electrophoresis is a technique for separation that relies on how quickly charged particles or ions migrate through a liquid in a capillary tube when an electric field is applied. It uses a small-diameter capillary, increasing the electric field intensity and dramatically enhancing separation effectiveness. An electric field is generated by injecting the sample into an electrolyte solution–filled capillary and then applying a voltage across it. Under the influence of an electric field, charged particles start to migrate. The migration rate depends on the charge, particle size, shape, and characteristics of the electrolyte solution.

Benefits of this method include high separation capability, minimal sample and reagent usage, and reduced apparatus costs compared to HPLC/GC. It may also be used to control HM quality for single marker analysis, fingerprinting, and metabolomic research. Low resolution for nonpolar or non-charged analytes unless combined with other partition-based separation methods. Lower sensitivity compared to chromatographic procedures due to small sample size. CE has been used to examine the fingerprints of a variety of medicinal herbs, including *Ginkgo biloba*, *Smilax glabra Roxb.*, *Carthamus tinctorius L.*, *Scutellaria baicalensis*, *Dendrobium candidum*, and *Glycyrrhiza uralensis* [26].

6.8.3 Chromatography

Once the sample has been applied or injected, the mobile phase carries the sample components through the system. The stationary phase is held in place by a support and remains stationary. Stronger interactions cause components to be held longer and flow through the system more slowly than weaker interactions, which cause components to spend more time in the mobile phase. This difference in travel rate leads to the separation of components as they move through the chromatographic system [24]. Target chemicals in HM matrices may be isolated into fractions with broad applicability, high resolution, selectivity, sensitivity, and completely automated operation. An HR-MS/MS detector may predict and determine the chemical structures of target peaks. Extraction and stability investigations for standards and samples are time-consuming and need costly, sophisticated instruments (LC-MS/MS). Before doing multivariate analysis, each sample requires peak alignment and retention time adjustment.

6.8.4 Direct MS

Mass spectrometry (MS) is an effective analytical method for quantifying known materials, locating novel chemicals within a sample, and clarifying various molecular structures and chemical compositions. Using a sample under investigation, a mass spectrometer produces various ions, separates them based on their unique mass-to-charge ratio (m/z), and then calculates the relative abundance of each ion type. There are three components of MS: the ion source, which makes gaseous ions from the material being studied; the mass analyzer, which separates the ions based on their m/z ratio into individual mass components; and the detector system, which finds the ions and records the relative abundance of each resolved ionic species.

Ambient MS methods may quickly separate analytes in complicated HM matrices based on mass per charge. Sample preparation is simple; some approaches even allow for chemical fingerprint imaging. The cost of an MS detector is higher than that of other detectors, and the ionization efficiency may differ depending on the techniques used and the sample matrices, leading to potential issues with reproducibility. Herbal compounds have been fingerprinted using direct mass spectroscopy from a range of plants such as *Allium sativum, Fritillaria* sp., *Panax quinquefolium* L., *Origanum* sp., *Gastro diaeelata, Cynanchum* sp., etc. [27].

6.8.5 Vibrational Spectroscopy

NIRS technology offers unmatched advantages in analysis speed because it simply needs to capture the NIR while creating samples to perform quality analysis without complex preprocessing. Chemometric approaches are necessary for separating, standardizing, and controlling complex HMs, which are lacking in vibrational spectroscopy methods. Using FTIR spectroscopy, the QC of several extracts of Sonchus arvensis, also referred to as Tempuyung in Indonesia, was carried out in conjunction with chemometrics. The QC of turmeric that was tampered with using the hazardous azo dye metanil yellow was conducted using both ATR-FTIR and Raman spectroscopy.

6.8.6 DNA Barcoding

Identifying species through DNA barcoding involves creating a DNA barcode library of known species and then comparing or classifying unknown barcode sequences with the library. Regarding HMPs, accomplishing these procedures involves five fundamental steps that carry significant importance, as mistakes or misunderstandings can occur at each stage. The process begins with (i) grinding the medicinal herbal product into a homogeneous powder, (ii) extracting the genomic DNA, (iii) using PCR to amplify the specific DNA barcode region, (iv) sequencing the amplified region, and (v) identifying the unknown sequence by comparing it to known standard reference material. It is possible to enable species-level authentication of medicinal samples, which makes it possible to identify adulteration from closely similar species. Plant genotyping using this technique can result in standardized medicinal crops. The technique is unable to identify adulterants from different plant parts within the same species and requires further data on the metabolite composition of HMs. Scientists have reported authentication of *Drynaria roosii*, a traditional Chinese herb, using DNA barcodes.

6.9 NEWER STRATEGIES FOR THE SAFETY EVALUATION OF HERBAL DRUGS

Herbal drugs, while often perceived as natural and safe, can pose significant risks if not appropriately evaluated. Herbal remedies and medicinal plants are widely used in therapy, even in the present era [28]. According to estimates from the WHO, 80% of people in developing nations use these kinds of treatments [29]. Traditional herbal therapy is widely utilized because of its lower cost, greater conformity to patients' beliefs, and ability to meet the need for more specialized care. Therefore, ensuring their safety is crucial before widespread use. While traditional methods have been employed for decades, traditional safety assessment methods may not be sufficient and often face challenges for these complex biological products and their potential for interactions with other substances.

One of the primary concerns in the safety evaluation of herbal drugs is the need for standardized analytical methods and quality-control procedures [30]. Reliable safety profiles are challenging to establish due to the complex chemical composition and heavy metal content, which may vary based on geographical origin, growing practices, and processing procedures. In addition, the underreporting of adverse events results from weak pharmacovigilance systems and poor communication about the use of herbal treatments between patients and healthcare practitioners. Therefore, sophisticated methods are required to enable quick, inexpensive, and more accurate techniques. Conventional authentication techniques consist of identification through morphology, microscopy, and chemical analysis [31]. However, in the last decade, several novel and innovative techniques have gained attention. Recent strategies for assessing the safety of HMs encompass a variety of methodologies, including pharmacovigilance, quality-control measures, and regulatory frameworks. Chemical fingerprinting using a metabolomics approach and DNA-based techniques for herbal authentication are the techniques implemented in the herbal drug industry's quality and safety assessment process [32].

6.9.1 Pharmacovigilance

It is a well-established field of study in the developed world that studies the side effects of drugs and how to identify, treat, record, report, and make regulatory decisions based on them. The pharmacological science of pharmacovigilance relates to identifying, evaluating, comprehending, and preventing adverse effects or other drug-related issues, especially those related to long- and short-term side effects of medications. Efficacy refers to the ability of an intervention to produce desirable, beneficial effects. Efficacious herbal medication has helped human health and the life expectancy of human beings safely. Preclinical efficacy research is concentrated on forecasting a drug's benefits, whereas safety research aims to assess the degree of risk connected with medication. As a result, clinical trials are carried out to evaluate novel medications and herbal preparations for safety and efficacy. The purpose of clinical studies is to disclose side effects, effectiveness, and benefit-harm-risk characteristics. They provide results that should yield trustworthy information for broader populations sharing the same attributes as the trial groups. However, clinical trials have shortcomings such as animal experiments having limited value in forecasting human safety, patients being selected based on a range of criteria, clinical trials having time and patient capacity restrictions, and so on.

Therefore, pharmacovigilance is a helpful tool and a new and emerging area to identify, assess, and handle ADRs and safety concerns, encompassing not only chemical drugs but also herbal, traditional, and CMs and biological products. HMs are widely used in healthcare, both in wealthy and poor nations. Post-marketing surveillance involves tracking the adverse effects of prescription drugs and herbal remedies while the general public uses them. Good Vigilance Practice defines pharmacovigilance as the post-marketing safety monitoring of a drug's side effects [33]. Therefore, pharmacovigilance is one of the most essential instruments for post-marketing surveillance because it guarantees the integrity of pharmaceuticals and related herbal remedies. Despite the great potential of traditional therapies, many of them are still unproven, and their uses are frequently not sufficiently disclosed on labels, leaving little information about possible adverse effects [34]. One prominent instance is the replacement of toxic Aristolochia species in traditional Chinese medicine, which has been linked to cases of kidney cancer recorded in Europe, China, and America, as well as severe renal toxicity [33]. These incidents highlight the diverse safety issues surrounding traditional HMs, prompting the establishment of pharmacovigilance studies that have significantly contributed to ethnobotanical and ethnopharmacological research, as well as to the safety monitoring of herb–drug interactions, which are critical components of comprehensive pharmacovigilance efforts [35].

6.9.2 Intrinsic Toxicity Evaluation

The evaluation of HMP toxicity is frequently associated with bioactive substances found in plant material, and complex combinations can pose unanticipated hazards. Various toxic endogenous compounds such as pyrrolizidine alkaloids, furan derivatives, anthraquinones, epoxy-diterpenoids, ginkgolide acids, and many more have toxic potential, and regulatory authorities have established concentration limits for known toxic compounds found in plants, necessitating effective detection methods [36].

Toxicological evaluation of HMs utilizes both *in vitro* and *in vivo* models, with OECD guidelines established for assessing acute and chronic toxicity. OECD refers to the *in vivo* repeated-dose toxicity study of the tested products over variable periods: OECD-407 (28 days), OECD-408 (90 days), and OECD-452 (12 months). *In vivo* studies are associated with ethical concerns, and a general trend exists to minimize the number of experimental animals. Therefore, ethical considerations lead to reduced animal testing, prompting the development of advanced *in vitro* models that better mimic human physiology. There has been an exponential evolution of new models from 2D and 3D cell cultures to microfluidic devices and miniature artificial organs (organ-on-chip systems). 3D cell culture offers more advantages over 2D-cell culture, better resembles *in vivo* conditions, and improves bioactivity prediction [37]. An alternative *in vivo* testing model, EpiAirway™ is a potential model for assessing inhalation toxicity. It is a credible method for evaluating the effects of herbal products on tissue integrity, cell viability, and the preservation of cilia. Ng'uni et al. used Episkin (reconstituted epidermis skin), another *in vitro* tissue model used in dermal toxicity studies [38], to evaluate the irritation potential of *Galenia africana*.

Another model for safety evaluation is invertebrate animal models, which include zebrafish (*Danio rerio*) and an insect model, *Galleria mellonella* (greater wax moth). The zebrafish assay is widely utilized as an effective and cost-efficient tool for toxicity screening. A key advantage of using zebrafish embryos is their transparency, which allows for detailed observation of developmental processes. This transparency, combined with the use of fluorescence and transgenic strains, enables real-time visualization of organogenesis, making zebrafish a valuable model for studying toxicity at different developmental stages. Different medicinal plants were assessed using zebrafish for toxicity screening, such as *Hypericum lanceolatum* [39], *Piper sarmentosum* [40], and *Urena lobata* [41]. In contrast, Mbarga et al. performed the toxicity evaluation of different medicinal plants from Cameroon, such as *Moringa oleifera*, *Vernonia amygdalina*, *Cinchona officinalis*, *Enantia chlorantha*, *Garcinia lucida*, *Cymbopogon citrates,* and *Azadirachta indica*. The findings revealed that *Galleria mellonella* can be a potential model for toxicity evaluation [42].

6.9.3 Genotoxicity and Carcinogenicity Evaluation

This study is crucial for safety assessment in pharmaceuticals and natural products, as genotoxic agents may act as carcinogens and mutagens in humans. The ICH guidelines recommended a comprehensive approach to genotoxicity testing, emphasizing that no single test is sufficient. In the context of HMPs, genotoxicity assessment is particularly significant, as many plants contain phytochemicals with potential genotoxic or carcinogenic properties. For instance, plants from the genera *Symphytum, Senecio, Crotalaria,* and *Heliotropism* produce pyrrolizidine alkaloids, which can form DNA adducts responsible for genotoxicity and tumor induction [43].

Genotoxicity is evaluated using a variety of *in vitro* and *in vivo* assays, and a battery of tests is typically advised to increase the accuracy of results. For a thorough genotoxicity assessment, the EMA and EFSA advise combining assays such as the Ames test, mouse lymphoma assay, and *in vivo* micronucleus test. Additional *in vivo* testing, such as the Comet assay or the mammalian erythrocyte micronucleus test, is necessary if the *in vitro* results are positive. Novel techniques have been developed, such as the *in vitro* CometChip assay, the *in vitro* MicroFlow micronucleus assay, ToxTRacker, Vitotox, and the TGx-DDI (a DNA damage inducing) transcriptomic biomarker assay, to improve the accuracy of *in vitro* genotoxicity assessments and lessen the dependence on *in vivo* testing. When paired with conventional techniques, these instruments have demonstrated efficacy in distinguishing between pertinent and extraneous positive outcomes in genotoxicity tests [44–46]. For instance, the CometChip was utilized to assess the mutagenic potential of plant extracts, whereas the Vitotox and TGx-DDI assays classified eugenol as a DNA-damaging agent at high concentrations.

6.9.4 Chemical Fingerprinting and Chemometric Evaluations

Polyherbal formulations, commonly used in TCM, aim to achieve synergistic effects by combining multiple plant species and often include animal parts and mineral substances. These multi-item concoction formulations contain highly diverse chemical compositions with hundreds of known and unknown compounds. Most current techniques for determining the authenticity and quality of plant species in products rely on chromatographic examination of one or two biomarkers or pharmacologically active ingredients. Nevertheless, this method is frequently insufficient for intricate blends like polyherbal formulations, where a thorough chemical fingerprint is necessary to avoid problems such as adulteration with similar species or prohibited compounds, which can seriously endanger customers [32].

Herbal profiling, especially for polyherbal products, is challenging. The integration of advanced chromatography techniques, such as UHPLC, reversed-phase ion-pairing HPLC, and size-exclusion chromatography, with MS tools (e.g., UV diode-array detectors and electrospray ionization MS), provides superior analytical capabilities compared to conventional methods. These hyphenated techniques offer rapid and effective separation, higher selectivity and precision, and the ability to eliminate instrumental interferences, enhancing the accuracy of retention time measurements. The resulting chromatographic fingerprints and spectral data can be analyzed using chemometrics, a discipline that applies mathematical and statistical methods to interpret complex chemical data. By employing chemometric techniques, intricate fingerprint patterns can be classified, and instrument errors can be corrected, allowing meaningful information to be extracted from the data. This advanced approach is essential for accurate QC and authentication of complex herbal products [47].

6.9.5 Omics-Based Toxicology

Evaluation using this strategy provides a cellular-level understanding of the interactions between toxic xenobiotics and biological systems, making it a valuable approach for assessing the toxicity of herbal products. Even in the early stages of drug discovery and development, the potential toxicity of a medicine can be predicted thanks to omics research. Toxico-proteomics, toxico-transcriptomics, and toxico-metabolomics are examples of key omics technologies that evaluate these interactions at the transcriptomic, proteomic, and metabolomic levels. This method improves our comprehension of the molecular mechanisms underlying the toxicity of herbal remedies [48]. As Table 6.6 illustrates, most studies use animal experimental models to evaluate the metabolic response of the interaction.

6.10 SAFETY HAZARDS OF HERBAL MEDICATIONS

Expansion of the usage of HMs has also led to an increase in the reporting of toxicity and adverse effects. Potential causes of these reactions would include harmful effects, overdose, hypersensitivity, toxic effects such as neurotoxicity, genotoxicity, and teratogenicity, as well as addiction or idiosyncratic reactions. According to the WHO, the primary cause of adverse reactions due to herbal drugs has been attributed to their substandard or poor quality and improper use, which makes the identification process of the genuine adverse effects more challenging. Another issue that arises with the adverse effects of drugs is clarity in the identification of HMs due to the need to use the scientific names of the drugs. Thus, the WHO Collaborating Centre for International Drug Monitoring has recommended using precise scientific binomial nomenclature for medicinal plants in order to prevent such uncertainty in the adverse response reports [49].

TABLE 6.6
Different Animal Experimental Models to Assess The Metabolic Response of Interactions

Herbal Sample	Analytical Methods	Aim of the Study
Aconitum spp. alkaloids	GC-TOF/MS	Investigation of modified metabolites linked to metabolic diseases that are triggered
Pinelliaternata	UPLC Q-TOF-MS	Examining overall toxicity (liver, kidney, and heart) by monitoring changes in serum metabolism
Sophora alopecuroides	^{1}H NMR	Analysis of the metabonomic profile
Coptidis rhizome	^{1}H NMR GC-MS	Examining the effects of gut microbiota changes on GI toxicity
Aurantio-obtusin	UPLC-QTOF-MS	Finding urine-related biomarkers linked to hepatotoxicity
Xanthii fructus	UPLC-QTOF-MS	Finding biomarkers for urine toxicity linked to changes in lipid metabolism
Dioscorea bulbiferaroot	^{1}H NMR	Hepatotoxicity prediction using urine metabolic disturbances
Polygonum multiflorum	UHPLC-MS HPLC-MS UPLC-QTOF-MS	Finding the routes and biomarkers of hepatotoxicity
Aconitum carmichaelii	LC-QTOF-MS	Assessing toxicity-related indicators in serum
Tripterygium wilfordii	GC-MS	Locating urine biomarkers to help explain toxicity from *Tripterygium wilfordii*

6.10.1 Carcinogenic Effects of Traditional HMs

Conventional drug development involves the use of a variety of *in vitro* and *in vivo* short-term genotoxicity assays, as well as long-term rodent carcinogenicity studies, to assess a medicine's carcinogenic potential. Carcinogenicity testing in rats and mice, the most time-consuming and expensive nonclinical safety studies, is necessary for new medication market approval. There has been little evidence of herbal treatments' potential for genotoxicity and little long-term carcinogenicity research.

6.10.2 Liver Toxicity Associated with HMs

The severity of drug-induced liver toxicity (DILI) varies from mild impairment that results in elevated levels of alanine aminotransferase without a spike in bilirubin levels and clinical evidence, which can be reversed by discontinuing treatment, to jaundice and apparent liver failure that may lead to liver transplantation or death. Two main categories can be used to further categorize DILI: While idiosyncratic DILI, a rare but severe form of liver injury that occurs only in predisposed individuals (perhaps related to immunological allergy processes), cannot be anticipated by premarketing research, preclinical and clinical studies can predict dose-dependent liver toxicity.

Herbal medications are just as capable of producing both types of DILI as conventional pharmaceuticals are. The difficult part is figuring out a direct cause-and-effect relationship between HM consumption and liver damage. Additionally, because self-medication is common, exercising caution when using herbal medications can be a difficult task. The widespread perception that natural products are intrinsically safe is reinforced by products of various plants that are utilized in conjunction with conventional pharmaceutical ingredients. As a result, people frequently forget to tell their doctors about this information. Furthermore, problems with low quality occur, including erroneous plant identification, incorrect part selection, adulteration, faulty labeling, inappropriate storage conditions, and fungal contamination.

6.10.3 HMs' Kinetic Interactions with Traditional Medications

Pharmacokinetic interactions that occur due to the simultaneous administration of HMs along with conventional medicines are a cause of significant concern. Herbal medication co-administration with drugs that have limited therapeutic indices adds even more safety risks. Some examples of herbal drugs that interact with conventional medicines involve pharmacokinetic parameters such as AUC, C_{max}, T_{max}, bioavailability, and metabolic activation, which are affected by the ability of the HMPs and their constituents to alter the action of drug-metabolizing enzymes and transmembrane transporters.

So complete knowledge about the herbal drug's ability to hinder or prompt the key drug-metabolizing enzymes should be known before it is used concomitantly with conventional drugs. According to the modern definition of evidence-based medicine, traditional use and expert opinion are at the bottom of the evidence ladder for pharmaceutical safety and efficacy. The most substantial evidence for safety and efficacy comes from unbiased systematic reviews and randomized clinical trials, either with or without meta-analyses.

6.11 CHALLENGES AND GUIDELINES OF CLINICAL TRIALS OF HERBAL DRUGS

6.11.1 Challenges

Researching herbal drugs is a challenging endeavor. Significant phase III trial design, regulatory requirements, product standardization, finances, and ethical considerations must be considered before filing an IND. The WHO published operational recommendations on regulatory standards for clinical investigations, including herbal products, in 2005. The modern medical diagnosis of HMs determines the inclusion criteria. Due to the disparities in modern medicine's understanding of HMs and their diagnostic criteria, it is challenging to generate homogeneous subject groups for each diagnosis of HMs. One of the key elements that significantly influences how well HM patients respond to treatment is patient participation. By combining physical therapy with psychological support, traditional medicine increases nonspecific aspects, contributing to practice success by leveraging placebo effects. Because HMs comprise several active components, administering and treating them is a complicated process. The patient's commitment to completing herbal therapy is what determines its success. The gold standard, randomized clinical trials with blinding, is supposed to address every significant issue that herbal medicine clinical trials encounter. Most RCTs are double blinded, meaning neither the subject nor the investigator is aware of the treatment allocation. However, because HMs necessitate a holistic strategy involving counseling, listening, explaining, lifestyle and dietary suggestions, and herbal prescriptions, double-blind studies are not practical, and RCTs encounter substantial obstacles during the clinical trials of herbal treatments. Choosing a control group is another one of these challenges. To demonstrate a particular effect of the herbal medication, there should be similarities between the intervention and control groups; therefore, the selection process should be conducted appropriately. Color, smell, duration, frequency of intake, treatment credibility, and the patient's physical posture during administration should all be taken into consideration via standardized controls. On the other hand, choosing a matching control for natural items with an odd flavor or aroma can be challenging. It has been acknowledged that occupational standardization is necessary. This is made clearer by the active role the therapist performs in HMP studies. Protocol violations may happen when practitioners try to deliver the optimum treatment result without considering the protocol. Since HM techniques are customized for each patient, they might not be applied universally as a treatment for a whole community. Furthermore, the results of the experiments are more likely to be influenced by this high sense of confidence. Initial assessments of many psychological factors, such as mood and personality, must be conducted. There is also the matter of drop-ins and sample contamination, which requires attention. Trial investigators should keep a close eye on their subjects to ensure that participants are not using herbal products that are not called for in the protocol.

6.11.2 Guidelines

Various polls indicate that HMs have become an essential component of public health because of their widespread usage. Clinical trials, however, must be carried out to expand their adoption and offer proof of their effectiveness. Using GCP at every level of clinical trials should fulfill the ethical and quality standards for clinical studies. European regulatory bodies use the following protocols in clinical trials involving HMs. Product validation, informed consent, safety information, product dosage form, and ethical issues were all examined in the process.

According to a WHO guideline, identifying the plant species is the initial step in ensuring the quality, safety, and efficacy of HMs. This requires botanical verification. The WHO states that a clinical trial should adhere to a protocol that involves securing informed consent from participants and clearly defining the efficacy endpoints. The origin of the patients under study and the recruitment details should be described. Bias should be avoided at all costs.

According to WHO guidelines, the safety assessment is not the major clinical research interest for herbal products. The lack of adverse effects from substances is generally considered evidence of their safety, and examining historical literature on their use can provide insight into their potential toxicity. If traditional medicine does not have adequate evidence and documentation, then clinical trial evidence is required. The WHO distinguishes between testing standards for traditional herbal remedies and new medications. Humans have used traditional HMs for millennia, and they may be considered time-tested. The testing requirements for traditional HMs, according to the WHO, differ from those for new drugs due to the centuries-long human usage of traditional HMs, which can be considered evidence of their time-tested efficacy. The dose used in the study should be strictly in accordance with the information provided in the relevant literature and as prescribed by traditional medicine practitioners [49].

6.11.3 Ethical Considerations in Clinical Trials with Herbal Products

The ethical norms guiding human participation in research apply equally to herbal treatments and studies employing these drugs. Obtaining approval, ensuring equitable subject selection, evaluating the pros and cons to favor the potential participant, and maintaining a sound experimental design are all essential. Concerns that primarily pertain to clinical studies using HMPs include the following:

- Adulteration of products
- Connections between herbal remedies and other substances
- Data on reproductive and organ toxicity
- Prior dose-finding

The unpredictability should be disclosed to everyone, especially during informed consent. People in various parts of the world have a strong belief that herbal drugs are safe and beneficial, which introduces bias; this can be reduced by careful attention to study design and appropriate selection of the control group. When feasible, consult the community from whom the medication originated during the research process. Share the results and benefits with them. Ethical and well-trained investigators are crucial for ensuring patient safety in research. Recruiting proficient physicians as investigators is essential for promptly recognizing and addressing any adverse events or deterioration of preexisting conditions. Ethics committees must approach herbal trials with the same vigilance as conventional treatment methods.

6.12 OUTLOOK AND CONCLUSION

The future of QC, standardization, and safety evaluation methodologies for HMs in clinical trials is critical for increasing their global acceptability and efficacy. As the herbal industry expands, solid foundations for these characteristics are crucial. The future of HMs in clinical trials hinges on developing robust QC, standardization, and safety evaluation strategies. As the demand for natural therapeutics rises, particularly in complementary and alternative medicine, ensuring the clinical efficacy and safety of HMs becomes paramount. Some strategies, such as safety and toxicological assessments, increased regulatory scrutiny, advancements in toxicogenomics, enhanced pharmacovigilance, and a focus on quality management systems, may become more effective in the future if more attention is paid to these important factors.

There will be an increase in the use of toxicogenomic screening tools. With these tools, possible toxicological interactions between herbs at different biological levels may be evaluated, which can aid in the early detection of harmful effects. This will entail keeping an eye out for adverse responses and adjusting the current synthetic drug monitoring methods to better suit the special characteristics of herbal products. Next-generation sequencing and omics technologies (genomics, proteomics, and metabolomics) may help in understanding how different people react to herbal remedies, leading to more individualized treatment plans. The detection of potentially harmful plants and compounds will be made easier by technological advances. This technological integration will enhance the understanding of herbal toxicokinetics and help predict toxicity due to herb–drug interactions. Better documentation, compliance with quality management systems, and GMP would help improve safety monitoring and speed up response times to safety issues. It will be crucial to do an ongoing study into the safety and effectiveness of herbal remedies. Regulatory agencies will probably set more stringent rules as demand for herbal products grows to ensure their efficacy and safety. To assess the efficacy and safety of natural medications, extensive clinical and scientific evidence is required. Herbal medications will progressively go through clinical studies to receive regulatory permission for widespread therapeutic usage as interest in integrative medicine grows. Traditional healers, researchers, and regulatory agencies must work together to close the knowledge gap between ethnopharmacology and modern medicine. In summary, the development of herbal drugs in the future will be primarily driven by technical developments and improved consumer knowledge, with a greater emphasis on safety, quality, and regulatory compliance [50].

ACKNOWLEDGMENTS

The authors thank the Department of Pharmaceuticals, Government of India for administrative and research support.

REFERENCES

[1] Kumari, R., & Kotecha, M. (2016). A review on the standardization of herbal medicines. *International Journal of Pharma Sciences and Research*, *7*(2), 97–106.

[2] Market Research Future. *Herbal Medicine Market.* (Accessed on Sept 2024). https://www.marketresearchfuture.com/reports/herbal-medicine-market-3250.

[3] Nanjappan, S. K., Paul, D., Ramani, R., Arumugam, S., & Chellappan, D. K. (2022). Profiling of trace elements and regulatory landscape of dietary herbal supplements. In *Environmental Challenges and Medicinal Plants: Sustainable Production Solutions under Adverse Conditions* (pp. 303–317). Springer.

[4] EMEA. *Herbal Medicine Products.* (Accessed onSept 2024). https://www.ema.europa.eu/en/human-regulatory-overview/herbal-medicinal-products.

[5] Rojas, P., Jung-Cook, H., Ruiz-Sánchez, E., Rojas-Tomé, I. S., Rojas, C., López-Ramírez, A. M., & Reséndiz-Albor, A. A. (2022). Historical aspects of herbal use and comparison of current regulations of herbal products between Mexico, Canada and the United States of America. *International*

Journal of Environmental Research and Public Health, *19*(23), 15690.

[6] Maldonado Miranda, J. J. (2021). Chapter 7—Medicinal plants and their traditional uses in different locations. In R. A. Bhat, K. R. Hakeem, & M. A. Dervash (Eds.), *Phytomedicine* (pp. 207–223). Academic Press.

[7] Chaachouay, N., & Zidane, L. (2024). Plant-derived natural products: A source for drug discovery and development. *Drugs and Drug Candidates*, *3*(1), 184–207.

[8] Rizvi, S. A. A., Einstein, G. P., Tulp, O. L., Sainvil, F., & Branly, R. (2022). Introduction to traditional medicine and their role in prevention and treatment of emerging and re-emerging diseases. *Biomolecules*, *12*(10).

[9] Barkat Abul, M., Goyal, A., Barkat, A. H., Salauddin, M., Pottoo, H. F., & Anwer, T. E. (2021). Herbal medicine: Clinical perspective and regulatory status. *Combinatorial Chemistry & High Throughput Screening*, *24*(10), 1573–1582.

[10] Izah, S. C., Ogidi, O. I., Ogwu, M. C., Salimon, S. S., Yusuf, Z. M., Akram, M., Raimi, M. O., & Iyingiala, A.-A. (2023). Historical perspectives and overview of the value of herbal medicine. In S. C. Izah, M. C. Ogwu, & M. Akram (Eds.), *Herbal Medicine Phytochemistry: Applications and Trends* (pp. 1–33). Springer International Publishing,

[11] Bhairam, M., Roy, A., Bahadur, S., Banafar, A., & Turkane, D. (2013). Standardization of herbal medicines: An overview. *Journal of Applied Pharmaceutical Research*, *1*(1), 14–21.

[12] Gupta, M., Chaudhary, P. H., Tawar, M. G., & Shrivastava, B. (2021). Need and scope of standardization of herbal medicines: A review. *International Journal of Green Pharmacy*, *15*, 346.

[13] Mahima, K., Sunil Kumar, K. N., Rakhesh, K. V., Rajeswaran, P. S., Sharma, A., & Sathishkumar, R. (2022). Advancements and future prospective of DNA barcodes in the herbal drug industry. *Frontiers in Pharmacology*, *13*, 947512.

[14] Pulimamidi, S. S., Naik, D. D., Yadav, M., Suryawanshi, K. G., Marathe, S. S., Jorvekar, S. B., Ponneganti, S., Ahmed, S., Hazarika, A., & Borkar, R. M. (2024). Seasonal dynamics of phytometabolites content in Assam tea, Camellia sinensis var. assamica by LC-MS/MS: Implications for quality. *Journal of Food Composition and Analysis*, *134*, 106546.

[15] Borges, C. V., Minatel, I. O., Gomez-Gomez, H. A., & Lima, G. P. P. (2017). Medicinal plants: Influence of environmental factors on the content of secondary metabolites. *Medicinal Plants and Environmental Challenges*, 259–277.

[16] Tanaka, N., & Kashiwada, Y. (2021). Phytochemical studies on traditional herbal medicines based on the ethnopharmacological information obtained by field studies. *Journal of Natural Medicines*, *75*, 762–783.

[17] Kushwaha, S. K., Kushwaha, N., Maurya, N., & Rai, A. (2010). Role of markers in the standardization of herbal drugs: A review. *Archives of Applied Science Research*, *2*(1), 225–229.

[18] Pant, J., Singh, R. M., & Marwah, H. (2021). Methods employed in quality control in ayurveda. *Journal of Medicinal Plants Studies*, *9*, 141–155.

[19] Kadir, D. H. (2021). Statistical evaluation of main extraction parameters in twenty plant extracts for obtaining their optimum total phenolic content and its relation to antioxidant and antibacterial activities. *Food Science & Nutrition*, *9*(7), 3491–3499.

[20] Won, H., Jeong, D. H., Shin, H.-S., Lee, J. H., Lee, J. P., Yang, J.-Y., Jung, K., Jeong, J., & Oh, J. H. (2021). Toxicological assessment of Bromochlorophene: Single and repeated-dose 28-day oral toxicity, genotoxicity, and dermal application in Sprague–Dawley rats. *Frontiers in Pharmacology*, *12*, 690141.

[21] Ekka, N. R., Namdeo, K. P., & Samal, P. K. (2008). Standardization strategies for herbal drugs: An overview. *Research Journal of Pharmacy and Technology*, *1*(4), 310–312.

[22] Pradhan, N., Gavali, J., & Waghmare, N. (2015). WHO (World Health Organization) guidelines for standardization of herbal drugs. *International Ayurvedic Medical Journal*, *3*(8), 2238–2243.

[23] Tambare, P., Tamboli, F. A., & More, H. N. (2011). Standardization of herbal drugs: An overview. *International Research Journal of Pharmaceutical*, *2*(12), 56–60.

[24] Patra, K. C., Pareta, S. K., Harwansh, R. K., & Kumar, K. J. (2010). Traditional approaches towards standardization of herbal medicines: A review. *Journal of Pharmaceutical Science and Technology*, *2*(11), 372–379.

[25] Nurani, L. H., Rohman, A., Windarsih, A., Guntarti, A., Riswanto, F. D. O., Lukitaningsih, E., Fadzillah, N. A., & Rafi, M. (2021). Metabolite fingerprinting using 1H-NMR spectroscopy and chemometrics for classification of three curcuma species from different origins. *Molecules*, *26*(24), 7626.

[26] Gong, D., Zheng, Z., Chen, J., Pang, Y., & Sun, G. (2021). Holistic quality evaluation of compound liquorice tablets using capillary electrophoresis fingerprinting combined with chemometric methods. *New Journal of Chemistry*, *45*(5), 2563–2572.

[27] Pereira, H. V., Pinto, F. G., Dos Reis, M. R., Garrett, T. J., Augusti, R., Sena, M. M., & Piccin, E. (2021). A fast and effective approach for the discrimination of garlic origin using wooden-tip electrospray ionization mass spectrometry and multivariate classification. *Talanta*, *230*, 122304.

[28] Woo, C. S. J., Lau, J. S. H., & El-Nezami, H. (2012). Herbal medicine: Toxicity and recent trends in assessing their potential toxic effects. In *Advances in Botanical Research* (Vol. 62, pp. 365–384). Elsevier.

[29] Islam, S. U., Dar, T. U., Khuroo, A. A., Bhat, B. A., Mangral, Z. A., Tariq, L., Tantray, W. W., & Malik, A. H. (2021). DNA barcoding aids in identification of adulterants of Trillium govanianum Wall. ex D. Don. *Journal of Applied Research on Medicinal and Aromatic Plants*, *23*, 100305.

[30] Wang, H., Chen, Y., Wang, L., Liu, Q., Yang, S., & Wang, C. (2023). Advancing herbal medicine: Enhancing product quality and safety through robust quality control practices. *Front Pharmacol*, *14*, 1265178.

[31] Jităreanu, A., Trifan, A., Vieriu, M., Caba, I.-C., Mârțu, I., & Agoroaei, L. (2023). Current trends in toxicity assessment of herbal medicines: A narrative review. *Processes*, *11*(1), 83.

[32] Pauzi, N. A. M., Cheema, M. S., Ismail, A., Ghazali, A. R., & Abdullah, R. (2022). Safety assessment of natural products in Malaysia: Current practices, challenges, and new strategies. *Reviews on Environmental Health*, *37*(2), 169–179.

[33] Fokunang, E. T., Fonmboh, D. J., Mballa, R. N., Nyuyki, A. B., Fokunang, L. B., Kaba, N., Abong, T. B., Duerr, R., Richard, E., & Ondoua, M.-T. A. (2020). Pharmacovigilance of natural herbal medicines research for efficacy, safety and quality assurance of phytomedicine products. *Journal of Complementary Alternative Medical Research*, 21–37.

[34] Paul, D., Surendran, S., Chandrakala, P., & Satheeshkumar, N. (2019). An assessment of the impact of green tea extract on palbociclib pharmacokinetics using a validated UHPLC–QTOF–MS method. *Biomedical Chromatography*, *33*(4), e4469.

[35] Rodrigues, E., & Barnes, J. (2013). Pharmacovigilance of herbal medicines: The potential contributions of ethnobotanical and ethnopharmacological studies. *Drug Safety*, *36*, 1–12.

[36] Wang, Y.-K., Li, W. Q., Xia, S., Guo, L., Miao, Y., & Zhang, B.-K. (2021). Metabolic activation of the toxic natural products from herbal and dietary supplements leading to toxicities [Review]. *Frontiers in Pharmacology*, *12*.

[37] Badr-Eldin, S. M., Aldawsari, H. M., Kotta, S., Deb, P. K., & Venugopala, K. N. (2022). Three-dimensional *in vitro* cell culture models for efficient drug discovery: Progress so far and future prospects. *Pharmaceuticals*, *15*(8), 926.

[38] Słoczyńska, K., Popiół, J., Gunia-Krzyżak, A., Koczurkiewicz-Adamczyk, P., Żmudzki, P., & Pękala, E. (2022). Evaluation of two novel hydantoin derivatives using reconstructed human skin model EpiskinTM: Perspectives for application as potential sunscreen agents. *Molecules*, *27*(6), 1850.

[39] Gence, L., Fernezelian, D., Bringart, M., Veeren, B., Christophe, A., Brion, F., Meilhac, O., Bascands, J. L., & Diotel, N. (2022). Hypericum lanceolatum lam. Medicinal plant: Potential toxicity and therapeutic effects based on a Zebrafish model. *Frontiers in Pharmacology*, *13*, 832928.

[40] Zainol Abidin, I. Z., Fazry, S., Jamar, N. H., Ediwar Dyari, H. R., Zainal Ariffin, Z., Johari, A. N., Ashaari, N. S., Johari, N. A., Megat Abdul Wahab, R., & Zainal Ariffin, S. H. (2020). The effects of Piper sarmentosum aqueous extracts on zebrafish (Danio rerio) embryos and caudal fin tissue regeneration. *Scientific Reports*, *10*(1), 14165.

[41] Purnomo, Y., Aini, N., & Noerhayati, E. (2022). Acute toxicity level of pulutan (Urena lobata) leaf extract on Zebrafish (Danio rerio) and its analysis by *in silico* study. *Research Journal of Pharmacy and Technology*, *15*(6), 2477–2482.

[42] Mbarga, M. J. A., Podoprigora, I. V., & Anyutoulou, K. L. D. (2021). Galleria mellonella (greater wax moth) as an eco-friendly *in vivo* approach for the assessment of the acute toxicity of medicinal plants: Application to some plants from Cameroon. *Open Veterinary Journal*, *11*(4), 651–661.

[43] Baldrick, P. (2021). Genotoxicity test battery: An assessment of its utility in early drug development. *Mutation Research/Genetic Toxicology and Environmental Mutagenesis*, *868–869*, 503388.

[44] Allemang, A., De Abrew, K. N., Shan, Y. K., Krailler, J. M., & Pfuhler, S. (2021). A comparison of classical and 21st century genotoxicity tools: A proof of concept study of 18 chemicals comparing *in vitro* micronucleus, ToxTracker and genomics-based methods (TGx-DDI, whole genome clustering and connectivity mapping). *Environmental and Molecular Mutagenesis*, *62*(2), 92–107.

[45] Verschaeve, L. (2015). Investigations of plant-derived products with the *in vitro* comet assay. Front. Genet. Conference Abstract: ICAW.

[46] Li, H.-H., Yauk, C. L., Chen, R., Hyduke, D. R., Williams, A., Frötschl, R., Ellinger-Ziegelbauer, H., Pettit, S., Aubrecht, J., & Fornace, A. J. (2019). TGx-DDI, a transcriptomic biomarker for genotoxicity hazard assessment of pharmaceuticals and environmental chemicals [Review]. *Frontiers in Big Data*, *2*.

[47] Bai, Y., & Zhang, H. (2024). The cluster analysis of traditional Chinese medicine authenticity identification technique assisted by chemometrics. *Heliyon*, *10*(18), e37479.

[48] Efferth, T., & Greten, H. J. (2012). Potential of 'Omics' technologies for implementation in research on phytotherapeutical toxicology. In *Advances in Botanical Research* (Vol. 62, pp. 343–363). Elsevier.

[49] Al-Worafi, Y. M. (2020). Herbal medicines safety issues. In *Drug Safety in Developing Countries* (pp. 163–178). Elsevier.

[50] Nanjappan, S., Paul, D., & Bolla, L. (2018). Assessing herb–drug interactions of herbal products with therapeutic agents for metabolic diseases: Analytical and regulatory perspectives. *Studies in Natural Products Chemistry*, *59*, 283–322.

7 Ethical Aspects of Clinical Trials of Herbal Medicines Across the Globe

Tanya Ralli[1], *Nitish Singh Jangwan*[2], *Swati Dobhal*[3], *Shradha Bisht*[4], *and Mamta F. Singh*[1]

[1]College of Pharmacy, COER University, Roorkee, Uttarakhand, India

[2]Department of Pharmacognosy and Phytochemistry, School of Pharmaceutical Sciences, Delhi Pharmaceutical Sciences and Research University, New Delhi, India

[3]School of Pharmaceutical Sciences and Technology, Sardar Bhagwan Singh University, Balawala, Dehradun, Uttarakhand, India

[4]Uttaranchal Institute of Pharmaceutical Sciences, Uttaranchal University, Dehradun, Uttarakhand, India

7.1 INTRODUCTION

Plants and herbs have been utilized by humanity for centuries and are employed worldwide for disease treatment and prevention. Plants contribute around 25% of worldwide prescribed contemporary pharmaceuticals [1]. The traditional healthcare system is increasingly popular globally, driven by public interest in herbal remedies and their remarkable acceptance for healing features with few or no adverse effects for various health issues [2]. Traditional medicine, particularly the utilization of medicinal plants, has been integral to human history globally, with knowledge transmitted between generations. In several regions of the world, including Asia and Europe, documentation can be traced to antiquity. Currently, herbal and traditional medicines are enhancing healthcare in several places globally [3]. Herbal medicines are utilized as both primary treatments and adjuncts to conventional pharmaceuticals by various populations. In numerous developing countries, traditional medicine constitutes the sole therapeutic option accessible to a significant portion of the population. Herbal remedies have been widely embraced in numerous industrialized nations and are increasingly significant in the European Union, Australia, North America, and the United Kingdom. The World Health Organization (WHO) reports that 90% of the people in Africa and 70% in India rely on herbal medicines for their healthcare requirements. The percentage of the population using herbal medicine for some major countries has been shown in Figure 7.1. Traditional medicine facilities exist in over 90% of general hospitals in China. Herbal medicine in Germany constitutes one of the five principal components of naturopathy, sometimes referred to as Kneipp therapies [1, 4].

The majority of people have the perception that since the herbal products are isolated from plants, they must be safe. However, as herbal products also possess pharmacological actions similar to synthetic drugs, they must possess one or more side effects [5–7]. Various side effects include interactions between drugs, toxicity due to overdose or at normal doses, and combining it with different products [8–10]. As can be seen with synthetic drugs, herbal medicines were not subjected to different phases of clinical trials for hundreds and thousands of years, which poses problems related to the risk of taking herbal drugs. Thus, to overcome the problems related to side effects, toxicity, and drug interactions, the clinical trials of herbal medicines need to be done in a systematic way. However, while making the guidelines for the regulation of herbal medicines, various unique ethical challenges associated with them must be considered. These unique challenges include biopiracy, cultural sensitivity, scientific validity, risk assessment, intellectual property issues, and informed consent. Furthermore, the ethical issues that emerge when clinical trials of herbal treatments are incorporated into the worldwide research agenda include intellectual property, safeguarding indigenous knowledge, cultural sensitivity, and the potential for exploitation in low- and middle-income countries (LMICs) [3, 11, 12].

This chapter highlights some major unique ethical challenges that hinder the development of regulatory guidelines for herbal medicines. Furthermore, the multifaceted aspects of collaboration in herbal medicine trials, highlighting the significance of ethical practices and their application in diverse circumstances, have also been discussed in this chapter.

7.2 ETHICAL PRINCIPLES IN CLINICAL RESEARCH AND GCP GUIDELINES

Normative ethics focuses on which moral norms to follow. Universal norms, such as prohibitions against killing and

DOI: 10.1201/9781003522065-7

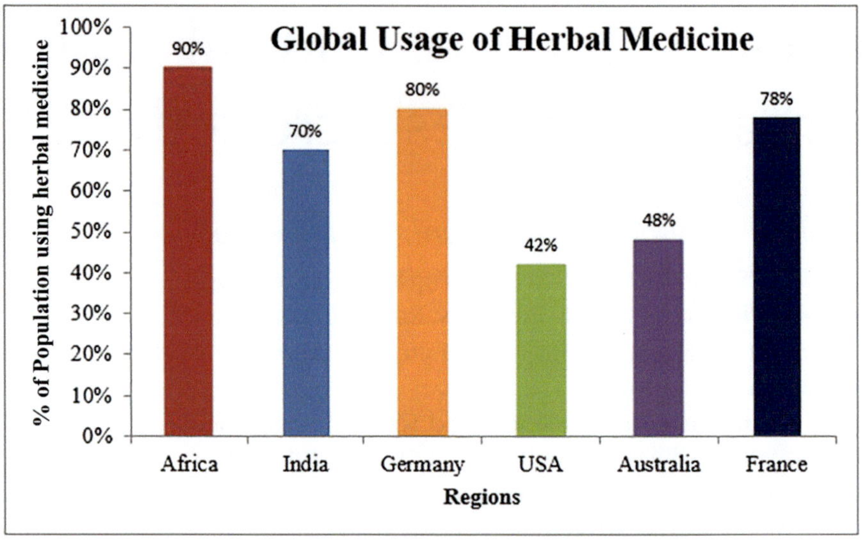

FIGURE 7.1 The global usage of herbal medicine in Africa, India, Germany, the United States, Australia, and France.

stealing, are part of common morality, while particular morality includes norms specific to culture and profession. Professional codes sometimes prioritize the profession's interests over broader moral concerns. Various evidences of unethical practices, such as medical interventions without informed consent and experiments conducted in concentration camps during World War II, along with significant medical advancements and societal changes, have driven the evolution of bioethics. Initially focused on professional conduct and codes, bioethics has expanded to encompass a broad range of areas, including research ethics, public health ethics, organizational ethics, and clinical ethics [13]. In order to ensure the protection of research participants and the integrity of the research process, ethics is essential in clinical research. It ensures that studies are conducted responsibly, with respect for participants, and with the aim of advancing knowledge in a manner that is both morally and legally sound.

Multiple past unethical human experiments highlighted the urgent need for establishing ethical principles in clinical research, but two significant events played a crucial role in establishing guidelines for clinical research. The first one occurred during World War II, when Nazi doctors conducted brutal experiments on concentration camp prisoners involving mutilation, pain, and often fatal outcomes. Another most infamous example of unethical research is the Tuskegee Syphilis Study from 1932 to 1972, which involved withholding treatment for syphilis from African-American men in Alabama [14]. The Nuremberg Code of 1947 was the first international document on research ethics involving humans, emphasizing the necessity of voluntary consent [15]. In 1964, the World Medical Association introduced the Declaration of Helsinki, which has been revised seven times, with the latest update in October 2013 in Fortaleza, Brazil [16]. The Belmont Report, released in 1979 by the National Commission for the Protection of Human Subjects for Biomedical and Behavioral Research in the United States, established three core ethical principles that are respect for persons, beneficence, and justice for human research [17]. In 1991, the U.S. Department of Health and Human Services (DHHS) introduced the Federal Policy for the Protection of Human Subjects, known as the "Common Rule," which was revised in 2017 [18]. The International Conference on Harmonization (ICH) issued the Good Clinical Practice Guidelines E6 (R1) in 1996, which were updated to E6 (R2) in 2016. Additionally, recommendations and guidelines relevant to bioethics were released by the National Bioethics Advisory Commission in the United States (2001), the Council for International Organizations of Medical Sciences (CIOMS) in Geneva (2002, revised in 2016), and the Nuffield Council on Bioethics in the United Kingdom (2002) [19]. Ethical guidelines are crucial for maintaining research integrity, safeguarding participants, and preventing exploitation. Government regulations mandate that all clinical trials receive institutional review board approval to ensure ethical standards and participant protection. A key ethical issue is ensuring participants are fully informed about the risks and the possibility of not benefiting personally from the research, as highlighted by the Hastings Center [20].

7.2.1 Basic Ethical Principles of Clinical Research Involving Human Participants

7.2.1.1 Beneficence and Non-Maleficence

This principle mandates that researchers should enhance benefits and reduce harms associated with their research. Studies should be designed in such a way as to have a beneficial risk-benefit ratio, and the risks associated with the research must be justified by the anticipated benefits.

The principle of non-maleficence underpins several moral rules such as not killing, causing pain, incapacitating, offending, or depriving others of essential goods. In practice, physicians must evaluate the benefits and burdens of treatments, avoid those that are excessively burdensome, and select the best course of action for their patients.

7.2.1.2 Justice

This principle mandates the fair distribution of the benefits and harms of research. Equals should be treated equally. It ensures that the selection of participants in research is fair and that no groups are unjustly excluded from the potential benefits.

7.2.1.3 Autonomy

This principle emphasizes respecting individual autonomy and protecting vulnerable participants by ensuring informed consent. This means that the participants should voluntarily choose to take part in research and have the choice to withdraw at any time. In addition to this, the participants must ensure that they understand the purpose of the research and their role in it [17]. Respecting autonomy requires physicians to provide patients with the necessary medical information and treatment options to enable self-determination. This principle supports informed consent, honesty, and confidentiality.

7.2.2 Overview of Good Clinical Practices

GCP is an international ethical and scientific quality standard for the design, conduct, performance, monitoring, auditing, recording, analysis, and reporting of clinical trials [21]. GCP originated from the ethical principles outlined in the Declaration of Helsinki in 1964. Over time, GCP has expanded to include detailed guidelines from regulatory bodies such as the ICH and various global regulatory agencies. Adhering to GCP regulations is a legal requirement in many countries and serves as an ethical obligation to safeguard the rights, safety, confidentiality, and well-being of research participants. Adhering to GCP indicates that researchers, sponsors, and investigators contribute to the continuous improvement of healthcare and the development of safe and effective treatments for patients worldwide. GCP forms the backbone of ethical and clinical research [21].

WHO formulated a scientific group in 1968 to establish principles for the clinical evaluation of drugs, including new products and new uses for existing ones. In 1975, another group was formed to focus on evaluating and testing drugs, leading to proposals for drug development guidelines. These efforts led to the development of the "Guidelines for Good Clinical Practices" for pharmaceutical trials, published in 1995, and also influenced other guidelines such as the International Conference of Harmonization (E6), "Good Clinical Practice: Consolidated Guideline" in 1996 [22]. The purpose of these principles was to safeguard research participants, ensure high-quality data collection, and encourage global collaboration to further medical science. Complying with GCP is an ethical responsibility to prioritize the well-being of participants and uphold the scientific integrity of clinical trials [23].

7.2.3 Key Principles of GCP

The principles of GCP are designed to be flexible and applicable across a wide range of clinical trials. Along with the ICH E8 guidelines, they encourage thoughtful planning and consideration of the unique aspects of each individual trial. This includes evaluating various trial characteristics such as design elements, the investigational product being assessed, the medical condition being targeted, participant characteristics, the trial settings, and the type of data collected. Consideration of these relevant factors is essential to ensure the quality of each trial. The principles aim to promote a more efficient and improved approach to trial design and conduct. In order to increase the reliability of test results, clinical trials should be designed in such a way that they protect the rights, safety, and well-being of study participants [24]. Table 7.1 outlines the various WHO principles of research involving humans. Clinical trial designs and processes should be proportionate to the risks inherent in the trial and the importance of the data being collected. Clinical trials must generate high-quality, reliable information to support decision-making. Systems for data capture, management, and analysis should be appropriately designed to mitigate risks and ensure data integrity. Key factors include system validation, data protection, IT security, and user management. Effective processes are essential for maintaining data integrity and protecting personal information. Sponsors and investigators should securely retain trial-related data for the required duration, ensuring it is accessible to regulatory authorities for verification and reconstruction of trial results [25].

7.3 UNIQUE ETHICAL CHALLENGES IN CLINICAL TRIALS OF HERBAL MEDICINES

Since ages, herbal drugs have been utilized widely, as they are considered to be safe. However, as these drugs are used as dietary supplements in the majority of countries, there are no strict regulatory guidelines for them. Hence, due to a lack of safety and efficacy data, these drugs possess several side effects and adverse effects [5–7]. Thus, to overcome this issue, the clinical trials of herbal drugs must be conducted in a systematic way, like those of synthetic drugs. Thus, this section highlights various ethical challenges that need to be overcome while implementing clinical trials in

TABLE 7.1
Brief Description of The Principles of GCP as Per Who Guidelines [13]

S. No.	Principle	Description
1.	Ethical principle	Research must be scientifically sound and follow ethical principles from the Declaration of Helsinki, with emphasis on respect for persons, beneficence, and justice.
2.	Safety of participants	Foreseeable risks, discomforts, and anticipated benefits for participants and society must be identified before initiating research. Research should only begin if the benefits clearly outweigh the risks, taking care of the rights, safety, and well-being of subjects.
3.	Informed consent	Informed consent must be obtained from every participant, indicating that the participants are well informed about the study protocol. Research should be scientifically justified and described in a clear, detailed protocol. The protocol should outline study organization, objectives, design, methodology, and statistical considerations.
4.	Independent Ethics Committee/ Institutional Review Board	Institutional Review Board (IRB) approval must be obtained before initiating research. Research must be conducted in compliance with the approved protocol.
5.	Scientific Validity	Clinical trials must be scientifically valid for their intended goals and should be based on robust scientific knowledge.
6.	Qualified individuals	Qualified and licensed personnel must be responsible for medical decisions, care of research subjects, and conduct of the trial.
7.	Quality of the trial	A valid quality system must be implemented for all aspects of the trial. Factors contributing to the quality of the trial must be identified. Quality by design (QbD) in clinical trials aims to proactively ensure trial quality by incorporating it into the study protocol and processes from the beginning.
8.	Risk management	Clinical trial processes should align with the inherent risk of the trials and the significance of the information collected. The trials should consider risks to the participants' rights, safety, and well-being, as well as the reliability of the results.
9.	Study protocol	A well-designed study protocol is essential for the protection of study participants and generating reliable results. The scientific objectives must be clearly stated in the protocol, and the trial processes should be feasible, avoiding unnecessary complexities and data collection. Trial processes should support the key study objectives.
10.	Reliable results	Clinical trials must produce high-quality, reliable information to support decision-making. Procedures for data collection, management, and analysis should be appropriately designed to mitigate risks and ensure data integrity.
11.	Data integrity	Information from clinical trials must be recorded, handled, and stored to ensure accurate reporting and verification. Effective measures are required for maintaining the integrity of results and the protection of personal information. Investigators should retain trial-related data for the required duration, ensuring it is accessible to regulatory authorities for verification and reconstruction of trial results.
12.	Confidentiality of data	Confidentiality of records or identifying objects must be protected according to regulatory requirements.
13.	Good Manufacturing Practice	Investigational products should be manufactured and stored according to the Good Manufacturing Practice and use.

herbal medicines research. Figure 7.2 illustrates the major ethical challenges associated with the clinical trial of herbal drugs.

7.3.1 Biopiracy

Biopiracy constitutes a significant ethical issue in international herbal medicine research. This pertains to the utilization of biological resources, including medicinal plants, for commercial gain without securing the requisite consent from the originating country or community. This is especially concerning when traditional knowledge regarding the therapeutic applications of these plants is involved. Biopiracy frequently transpires during herbal medicine trials, wherein indigenous knowledge is appropriated, exploited, and patented without appropriate acknowledgment or remuneration to the knowledge custodians [26, 27].

Despite global acknowledgment and legal structures advocating for the preservation of traditional knowledge, indigenous people persist in encountering biopiracy,

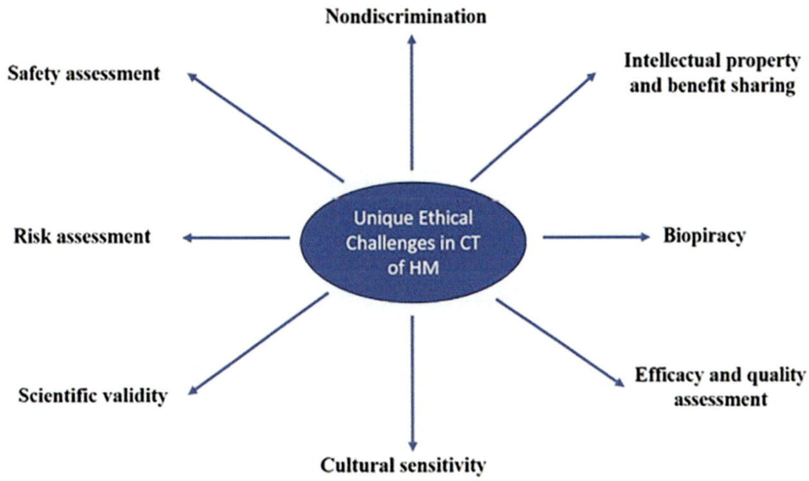

FIGURE 7.2 Unique ethical challenges related to clinical trials of herbal medicines.

wherein commercial entities use their resources without consent. This unethical behavior, motivated by profit-driven multinational businesses, jeopardizes the traditional legacy and dignity of indigenous peoples while raising significant ethical concerns over cultural appropriation and economic disparity. By restricting recognition, respect, and equitable remuneration for these people, biopiracy exacerbates poverty and intensifies socioeconomic inequality. The term "biopiracy," introduced by Pat Mooney in 1993, denotes the use of genetic resources and indigenous knowledge without appropriate acknowledgment. Occurrences of biopiracy, particularly within traditional medicinal systems, are prevalent. A 2013 article by the *Times of India* claimed that 2,000 worldwide patents for traditional Indian medicine were filed annually, predominantly by countries such as the United States, Japan, and China, indicating significant misuse of indigenous knowledge. These statistics are probably understated, highlighting the pressing necessity for enhanced legal protections and international rules to protect indigenous intellectual property [26].

Global ethical concerns necessitate the establishment of stringent legal frameworks to avert biopiracy and safeguard the rights of indigenous populations. The Convention on Biological Diversity (CBD) and the Nagoya Protocol are significant international accords designed to tackle these issues; yet enforcement is sometimes inadequate, especially in nations with constrained resources or insufficient political will. Researchers and organizations conducting herbal medicine trials must comply with international norms mandating prior informed consent and equitable benefit-sharing to avert the exploitation of indigenous communities [26, 28, 29].

7.3.2 Scientific Validity

The clinical trials conducted must be designed in such a way that the results generated are scientifically valid and produce importance in society. The clinical trials of herbal drugs pose problems related to the design of the study and the methodology adopted for obtaining the results [30].

7.3.2.1 Study Design and Methodology Adopted

Due to the problems in the study design and the methodology adopted for estimating the safety, efficacy, and quality of herbal medicines, the results are mainly treated as uncertain [31]. Out of all the methods, randomized controlled trials (RCT) always give results that can often be treated as valid in scientific terms. However, these types of trials are not applicable in cases where different individuals are receiving different types of herbal medicines simultaneously [32]. To overcome these shortcomings, different novel techniques must be incorporated into the clinical trials of herbal medicine. The clinical trials of herbal medicines must be considered in three groups: (a) standardized herbal formulation, (b) customized herbal formulation, and (c) the placebo [33]. Moreover, there is also an issue in preparing the placebo of herbal medicines, as herbal preparations have very peculiar taste and odor, which makes it very difficult to prepare a placebo that has similar organoleptic properties to the herbal preparations [34].

Furthermore, since the herbal medicines are isolated from different plants, which grow in different locations, they might possess varying amounts of active constituents, which in turn results in varying quality standards [33]. Thus, without proper quality control, the results obtained from clinical trials would surely be wrong. Moving further, the identification of an active herbal constituent from plenty of bioactive molecules in a single herbal preparation would also be very difficult. Indeed, it would be very difficult to understand which bioactive molecule is involved in the pharmacological action of the herbal preparation. Hence, to overcome these problems, new strategies like the "bioactive equivalence oriented feedback screening method" [35] and omics techniques have been proposed, which help in the identification of the bioactive marker [36, 37].

7.3.3 Risk Assessment

Before starting any clinical trial, risk-benefit analysis must be done adequately by checking the available data on the toxicity and preclinical trials. As human subjects are involved in clinical trials, it is necessary to first check the risk associated with the drug. Any clinical trial is ethical if the risk involved is much less than the anticipated benefit of the drug. There is a perception among the majority of people that, as herbal products are isolated from plants, they must be safe [5–7]. Hence, many people do not report the side effects observed due to the consumption of herbal drugs. This leads to a false risk-benefit analysis in the case of herbal or traditional medicines.

The WHO has developed guidelines related to the monitoring of safety data of herbal medicines, but several challenges still exist in its reporting, which leads to nonauthentic safety data [38]. The major issues in the reporting of pharmacovigilance data of herbal drugs include (a) many different drugs are known by the same local name in different areas and (b) it is very difficult to differentiate herbal medicines from their close relatives. These factors, in turn, lead to nonreporting or false reporting of data related to herbal drugs [39]. Hence, to overcome these issues, novel strategies i.e., DNA barcoding and omics techniques, were introduced, facilitating greater contribution of safety data in pharmacovigilance [37, 40]

7.3.4 Safety Assessment

Similar to synthetic drugs, it is very important to identify the adverse effects and side effects that occur with the administration of herbal preparations. In the case of herbal medicines, since they are a combination of multiple active molecules, it would be very difficult to understand which side effect or adverse effect is occurring due to which particular biomarker. This problem is not seen in the case of synthetic molecules. Thus, it falls under the unique ethical challenges in the clinical trials of herbal medicines. Another challenge observed in herbal medicines is the presence of quacks because there are no proper regulations for them [23, 41].

7.3.5 Quality and Efficacy Assessment

Most of the incidents observed were due to the administration of poor and adulterated quality herbal drugs. Thus, while preparing herbal preparations, the pharmacopoeias of different countries stating the quality standards of herbal drugs must be followed adequately. Moreover, the proper regulations of Good Manufacturing Practice (GMP) and Good Agricultural and Collection Practices (GACP) must be followed [42].

7.3.6 Nondiscrimination

In many areas, the first line of treatment is always herbal medicine due to either the nonavailability of synthetic drugs or cultural beliefs. However, because of their cultural beliefs, it is not advisable that these communities of people do not receive facilities for diagnosis, treatment, etc. Therefore, there should be no discrimination in providing modern medicines to these individuals. Both modern and traditional medicine can coexist and work together in giving better results [43].

7.3.7 Intellectual Property and Benefit-Sharing

A significant ethical issue pertains to intellectual property (IP) rights. When pharmaceutical corporations or researchers patent herbal remedies derived from indigenous knowledge, the original knowledge holders frequently receive no entitlement to the revenues created from these innovations. Many traditional treatments utilized for generations by indigenous populations may be copyrighted after they undergo scientific validation through clinical research. Nonetheless, the indigenous people who cultivated and preserved this knowledge are frequently marginalized from the financial and legal advantages associated with patent ownership [44–46].

The World Intellectual Property Organisation (WIPO) recognizes traditional knowledge as an essential repository of skills and practices fundamental to cultural and spiritual identity, emphasizing its persistence across generations. This knowledge is jeopardized by influential entities that utilize and copyright it without recognizing the indigenous groups that safeguard it. Multinational firms frequently exploit indigenous knowledge without equitable remuneration, underscoring the conflict between IP rights and commercial objectives. This presents ethical dilemmas around cultural appropriation and the potential erosion of centuries-old wisdom for commercial gain. This exploitation denies indigenous populations their proper share, increasing poverty and inequality. In addition to IP, the protection of traditional knowledge pertains to equity, cultural reverence, and global social justice, hence categorizing it as a core human rights concern [26]. This inequity in IP rights prompts apprehensions regarding intellectual colonialism, wherein the knowledge and resources of one group are used by a more dominant entity, frequently without equitable remuneration. The establishment of efficient benefit-sharing procedures is crucial to guarantee that the revenues from herbal medicine trials are reinvested in the communities that contributed the original knowledge. This may manifest as cash remuneration, investment in local healthcare facilities, or granting communities access to novel treatments developed from their traditional remedies. Ethical principles must emphasize the creation of procedures to guarantee that indigenous knowledge holders receive equitable benefits from their own knowledge [28, 47, 48].

7.4 CHALLENGES IN INFORMED CONSENT

Informed consent is crucial not only in clinical research but also for all research involving human subjects, reflecting the ethical principles of beneficence and nonmaleficence. It ensures that patients understand their rights, the purpose of the study, procedures, and the potential risks and benefits of the proposed intervention, enabling them to make informed decisions [49]. Informed consent is an ongoing process that includes providing relevant information to potential participants, ensuring their competence, and making the information easily understandable. It also emphasizes the voluntariness of participation, protecting individuals' freedom of choice and respecting their autonomy [50]. As a requisite, participants must have the capacity to understand the research and make an informed decision to provide consent voluntarily, without any coercion or undue inducements. It is required to obtain consent before starting any study-related procedures and to ensure the privacy and confidentiality of participants at all times. If an individual cannot give consent, the consent of a legally authorized representative is required. For differently abled participants, methods like Braille should be used to enhance understanding. If a participant is willing to participate but cannot sign or give a thumbprint, verbal consent may be obtained with the approval of the ethical committee in the presence of an impartial witness. However, verbal consent should only be used in exceptional circumstances. In the case of illiterate participants, the consent process must be witnessed by an impartial, literate individual who is not related to the participant and has no relation to the research. Each participant should sign the informed consent form after receiving, comprehending, and voluntarily agreeing to the information presented. If a participant cannot sign, a thumb impression must be obtained [50]. Despite meeting the necessary requirements for obtaining informed consent, challenges remain that make the process complex.

7.4.1 Key Challenges Involved in Informed Consent

7.4.1.1 Language Barrier

Individuals signing consent forms are assumed to understand the information, but evaluating their comprehension is difficult, often leading to misunderstanding, especially due to language barriers. In multilingual countries like the United States and the United Kingdom, patients with limited English proficiency may find it difficult to grasp medical information, resulting in lower satisfaction and nonadherence to treatment. Many times, patients sign informed consent forms without fully understanding the content, which can lead to their withdrawal from the trial at later stages of the study [49]. In order to solve this problem, healthcare providers are required to provide language assistance, ensuring that patients are fully informed about their treatment options [51].

7.4.1.2 Lack of Health Literacy

Health literacy is the ability of an individual to understand, obtain, critically assess, and apply health-related information. Variations in health literacy can impact patients' understanding of clinical trial informed consent forms. Patients with lower literacy levels often struggle to comprehend the material. Standard informed consent forms are frequently written at a level that exceeds the average literacy level of the population, creating a gap between patients' health literacy and their ability to fully understand information about clinical trials [52]. In this case, patients would rather rely on the trust of healthcare providers than truly understand the information presented. It may further lead to impaired decision-making capacity, as patients may not make informed choices that align with their values and preferences [53].

7.4.1.3 False Expectations

Even in the absence of language barriers or religious conflicts, misunderstanding can still occur due to unrealistic expectations about the results of the experiment. Some patients may fear being treated as mere "test subjects," while others might hesitate to participate because of concerns over past incidents of clinical trial fraud and misconduct [49].

7.4.1.4 Consent in Emergency Situations

In situations that are critical by nature and require urgency, there may be little to no time for patients to engage in detailed discussions and make informed decisions. During emergencies, the main priority is to provide medical care, which results in delayed or incomplete documentation of the consent process. Limited information about the patient, his values and goals, and his preexisting medical conditions is a major concern in emergency situations. Due to the lack of medical history, providers will not be able to make informed decisions about the course of action [54].

7.4.1.5 Informed Consent in Pediatric Population

Ethical medical care prioritizes patients' autonomy, but for children or those with developmental delays, other principles such as beneficence, non-maleficence, and dignity guide care. While adults can make medical decisions, minors may be capable of understanding and engaging in medical decision-making. Children in the age group of 7–8 years can understand the risks and benefits of treatments, and many adolescents can make informed decisions similar to adults. Adolescents typically expect to be included in decisions, although legal expectations allow minors to consent to certain treatments without parental consent. It is important to consider a child's developing autonomy and avoid coercion. Respecting a child's refusal or reluctance to assent is important, especially when the intervention is not urgent or essential to their health [55].

7.4.1.6 Religious Influences

The informed consent process allows participants to decide whether to accept or refuse recommended medical treatment. However, their decision to participate in research may sometimes be influenced by their religious beliefs. It is common for the methods used in an experiment to conflict with the behavioral guidelines established by a participant's religion [49].

7.4.1.7 Patient Perception

Many patients perceive clinical trials as an added burden, preferring conventional treatments and fearing unknown side effects of new ones. This makes obtaining informed consent difficult, as too much emphasis on potential risks may unnecessarily alarm them and discourage consideration of potentially lifesaving or life-improving options [49].

7.4.2 Evolving Regulatory Framework

Advancements in medical technology, along with increased patient autonomy and evolving societal norms, require healthcare providers to regularly update their informed consent practices to meet ethical and legal standards. This includes tailoring consent to individual preferences and values, promoting shared decision-making, appointing surrogate decision-makers when necessary, and ensuring clear communication about data privacy. It is also crucial to address therapeutic misconceptions and consider cultural backgrounds and communication preferences to ensure patients fully understand the distinction between research and treatment [56].

7.5 GLOBALIZATION AND ETHICAL CONCERNS

Plants and herbs have been utilized by humanity for centuries and are employed worldwide for disease treatment and prevention. Plants contribute to around 25% of worldwide prescribed contemporary pharmaceuticals [1]. The traditional healthcare system is increasingly popular globally, driven by public interest in herbal remedies and their remarkable acceptance for healing features with few or no adverse effects for various health issues [2]. Traditional medicine, particularly the utilization of medicinal plants, has been integral to human history globally, with knowledge transmitted between generations. In several regions of the world, including Asia and Europe, documentation can be traced to antiquity. Currently, herbal and traditional medicines are enhancing healthcare in several places globally [3]. Herbal medicines are utilized as both primary treatments and adjuncts to conventional pharmaceuticals by various populations. In numerous developing countries, traditional medicine constitutes the sole therapeutic option accessible to a significant portion of the population. Herbal remedies have been widely embraced in numerous industrialized nations and are increasingly significant in the European Union, Australia, North America, and the United Kingdom. Sixty percent of the global population currently uses herbal or conventional therapy as primary therapeutic agents for the treatment of fever associated with malaria. Eighty percent of Africans, 30–50% of Chinese, 48% of Australians, 70% of Canadians, 80% of Germans, 42% of Americans, 39% of Belgians, and 76% of French individuals prefer herbal or alternative therapies as their primary treatment for various maladies. The WHO reports that 90% of people in Africa and 70% in India rely on herbal medicines for their healthcare requirements. Traditional medicine facilities exist in over 90% of general hospitals in China. Herbal medicine in Germany constitutes one of the five principal components of naturopathy, sometimes referred to as Kneipp therapies [1, 4]. The majority of individuals in developing countries are projected to depend on traditional remedies as their principal form of healthcare, with the utilization of herbs regarded as a vital component of their cultural heritage [57, 58]. Regardless of their rationale, customers possess the right to obtain herbal treatments that are both safe and effective. Moreover, scientific investigations must be conducted to produce evidence-based data on efficacy, effectiveness, dosage, adverse effects, and contraindications [1].

The global expansion of clinical trials for herbal treatments has introduced various ethical dilemmas, especially around IP, safeguarding indigenous knowledge, cultural sensitivity, and the potential for exploitation in LMICs. Traditional herbal remedies garner global interest for their possible medicinal advantages, prompting clinical research to be conducted in many countries. Nonetheless, the ethical guidelines governing these studies, particularly in relation to vulnerable populations or indigenous cultures, are not consistently maintained. This section examines the particular ethical issues that emerge when clinical trials of herbal treatments are incorporated into the worldwide research agenda [3, 11, 12].

7.5.1 Exploitation of Traditional Knowledge in Global Trials

The exploitation of indigenous and traditional knowledge represents a significant ethical concern in the globalization of herbal medicine trials. Numerous herbal remedies are based on knowledge systems developed over millennia by indigenous and local populations. These groups frequently lack legal safeguards for their traditional medicinal knowledge, rendering it susceptible to exploitation by researchers or pharmaceutical companies from affluent nations that aim to profit from it without providing adequate recompense or acknowledgment [26, 59, 60]. This issue is known as bioprospecting or biopiracy—the commercial exploitation of indigenous biological resources or traditional knowledge without appropriate recognition or equitable benefit-sharing with the local communities of origin. The Convention on Biological Diversity (CBD) [61] and the Nagoya Protocol underscore the necessity for prior informed consent (PIC) [62] and access and benefit-sharing (ABS) agreements to avert the [63] exploitation of traditional knowledge. The Biodiversity Act of 2002 conforms to the United Nations (UN) CBD, highlighting sustainable utilization, conservation, and equitable benefit-sharing of biological resources, accompanied by stringent penalties for infractions. The Protection of Plant Varieties and Farmers' Rights Act, 2001, facilitates the safeguarding of plant varieties and the rights of farmers, thereby fostering agricultural advancement and encouraging research investment. The National Biodiversity Authority, founded in 2003, guarantees equitable utilization and preservation of biodiversity. The UN Convention acknowledges the significance of indigenous populations in safeguarding genetic resources. The Traditional Knowledge Digital Library (TKDL) safeguards indigenous knowledge, whereas the WIPO's Intergovernmental Committee on Intellectual Property and Genetic Resources, Traditional Knowledge, and Folklore (IGRTKF) promotes worldwide discussion on the protection of traditional knowledge. Ultimately, the International Treaty on Plant Genetic Resources addresses food security, biodiversity, and IP rights, fostering ABS [63]. These agreements guarantee that local populations are apprised of the utilization of their expertise and receive equitable remuneration for it. In practice, these systems are frequently neglected or poorly executed, raising worries that indigenous populations are being exploited for their resources and knowledge without receiving corresponding advantages [48, 63].

7.5.2 Cultural Sensitivity in International Research

Indigenous communities are essential stewards of traditional knowledge, intricately connected to their cultural heritage, identity, and lifestyle. This information, transmitted through generations, embodies their extensive past and shapes their relationships with the environment and civilization. It encompasses fields including medicine, agriculture, biodiversity, craftsmanship, and climate analysis. Disciplines such as Ayurveda, Siddha Vaidya, Unani, Tai Chi, and Acupuncture, originating from India and China, have significantly impacted modern science and are acknowledged worldwide. The profound ecological knowledge of indigenous peoples, cultivated over generations, is crucial for sustainable resource management; its extraction without recognition undermines their cultural heritage and impedes sustainability [26]. Ethical considerations involving cultural sensitivity emerge when conducting clinical studies in varied cultural contexts. Herbal medicines are frequently integral to the cultural and spiritual traditions of indigenous societies. Herbal remedies are often seen not just as biological treatments but also as integral to a holistic perspective on health and well-being. Consequently, scientific research aimed at evaluating the efficacy of these therapies must consider their cultural importance [64].

A report by the state-affiliated Institute of Information Technology, as cited by Business Standard, revealed a substantial reduction in traditional knowledge among tribal communities in Kerala. The Kuruchiya and Kuruba tribes have lost more than 50% of their indigenous knowledge, while other tribes such as the Cholanaikar, Malanpandaram, Kattunaikar, and Kaani have experienced comparable declines. The extraction and commercialization of resources from indigenous territories frequently transpire without equitable compensation, intensifying economic inequalities and restricting access to vital knowledge and resources. The disparity is apparent in the "Jeevani Case," when the Kani tribe's expertise regarding the *Arogyapacha* plant facilitated the creation of the Jeevani medicament. Disputes regarding raw materials, exploitation, and the imprisonment of tribal collectors have alienated the community, underscoring the persistent issue of biopiracy that provides neither reciprocity nor recognition to indigenous groups [26].

Extracting a specific plant from its ceremonial or spiritual context for scientific examination may be perceived as disrespectful or detrimental. This is particularly applicable when the trials are executed without community participation or engagement. Researchers must construct their trials to honor local customs and beliefs, collaborating closely with communities to ensure that traditional therapeutic techniques are not compromised by the scientific approach. Ethical criteria must mandate that clinical trials consider the holistic health approaches prevalent in societies utilizing herbal remedies, rather than enforcing exclusively biomedical frameworks [65, 66].

7.5.3 Regulatory Discrepancies in Global Clinical Trials

Research on herbal pharmaceuticals encounters numerous barriers, including financial, ethical, and regulatory obstacles, along with complications in study design and product standardization. In 2005, the WHO released recommendations to facilitate clinical studies of herbal products. A hurdle lies in delineating inclusion and exclusion criteria owing to disparities between contemporary and herbal medicine diagnoses. Jonas and Linde's "double classification method" resolves this issue by diagnosing people according to contemporary criteria and subsequently categorizing them using traditional approaches. Herbal therapies significantly depend on patient involvement and psychosocial elements, complicating randomization and blinding in clinical studies. Single-blinding is frequently employed due to the multifaceted nature of herbal therapies. Controls must be standardized for variables including color, odor, duration, frequency of administration, and the credibility of the treatment to the patient and the physical context in which treatments are delivered [67, 68]. Identifying suitable controls that correspond to the intervention group in studies can be challenging, particularly for products such as ginger, which possess a distinctive odor [69]. Ensuring standardization among practitioners is essential, and treatment guidelines can mitigate variability. Moreover, herbal remedies are tailored to individuals, complicating the standardization of treatments for communities, and foundational psychological evaluations are essential to consider belief and faith. The investigators must diligently observe their study subjects for any attempts to procure herbal products beyond the established procedure [70]. The globalization of clinical trials for herbal medicines highlights substantial issues associated with regulatory inconsistencies among nations. Although high-income nations possess comprehensive regulatory frameworks for clinical trials, several LMICs lack sufficient ethical monitoring. This has resulted in a practice termed "ethics dumping," wherein studies that fail to satisfy ethical norms in affluent nations are executed in less affluent countries with less stringent regulatory frameworks. Such trials may take advantage of the weak regulatory framework in LMICs, endangering vulnerable populations [70].

Ethics dumping presents issues of participant safety, informed consent, and the equitable conduct of trials in low-resource environments. Participants in these locations may lack a comprehensive understanding of the risks and benefits associated with clinical trial participation, especially if they possess low health literacy or if the informed consent process is inadequately translated into their native languages. Researchers and sponsors of international clinical trials must guarantee adherence to ethical standards irrespective of trial locations, and they must strive to standardize ethical norms across borders to avert exploitation [71].

7.5.4 Ethical Dilemmas in Low- and Middle-Income Countries

Trials undertaken in LMICs pose numerous ethical challenges, especially concerning the involvement of vulnerable groups in research. Justice is a fundamental ethical tenet in clinical research, necessitating the equitable distribution of burdens and benefits associated with the research. In numerous instances, clinical trials in LMICs primarily advantage persons in affluent countries, providing few direct benefits to the local communities involved. This raises concerns over the equitable distribution of benefits and the possible exploitation of resource-deprived populations to the advantage of the global pharmaceutical sector [72, 73]. Moreover, the informed consent procedure is especially challenging in LMICs, where linguistic obstacles, limited literacy levels, and cultural disparities hinder participants' comprehensive understanding of the trial's risks and benefits. Researchers must transcend the mere acquisition of signatures and ensure that participants are thoroughly informed about the trial's specifics. Ethical guidelines must mandate that researchers employ culturally suitable and accessible approaches to elucidate the goals, risks, and benefits of the trial [74].

7.6 COLLABORATION, ETHICAL PRACTICES, AND INTEGRATION OF ETHICAL CONSIDERATIONS

Clinical trials of herbal therapy encounter distinct problems arising from the convergence of traditional knowledge, scientific rigor, and ethical issues. Collaboration among various stakeholders, including researchers, regulatory agencies, indigenous people, and healthcare professionals, is crucial to tackle these difficulties. Collaborative methodologies guarantee the incorporation of ethical norms across the research progression, from trial design to the dissemination of findings. This section examines the multifaceted aspects of collaboration in herbal medicine trials, highlighting the significance of ethical practices and their application in diverse circumstances.

7.6.1 Community Engagement in Ethical Research

Overcoming cultural boundaries is crucial in international clinical research, as varying attitudes might hinder the informed consent procedure. Community interaction is essential for advancing ethical behaviors, enhancing mutual understanding, and honoring cultural settings. By acknowledging communities as stakeholders, researchers establish mutual accountability, thereby enhancing the relevance and quality of investigations. This methodology facilitates recruitment and fosters confidence between researchers and local communities, creating equitable relationships essential for ethical research. Involving local and indigenous groups in the study process is a paramount ethical consideration in herbal medicine trials. Numerous herbal medicines are founded on ancient knowledge systems transmitted over generations. In conducting clinical trials incorporating traditional treatments, it is imperative that researchers honor the intellectual and cultural ownership of the communities that have cultivated and safeguarded this knowledge [75].

Organizations such as the WHO and the CIOMS highlight the significance of community involvement in research ethics. A participatory method is advocated, engaging communities in the conception, execution, and dissemination of research. This promotes reciprocal learning, trust, and local ownership of research initiatives, while emphasizing the necessity to honor cultural circumstances and modify research methodologies accordingly. Successful community participation necessitates strategies adapted to local needs. Engaging community representatives in evaluative procedures, fostering enduring connections, and sustaining transparent communication for conversation and feedback are essential. Utilizing local networks and technological platforms can significantly improve outreach and participation initiatives. Community participation transcends just formal consent; it entails cultivating trust and fostering open communication with community leaders, traditional healers, and other essential stakeholders. Researchers must adhere to Free, Prior, and Informed Consent (FPIC), necessitating that community members are informed of the trial's objectives, risks, and potential benefits in a culturally sensitive manner [76]. This may involve conducting consultations in local languages, employing culturally relevant communication techniques, and allowing the community adequate time to evaluate the material prior to granting consent. FPIC entails a continuous procedure in which the community remains informed throughout the study [77]. Moreover, it is important to guarantee that the advantages of the research are fairly distributed among the participating community. This may entail granting access to the study results, ensuring the community benefits from any prospective commercial applications of the herbal treatments, or even investing in local healthcare facilities. In the absence of these considerations, there exists a potential for exploitation, when communities impart priceless traditional knowledge without obtaining any reciprocal advantages [30].

7.6.2 Interdisciplinary Collaboration in Herbal Medicine Trials

Interdisciplinary collaboration is crucial in clinical practice and in the generation of high-quality, practical research. Nevertheless, limited frameworks are available to promote successful communication among academics from diverse backgrounds with shared objectives. Collaboration relies on the premise that individuals contribute distinct skills or abilities, which, when exchanged, augment collective capacity, enhance efficacy, and optimize resource alignment. Interdisciplinary collaboration is widely acknowledged as a vital component in research and healthcare delivery [78]. Herbal medicine trials require an interdisciplinary approach due to their integration of contemporary scientific research and traditional knowledge systems. Ethnobotanists, who examine the interplay between humans and flora, can offer significant insights into the cultural background of herbal treatments and assist in maintaining respect for traditional practices during the trial. Pharmacologists have knowledge in medication development and safety evaluation, ensuring the scientific integrity of the trials is upheld [79]. Interdisciplinary collaboration is crucial for balancing the scientific rigor necessary for clinical studies with the need to honor cultural customs. Ethnobotanists can convert traditional knowledge into a scientifically testable format, while pharmacologists and medical researchers guarantee that herbal therapies undergo stringent safety and efficacy assessments. Furthermore, ethicists are essential in assisting researchers in addressing intricate ethical issues, including the sharing of traditional knowledge and the safeguarding of IP rights for indigenous groups [80]. A multidisciplinary approach guarantees that trials are crafted with cultural sensitivity and scientific precision, resulting in outcomes that are significant for both local communities and the worldwide scientific community [81].

7.6.3 Capacity Building and Education for Ethical Conduct

Capacity building is an essential component of ethical practice in worldwide herbal medicine research, especially in LMICs, where resources for conducting and managing clinical trials may be constrained. Locally conducted health research in LMICs is crucial for tackling global health issues. Nonetheless, despite global efforts, health research capacity in LMICs remains insufficient, and development projects frequently lack coherence. Moreover, global health research funding has offered minimal assistance in enhancing research ethics review procedures in these nations. Nations possessing inadequate ethics review frameworks, restricted knowledge, and diminished administrative skills have been especially susceptible during health crises [82, 83]. In recent decades, programs to improve the research knowledge and skills of investigators in LMICs have been established. Enhancing this capacity will enable local researchers to effectively lead research initiatives, establish priorities, and influence policy modifications that can enhance health outcomes [84]. Developing local research capacity entails educating local researchers, healthcare practitioners, and community members in ethical research methodologies. This fosters a sustainable research environment, enabling local stakeholders to perform their own clinical trials while ensuring the maintenance of ethical standards even after the international collaborators have left [85]. Capacity building involves improving the infrastructure required for clinical trials, including the establishment of ethics review boards and the improvement of laboratory facilities. Enhancing local capacity to conduct and supervise clinical trials ensures that these studies are morally robust and foster the sustainable advancement of local healthcare and research infrastructures [82].

7.6.4 Ethical Approval and Oversight in Collaborative Research

The history of clinical research is characterized by numerous instances of participant exploitation or damage to further scientific objectives. Consequently, ethical standards, norms, and laws have been established over time to protect the rights and welfare of study participants. Ethical oversight is essential for ensuring that clinical trials are done in a way that safeguards the rights, safety, and welfare of participants. In the realm of global herbal medicine trials, ethical monitoring must consider both international norms and local ethical standards. Ethics committees or IRBs are essential for evaluating trial protocols to guarantee adherence to GCP standards and the protection of participants' rights [86]. The safeguarding of participants and the acquisition of high-quality, significant data are fundamental principles directing trial monitoring. IRBs are essential in guaranteeing that appropriate procedures are implemented to protect participants' rights and welfare, both before the commencement of a study and through ongoing evaluations. They conduct ethical evaluations of research at designated locations, as required by U.S. legislation for organizations that receive government funding. In resource-limited settings (RLS), constrained financial and human resources, weak ethics training, and deficient standard operating procedures frequently impede the efficacy of these reviews [87].

Oversight of clinical trials is crucial for safeguarding research participants and maintaining the integrity of the collected data. Key stakeholders, such as the trial sponsor, local IRBs, and independent monitoring committees, possess overlapping duties in this process. Uniformity among these groups is essential for the seamless implementation of trials, although it may prove difficult in RLS. Augmenting capacity and delivering training in these environments can enhance oversight and lead to improved clinical trial results and medical management [88]. Ethical oversight frequently necessitates cooperation between local and international review boards. This is especially crucial when studies occur in indigenous communities, where the ethical considerations may markedly differ from those in standard pharmacological trials. IP and biopiracy issues require meticulous consideration. Biopiracy denotes the appropriation of indigenous knowledge without sufficient remuneration or recognition. Engaging with local ethics committees guarantees the resolution of these issues and the advantageous impact of the research on the local community [26]. Moreover, ethical oversight must be a continuous process throughout the entire study. Researchers are tasked with overseeing the clinical trial to ensure that any adverse outcomes or ethical concerns are swiftly addressed. This degree of attentiveness sustains confidence between researchers and participants, which is essential for the trial's success [86].

7.6.5 Building Trust with Indigenous Communities

Trust is frequently highlighted as a fundamental factor in collaborative research. Researchers are urged to establish trust with participants and communities through active involvement and informed consent, ensuring voluntary participation with a comprehensive grasp of risks and benefits. In bioethics, trust is regarded as one party's acknowledgment of vulnerability, expecting goodwill from the other. Nonetheless, trust-building needs ongoing effort and sustained commitment. The function of trust in researcher-participant relationships is little examined, despite many studies on trust in interpersonal contexts. Trust-building requires one side to willingly embrace vulnerability, anticipating that the more powerful counterpart would demonstrate kindness and exceed fundamental commitments to provide protection. In research, gaining confidence frequently necessitates participants giving up control, depending on the ethical integrity of researchers to act morally and avoid exploitation. While the informed consent process is essential for establishing trust, it frequently places the responsibility on participants to place their faith in researchers, which may entail the danger of betrayal [89]. Trust is a fundamental component of ethical research, especially when engaging with indigenous or underprivileged groups. Trust may diminish when people see exploitation or marginalization during the research process, resulting in opposition to further research endeavors. Researchers must promote transparency and open communication throughout the trial procedure to establish confidence. This entails transparency regarding the research objectives, the utilization of data, and the advantages the community will derive from the findings [84].

Establishing robust links between researchers and indigenous people is crucial for executing ethical and mutually advantageous indigenous health research. Due to the historical context of mistrust, exploitation, and unethical behaviors with indigenous populations, collaborative research must be grounded in trust. Sustained interactions with indigenous groups are essential for establishing trust. Rather than engaging with communities on a one-time basis for a single trial, researchers should aim to establish ongoing partnerships that extend beyond the immediate research project. In doing so, they exhibit a dedication to the community's welfare and acknowledge its knowledge and accomplishments [90]. Benefit-sharing constitutes a crucial element in trust establishment. Communities that provided their knowledge for herbal medicine studies must receive equitable rewards when these trials result in commercial products. This may encompass financial remuneration, enhancements in healthcare, or joint ownership of IP [91].

7.6.6 Developing a Global Ethical Framework

Investigations into traditional herbal remedies present intricate ethical dilemmas, especially regarding public health. This research must comply with the same ethical requirements as any studies involving human subjects. An ethical framework previously established by Emanuel et al. and adapted for international research provides a valuable foundation for considering the ethics of international traditional herbal medicine research [92]. The framework underscores several fundamental principles. A collaborative relationship includes reciprocal respect among equal partners, emphasizing capacity building within the host nation. Social value mandates that research must yield generalizable information or health enhancements, providing distinct advantages to particular communities. Scientific validity guarantees that the research design is suitable for the specific local environment. Equitable subject selection necessitates that participants be selected according to scientific criteria rather than convenience or susceptibility. The favorable risk-benefit ratio underscores that the benefits to people and their communities must surpass risks. Furthermore, independent review safeguards research integrity by guaranteeing a fair evaluation of risks and benefits. Informed consent must be culturally acceptable, and respect for subjects necessitates a strategy for communicating data and monitoring adverse occurrences, while also protecting participants' right to withdraw from the study [93]. The global expansion of herbal medicine trials necessitates the establishment of a robust ethical framework applicable across diverse nations and cultures. A framework must integrate international norms while respecting local cultures and knowledge systems. Although organizations such as WHO have progressed in formulating criteria for the ethical implementation of herbal medicine studies, there is still much work to be done [94].

A comprehensive ethical framework must consider IP rights, biopiracy, benefit-sharing, and informed consent, while being attuned to the requirements of indigenous populations. It should also offer help in navigating the frequently contradictory ethical requirements of various countries, especially in cases involving multinational collaborations in trials. This approach would safeguard participants' rights while ensuring that herbal medicine research is conducted with equity, transparency, and cultural respect [26, 27].

7.7 CONCLUSIONS

In conclusion, regulating the safety and efficacy of herbal drugs in an ethical way requires a multifaceted approach that addresses both scientific and cultural considerations. To ensure the integrity of herbal medicine research, regulatory authorities are developing comprehensive guidelines, including GMP and GCP, which provide a foundation for safe and ethical clinical trials. However, these guidelines must also incorporate safeguards for issues such as IP rights, biopiracy, benefit-sharing, and informed consent, with a particular focus on the needs and rights of indigenous populations. By aligning these ethical and regulatory frameworks with global standards, while respecting the diverse cultural contexts of herbal medicine, we can ensure that research is conducted in a way that promotes equality, transparency, and respect for all stakeholders involved.

REFERENCES

[1] Husain, G.M., Khan, M.A., Urooj, M. & Kazmi, M.H. (2017). Pharmacodynamic Evaluation: Herbal Medicine. In Hock, F.J. & Gralinski, M.R., Drug Discovery and Evaluation: Methods in Clinical Pharmacology. Springer International Publishing, 1(2), 483–497.

[2] Sachin, S., Prince Ahad, M., Nishant, K., Apporva, C., Jasreen, U., Shilpa & Anmoldeep, K. (2022). Traditional and Herbal Medicines: Opportunities and Challenges. Pharmacognosy Research, 14(2), 107–114.

[3] Knoess, W. & Wiesner, J. (2019). The Globalization of Traditional Medicines: Perspectives Related to the European Union Regulatory Environment. Engineering, 5(1), 22–31.

[4] Joos, S., Glassen, K. & Musselmann, B. (2012). Herbal Medicine in Primary Healthcare in Germany: The Patient's Perspective. Evidence-Based Complementary and Alternative Medicine, 2012(1), 294638–294648.

[5] Okoronkwo, I., OnyiaPat, J.L., Okpala, P.U., Agbo, M.A. & Ndu, A. (2014). Patterns of Complementary and Alternative Medicine Use, Perceived Benefits, and Adverse Effects among Adult Users in Enugu Urban, Southeast Nigeria. Evidence-Based Complementary and Alternative Medicine, 2014(1), 239372–239377.

[6] White, A., Boon, H., Alraek, T., Lewith, G., Liu, J., Norheim, A., Steinsbekk, A., Yamashita, H. & Fonnebo, V. (2013). Reducing the Risk of Complementary and Alternative Medicine (CAM): Challenges and Priorities. European Journal of Integrative Medicine, 6(1), 404–408.

[7] Werner, S.M. (2014). Patient Safety and the Widespread Use of Herbs and Supplements. Frontiers in Pharmacology, 5(1), 142–143.

[8] Posadzki, P., Watson, L. & Ernst, E. (2013). Contamination and Adulteration of Herbal Medicinal Products (HMPs): An Overview of Systematic Reviews. European Journal of Clinical Pharmacology, 69(3), 295–307.

[9] Curtis, P. & Gaylord, S. (2005). Safety Issues in the Interaction of Conventional, Complementary, and Alternative Health Care. Complementary Health Practice Review, 10(1), 3–31.

[10] Smith, C.A., Priest, R., Carmady, B., Bourchier, S.J. & Bensoussan, A. (2010). The Ethics of Traditional Chinese and Western Herbal Medicine Research: Views of Researchers and Human Ethics Committees in Australia. Journal of Evidence-based Complementary and Alternative Medicine, 2011(1), 256915–256921.

[11] Ahlawat, J., Verma, N. & Sehrawat, A.R. (2013). Globalization of Herbal Drugs: A Bliss and Concern. International Journal of Science and Research, 3(11), 466–474.

[12] Leonti, M. (2013). Traditional Medicines and Globalization: Current and Future Perspectives in Ethnopharmacology. Frontiers in Pharmacology, 4(1), 1–13.

[13] Varkey, B. (2021). Principles of Clinical Ethics and Their Application to Practice. Journal of Medical Principles Practice, 30(1), 17–28.

[14] MREC. Overview of Ethics of Clinical Research, Ethical Principles & Requirements. Retrieved 10th October 2024 from https://www.crc.gov.my/wp-content/uploads/documents/intranet/GCP/03_Overview_of_ethics_of_Clinical_Research.pdf.

[15] Tribunal, I.M. (1949). Trials of War Criminals Before the Nuremberg Military Tribunals Under Control Council Law No. 10. (2nd ed., Vol. 2). US Government Printing Office.

[16] Association, W.M. World Medical Association Declaration of Helsinki: Ethical Principles for Medical Research Involving Human Participants. Retrieved 19th October 2024 from https://www.wma.net/policies-post/wma-declaration-of-helsinki/.

[17] OHRP. The Belmont Report. Ethical Principles and Guidelines for the Protection of Human Subjects of Research. Retrieved 17th September 2024 from https://www.bing.com/search?pglt=299&q=The+Belmont+Report.+Ethical+principles+and+guidelines+for+the+protection+of+human+subjects+of+research&cvid=760174822f6b4413b10b9aabfca208c2&gs_lcrp=EgRlZGdlKgYIABBFGDkyBggAEEUYOdIBBzMxNWowajGoAgCwAgA&FORM=ANNTA1&PC=U531.

[18] Register, F. Federal Policy for the Protection of Human Subjects. Final Rule. Retrieved 12th October 2024 from https://www.federalregister.gov/documents/2017/01/19/2017-01058/federal-policy-for-the-protection-of-human-subjects.

[19] CIOMS. International Ethical Guidelines for Biomedical Research Involving Human Subjects. Retrieved 21st September 2024 from https://cioms.ch/wp-content/uploads/2021/03/International_Ethical_Guidelines_for_Biomedical_Research_Involving_Human_Subjects_2002.pdf.

[20] University, W.F. Ethics in Clinical Research: Foundations and Current Issues. Retrieved 5th September 2024 from https://school.wakehealth.edu/education-and-training/graduate-programs/clinical-research-management-ms/features/ethics-in-clinical-research.

[21] Vijayananthan, A. & Nawawi, O. (2008). The Importance of Good Clinical Practice Guidelines and Its Role in Clinical Trials. Biomedical Imaging and Intervention Journal, 4(1), 1–4.

[22] WHO. Handbook for Good Clinical Research Practice (GCP): Guidance for Implementation. Retrieved 7th October 2024 from https://iris.who.int/bitstream/handle/10665/43392/924159392X_eng.pdf.

[23] WHO. WHO Guidelines on Developing Consumer Information on Proper Use of Traditional, Complementary and Alternative Medicine. Retrieved 15th September 2024 from https://www.who.int/publications/i/item/9241591706.

[24] ICH. ICH-E6 Good Clinical Practice (GCP). Retrieved 16th September 2024 from https://database.ich.org/sites/default/files/ICH_E6-R3_GCP-Principles_Draft_2021_0419.pdf.

[25] Briana, R. Principles of Good Clinical Practice. Retrieved 19th September 2024 from https://www.longdom.org/open-access/principles-of-good-clinical-practice-102457.html.

[26] Raj, A. & Shree, S. (2023). The Silent Larceny: Exploitation of Traditional Knowledge by Way of Biopiracy. SSRN, 1(1), 1–9.

[27] Gatt, A.R., Vella Bonanno, P. & Zammit, R. (2024). Ethical Considerations in the Regulation and Use of Herbal Medicines in the European Union. Frontiers in Medical Technology, 6(1), 1–14.

[28] Wynberg, R. (2023). Biopiracy: Crying Wolf or a Lever for Equity and Conservation? Research Policy, 52(2), 104674–104684.

[29] Mackey, T.K. & Liang, B.A. (2012). Integrating Biodiversity Management and Indigenous Biopiracy Protection to Promote Environmental Justice and Global Health. American Journal of Public Health, 102(6), 1091–1095.

[30] Emanuel, E., Wendler, D., Killen, J. & Grady, C. (2004). What Makes Clinical Research in Developing Countries Ethical? The Benchmarks of Ethical Research. The Journal of Infectious Diseases, 189(5), 930–937.

[31] Chen, J., Huang, J., Li, J.V., Lv, Y., He, Y. & Zheng, Q. (2017). The Characteristics of TCM Clinical Trials: A Systematic Review of ClinicalTrials.gov. Evidence Based Complementary Alternative Medicine, 2017(1), 94614–94615.

[32] Akobeng, A.K. (2005). Understanding Randomised Controlled Trials. Archives of Disease in Childhood, 90(8), 840–844.

[33] Bensoussan, A., Talley, N.J., Hing, M., Menzies, R., Guo, A. & Ngu, M. (1998). Treatment of Irritable Bowel Syndrome with Chinese Herbal Medicine: A Randomized Controlled Trial. Jama, 280(18), 1585–1589.

[34] Fung, F.Y. & Linn, Y.C. (2015). Developing Traditional Chinese Medicine in the Era of Evidence-Based Medicine: Current Evidences and Challenges. Evidence-Based Complementary and Alternative Medicine, 2015(1), 425037–425045.

[35] Liu, P., Yang, H., Long, F., Hao, H.P., Xu, X., Liu, Y., Shi, X.W., Zhang, D.D., Zheng, H.C., Wen, Q.Y., Li, W.W., Ji, H., Jiang, X.J., Zhang, B.L., Qi, L.W. & Li, P. (2014). Bioactive Equivalence of Combinatorial Components Identified in Screening of an Herbal Medicine. Pharmaceutical Research, 31(7), 1788–1800.

[36] Ngo, L.T., Okogun, J.I. & Folk, W.R. (2013). 21st Century Natural Product Research and Drug Development and Traditional Medicines. Natural Products Report, 30(4), 584–592.

[37] Ulrich, M., G., Zeitler, H., Jobst, D., Panek, D., Vetter, H. & Wagner, H. (2007). Application of the "-Omic-" Technologies in Phytomedicine. Phytomedicine, 14(1), 70–82.

[38] Rodrigues, E. & Barnes, J. (2013). Pharmacovigilance of Herbal Medicines: The Potential Contributions of Ethnobotanical and Ethnopharmacological Studies. Drug Safety, 36(1), 1–12.

[39] Chugh, N.A., Bali, S. & Koul, A. (2018). Integration of Botanicals in Contemporary Medicine: Road Blocks, Checkpoints and Go-Ahead Signals. Integrative Medicine Research, 7(2), 109–125.

[40] Ouedraogo, M., Baudoux, T., Stévigny, C., Nortier, J., Colet, J.M., Efferth, T., Qu, F., Zhou, J., Chan, K., Shaw, D., Pelkonen, O. & Duez, P. (2012). Review of Current and "Omics" Methods for Assessing the Toxicity (Genotoxicity, Teratogenicity and Nephrotoxicity) of Herbal Medicines and Mushrooms. Journal of Ethnopharmacology, 140(3), 492–512.

[41] WHO. WHO Guidelines on Safety Monitoring of Herbal Medicines in Pharmacovigilance Systems. Retrieved 10th September 2024 from https://www.who.int/publications/i/item/9241592214.

[42] WHO. WHO Guidelines on Good Agricultural and Collection Practices (GACP) for Medicinal Plants. Retrieved 10th September 2024 from https://www.who.int/publications/i/item/9241546271.

[43] Yuan, H., Ma, Q., Ye, L. & G, P. (2016). Traditional Medicine and Modern Medicine from Natural Products. Molecules, 21(5), 559–567.

[44] Gidey, M., Asfaw, Z. & Giday, M. (2016). Value Chain Analysis of Medicinal Plants and the Associated Challenges. Journal of Medicinal Plants Studies, 4(1), 45–55.

[45] Ranjan, G., Bjarne, G. & Steven, M.F. (2005). Natures Medicines: Traditional Knowledge and Intellectual Property Management. Case Studies from the National Institutes of Health (NIH), USA. Current Drug Discovery Technologies, 2(4), 203–219.

[46] James, T. (2022). Traditional Medicine and Intellectual Property Rights Law and Policy Perspectives. Forum of Indian Traditional Medicine, 1(1), 1–11.

[47] Adhikari, K. & Lawson, C. (2018). Biodiversity, Genetic Resources and Intellectual Property: Developments in Access and Benefit Sharing. Routledge, 1(1), 1–280.

[48] Dutfield, G. (2017). Traditional Knowledge, Intellectual Property and Pharmaceutical Innovation: What's Left to Discuss. In David, M. & Halbert, D., The SAGE Handbook of Intellectual Property. SAGE Publisher, 1(1), 649–664.

[49] Nijhawan, L.P., Janodia, M.D., Muddukrishna, B.S., Bhat, K.M., Bairy, K.L., Udupa, N. & Musmade, P.B. (2013). Informed Consent: Issues and Challenges. Journal of Advanced Pharmaceutical Technology & Research 4(3), 134–140.

[50] ICMR. National Ethical Guidelines for Biomedical and Health Research Involving Human Participants. Retrieved 10th October 2024 from https://ethics.ncdirindia.org/asset/pdf/ICMR_National_Ethical_Guidelines.pdf.

[51] Pietrzykowski, T. & Smilowska, K. (2021). The Reality of Informed Consent: Empirical Studies on Patient Comprehension: Systematic Review. Trials, 22(1), 57–65.

[52] Burks, A.C. & Keim, M.J. (2019). Health Literacy and Informed Consent for Clinical Trials: A Systematic Review and Implications for Nurses. Nursing: Research and Reviews, 9(3), 1–40.

[53] Perrenoud, B., Velonaki, V.S., Bodenmann, P. & Ramelet, A.S. (2015). The Effectiveness of Health Literacy Interventions on the Informed Consent Process of Health Care Users: A Systematic Review Protocol. JBI Database of Systematic Reviews and Implementation Reports, 13(10), 82–94.

[54] Moskop, J.C. (1999). Informed Consent in the Emergency Department. Emergency Medicine Clinics of North America, 17(2), 327–340.

[55] Samuels, J.F., Bienvenu, O.J., Grados, M.A., Cullen, B., Riddle, M.A., Liang, K.Y., Eaton, W.W. & Nestadt, G.

(2008). Prevalence and Correlates of Hoarding Behavior in a Community-Based Sample. Behaviour Research and Therapy, 46(7), 836–844.

[56] Bazzano, L.A., Durant, J. & Brantley, P.R. (2021). A Modern History of Informed Consent and the Role of Key Information. The Ochsner Journal, 21(1), 81–85.

[57] Sachin, S., Prince Ahad, M., Nishant, K., Apporva, C., Jasreen, U., Shilpa & Anmoldeep, K. (2022). Traditional and Herbal Medicines: Opportunities and Challenges. Pharmacognosy Research, 14(2), 107–114.

[58] Coulter, I.D. & Willis, E.M. (2004). The Rise and Rise of Complementary and Alternative Medicine: A Sociological Perspective. The Medical Journal of Australia, 180(11), 587–589.

[59] Kasih, D.P.D., Dharmawan, N.K.S., Putra, I.B.W., Sudiarawan, K.A. & Rakhima, A.S. (2021). The Exploitation of Indigenous Communities by Commercial Actors. Journal of Ethnic and Cultural Studies, 8(4), 91–108.

[60] Kuruk, P. (2020). Exploitation of Traditional Knowledge of Indigenous and Local Communities. In Adebola, T., Traditional Knowledge, Genetic Resources, Customary Law and Intellectual Property: A Global Primer. Edward Elgar Publishing, 1(1), 1–512.

[61] Lachenmeier, D.W. & Montagnon, C. (2024). Convention on Biological Diversity (CBD) and the Nagoya Protocol: Implications and Compliance Strategies for the Global Coffee Community. Foods, 13(2), 254–260.

[62] Schebesta, H. (2021). The Potential of Private Standards for Valorizing Compliance with Access and Benefit Sharing Obligations of Genetic Resources and Traditional Knowledge. Agronomy, 11(9), 1823–1835.

[63] Heinrich, M., Scotti, F., Andrade-Cetto, A., Berger-Gonzalez, M., Echeverría, J., Friso, F., Garcia-Cardona, F., Hesketh, A., Hitziger, M. & Maake, C. (2020). Access and Benefit Sharing under the Nagoya Protocol—Quo Vadis? Six Latin American Case Studies Assessing Opportunities and Risk. Frontiers in Pharmacology, 11(1), 765–780.

[64] Gatt, A.R., Vella Bonanno, P. & Zammit, R. (2024). Ethical Considerations in the Regulation and Use of Herbal Medicines in the European Union. Frontiers in Medical Technology, 6(1), 1–14.

[65] Ezekwesili-Ofili Josephine, O. & Okaka Antoinette Nwamaka, C. (2019). Herbal Medicines in African Traditional Medicine. In Philip, F.B., Herbal Medicine. IntechOpen, 1(1).

[66] Chebii, W.K., Muthee, J.K. & Kiemo, K. (2020). The Governance of Traditional Medicine and Herbal Remedies in the Selected Local Markets of Western Kenya. Journal of Ethnobiology and Ethnomedicine, 16(1), 39–48.

[67] Dhiman, A., Nanda, A., Ahmad, S. & Narasimhan, B. (2011). *In Vitro* Antimicrobial Activity of Methanolic Leaf Extract of Psidium Guajava L. Journal of Pharmacy and Bioallied Sciences, 3(2), 226–229.

[68] Buchanan, D.R., White, J.D., O'Mara, A.M., Kelaghan, J.W., Smith, W.B. & Minasian, L.M. (2005). Research-Design Issues in Cancer-Symptom–Management Trials Using Complementary and Alternative Medicine: Lessons From the National Cancer Institute Community Clinical Oncology Program Experience. Journal of Clinical Oncology, 23(27), 6682–6689.

[69] Critchley, J., Zhang, Y., Suthisisang, C.C., Chan, T. & Tomlinson, B. (2000). Alternative Therapies and Medical Science: Designing Clinical Trials of Alternative/Complementary Medicines—Is Evidence-Based Traditional Chinese Medicine Attainable? The Journal of Clinical Pharmacology, 40(5), 462–467.

[70] Parveen, A., Parveen, B., Parveen, R., Ahmad, S. & Sciences, B. (2015). Challenges and Guidelines for Clinical Trial of Herbal Drugs. Journal of Pharmacy, 7(4), 329–333.

[71] Schroeder, D., Chatfield, K., Singh, M., Chennells, R. & Herissone-Kelly, P. (2019). Ethics Dumping and the Need for a Global Code of Conduct. In Schroeder, D., Chatfield, K., Singh, M., Chennells, R. & Herissone-Kelly, P., Equitable Research Partnerships: A Global Code of Conduct to Counter Ethics Dumping. Springer International Publishing, 1(1), 1–4.

[72] Bittker, B.M. The Ethical Implications of Clinical Trials in Low- and Middle-Income Countries. Retrieved 20th September 2024 from https://www.americanbar.org/groups/crsj/publications/human_rights_magazine_home/the-truth-about-science/the-ethical-implications-of-clinical-trials/

[73] Lahey, T. (2013). The Ethics of Clinical Research in Low- and Middle-Income Countries. In Bernat, J.L. & Beresford, H.R., Handbook of Clinical Neurology. Elsevier, 1(118), 301–313.

[74] Al Shamsi, H., Almutairi, A.G., Al Mashrafi, S. & Kalbani, T.A. (2020). Implications of Language Barriers for Healthcare: A Systematic Review. Oman Medical Journal, 35(2), 1–7.

[75] Gupta, S., Yadav, M.K., Thangamani, D., Vidhya, C., Kalaimani, P.S., Prabhavathi, S.J. & Vinuradha, R. (2023). Herbal Medicines: Bridging Traditional Knowledge with Modern Pharmacology. Journal of Biochemical Cellular Archives, 23(1), 10–21.

[76] George, D.R., Kuiken, T. & Delborne, J.A. (2019). Articulating 'Free, Prior and Informed Consent' (FPIC) for Engineered Gene Drives. Proceedings of the Royal Society B, 286(1917), 1–8.

[77] Mitchell, T., Arseneau, C., Thomas, D. & Smith, P. (2019). Towards an Indigenous-Informed Relational Approach to Free, Prior, and Informed Consent (FPIC). International Indigenous Policy Journal, 10(4), 1–28.

[78] VanWormer, A., Lindquist, R., Robiner, W. & Finkelstein, S. (2012). Interdisciplinary Collaboration Applied to Clinical Research: An Example of Remote Monitoring in Lung Transplantation. Dimensions of Critical Care Nursing, 31(3), 202–210.

[79] Yuan, H., Ma, Q., Ye, L. & Piao, G. (2016). The Traditional Medicine and Modern Medicine from Natural Products. Molecules, 21(5), 559–567.

[80] Garnatje, T., Penuelas, J. & Valles, J. (2017). Ethnobotany, Phylogeny, and Omics for Human Health and Food Security. Trends in Plant Science, 22(3), 187–191.

[81] Wang, H., Chen, Y., Wang, L., Liu, Q., Yang, S. & Wang, C. (2023). Advancing Herbal Medicine: Enhancing Product Quality and Safety through Robust Quality Control Practices. Frontiers in Pharmacology, 14(1), 1–15.

[82] Sina, B. & Tierney, J. (2024). In Practice: Capacity Building for Research Ethics Review in Low- and Middle-Income Countries. In Sorenson, R.A., Principles and Practice of Emergency Research Response. Springer International Publishing, 1(1), 883–893.

[83] Franzen, S.R.P., Chandler, C. & Lang, T. (2017). Health Research Capacity Development in Low and Middle Income Countries: Reality or Rhetoric? A Systematic

Meta-Narrative Review of the Qualitative Literature. BMJ Open, 7(1), 123–132.

[84] Cassell, H.M., Rose, E.S., Moon, T.D., Bello-Manga, H., Aliyu, M.H. & Mutale, W. (2022). Strengthening Research Capacity through an Intensive Training Program for Biomedical Investigators from Low- and Middle-Income Countries: The Vanderbilt Institute for Research Development and Ethics (VIRDE). BMC Medical Education, 22(1), 97–105.

[85] Franzen, S.R.P., Chandler, C., Siribaddana, S., Atashili, J., Angus, B. & Lang, T. (2017). Strategies for Developing Sustainable Health Research Capacity in Low and Middle-Income Countries: A Prospective, Qualitative Study Investigating the Barriers and Enablers to Locally Led Clinical Trial Conduct in Ethiopia, Cameroon and Sri Lanka. BMJ Open, 7(10), 17–24.

[86] McNair, L. (2022). Ethical and Regulatory Oversight of Clinical Research: The Role of the Institutional Review Board. Experimental Biology and Medicine, 247(7), 561–566.

[87] Kadam, R. & Karandikar, S. (2012). Ethics Committees in India: Facing the Challenges. Perspectives in Clinical Research, 3(2), 1–9.

[88] Godfrey, C., Payton, M., Tasker, S., Proestel, S. & Schouten, J.T. (2014). Ensuring Participant Safety and Trial Integrity with Clinical Trials Oversight. Journal of Acquired Immune Deficiency Syndromes, 65(1), S40–S43.

[89] Folayan, M.O. & Haire, B. (2023). What's Trust Got to Do with Research: Why Not Accountability? Frontiers in Research Metrics Analytics, 8(1), 123–142.

[90] Da, S.M., Flood, C.M., Goldenberg, A. & Singh, D. (2022). Regulating the Safety of Health-Related Artificial Intelligence. Health Policy, 17(4), 63–77.

[91] Bansal, S., Sarker, T., Yadav, A., Garg, I., Gupta, M. & Sarvaiya, H. (2024). Indigenous Communities and Sustainable Development: A Review and Research Agenda. Global Business and Organizational Excellence, 43(4), 65–87.

[92] Emanuel, E.J., Wendler, D., Killen, J. & Grady, C. (2004). What Makes Clinical Research in Developing Countries Ethical? The Benchmarks of Ethical Research. The Journal of Infectious Diseases, 189(5), 930–937.

[93] Miller, F.G., Emanuel, E.J., Rosenstein, D.L. & Straus, S.E. (2004). Ethical Issues Concerning Research in Complementary and Alternative Medicine. JAMA, 291(5), 599–604.

[94] Chatfield, K., Salehi, B., Sharifi-Rad, J. & Afshar, L. (2018). Applying an Ethical Framework to Herbal Medicine. Evidence-Based Complementary and Alternative Medicine, 2018(1), 19–39.

8 Regulatory Aspects of Clinical Trials of Herbal Medicines in the United States

Pankaj Dagur[1], *Naresh Rangra*[2], *Motseoa Grace Molefi*[1], *Ankita Jadhav*[3], *and Manik Ghosh*[4]
[1]School of Pharmacy, Sharda University, Greater Noida, Uttar Pradesh, India
[2]Department of Pharmaceutical Chemistry, Chitkara University, Punjab, India
[3]Department of Pharmaceutics, Shirpur Education Society's R.C. Patel Institute of Pharmaceutical Education & Research, Shirpur, Maharashtra, India
[4]Department of Pharmaceutical Sciences and Technology, Birla Institute of Technology, Mesra, Ranchi, Jharkhand, India

8.1 INTRODUCTION

Herbal medicines, derived from plants and natural substances, have been integral to healthcare for centuries [1]. In the United States, the popularity of herbal products has surged in recent decades, driven by consumer interest in natural health remedies and alternative therapies. The market for herbal medicines is substantial, accounting for billions in revenue, with products used for everything from general wellness to the treatment of specific conditions like arthritis and anxiety [2]. Despite this, the regulatory framework for herbal medicines in the United States remains complex (Table 8.1), with a need for stringent control to ensure safety and efficacy [3]. This rise in usage has also raised concerns regarding the potential for adverse reactions, contamination, or inconsistencies in product quality, emphasizing the importance of robust regulation.

In contrast to conventional pharmaceuticals, herbal medicines are often regulated as dietary supplements under the DSHEA of 1994. This categorization means they do not require premarket approval by the Food and Drug Administration (FDA) [4]. However, herbal products used in clinical trials, especially those claiming therapeutic benefits, must meet more rigorous standards akin to those applied to drugs. The evolving role of herbal medicines in modern clinical practice, alongside the rising expectations for scientific validation, underscores the need for clear regulatory oversight.

The integration of herbal medicines into mainstream healthcare brings challenges in ensuring product safety, efficacy, and quality. Regulatory compliance is critical not only to protect public health but also to gain the trust of healthcare professionals and consumers. Compliance with US FDA regulations is mandatory for companies involved in the research and development of herbal products intended for clinical use. Regulatory frameworks such as Good Manufacturing Practices (GMP) ensure that herbal products are produced consistently and controlled to quality standards. Furthermore, the FDA's guidance on clinical trials, including investigational new drug (IND) applications, outlines the necessary steps for conducting trials with herbal medicines, requiring extensive preclinical and clinical data to ensure safety before human trials commence [5].

Noncompliance with these regulations can result in product recalls, penalties, or reputational damage. The FDA monitors adverse effects reported by users of herbal products, and the manufacturer is responsible for reporting serious adverse events [6]. Therefore, adherence to regulatory standards is essential for companies seeking to introduce new herbal products into the market, and it is vital for ensuring that these products meet the expected safety and efficacy criteria [7].

This chapter aims to provide an in-depth exploration of the regulatory landscape surrounding herbal medicines in the United States, with a focus on clinical trials. Key objectives include the following:

1. Analyzing the role of the FDA and other regulatory bodies in overseeing the use of herbal medicines in clinical settings.
2. Reviewing the current regulations governing the approval, marketing, and clinical testing of herbal products.
3. Exploring the challenges faced by herbal medicine manufacturers in meeting regulatory requirements.
4. Discussing the future direction of herbal medicine regulation in the United States, particularly concerning clinical trials.

8.1.1 Classification of Herbal Medicines

8.1.1.1 Dietary Supplements versus Drugs

In the United States, the classification of herbal medicines is primarily determined by their intended use and the claims made by the manufacturer. Herbal products are most commonly classified as dietary supplements under the Dietary

DOI: 10.1201/9781003522065-8

TABLE 8.1
Overview of U.S. Regulatory Agencies for Herbal Medicines

Agency	Role/Responsibility	Relevant Regulations/Guidelines
FDA (Food and Drug Administration)	Oversees the safety and efficacy of drugs, including herbal medicines used as dietary supplements or new drugs	DSHEA, FD&C Act, IND Guidelines
FTC (Federal Trade Commission)	Monitors the advertising and marketing of dietary supplements	FTC Advertising Guidelines for Dietary Supplements
NIH (National Institutes of Health)	Conducts and supports clinical research, including for herbal medicines	NIH Complementary and Alternative Medicine Program
USDA (United States Department of Agriculture)	Regulates the cultivation and processing of plants used in herbal medicines	Organic Certification Standards, Good Agricultural Practices (GAP)

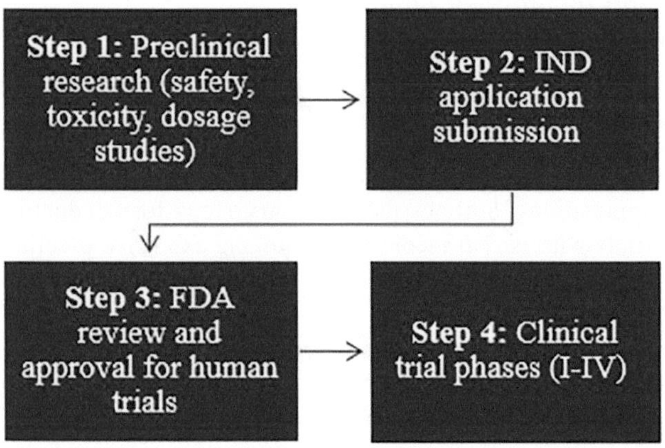

FIGURE 8.1 Investigational new drug.

Supplement Health and Education Act (DSHEA) of 1994, which allows for broad consumer access but imposes fewer regulatory requirements than drugs [4]. Dietary supplements are defined as products taken orally that contain a "dietary ingredient" intended to supplement the diet. These ingredients may include vitamins, minerals, herbs, amino acids, and enzymes [4].

The distinction between dietary supplements and drugs is critical for regulatory oversight. Dietary supplements are not required to undergo the rigorous premarket approval process that drugs must adhere to. Instead, manufacturers are responsible for ensuring the safety of their products and for making accurate, nonmisleading label claims. Dietary supplements can make structure/function claims (e.g., "supports immune function") but cannot claim to treat, cure, or prevent diseases [8].

In contrast, herbal medicines classified as drugs are subject to stricter FDA regulations under the Federal Food, Drug, and Cosmetic (FD&C) Act. If an herbal product is intended to diagnose, treat, cure, or prevent any disease, it is considered a drug, and manufacturers must provide scientific evidence of its safety and efficacy through clinical trials [9]. This requires submission of an IND application (Figure 8.1) and, eventually, a New Drug Application (NDA) for FDA approval before the product can be marketed as a drug [5].

The classification as either a dietary supplement or a drug significantly impacts the regulatory pathway the product must follow, and it often depends on how the product is marketed, its intended use, and the claims made by the manufacturer.

8.1.1.2 New Dietary Ingredient Notifications

The FDA requires manufacturers to submit a New Dietary Ingredient (NDI) notification if they intend to introduce a dietary supplement that contains an ingredient not marketed in the United States before October 15, 1994. This regulation is intended to ensure that new ingredients, including those derived from herbal sources, are safe for consumer use [10].

The NDI notification process requires manufacturers to provide evidence that the new ingredient is reasonably expected to be safe when used as directed. This evidence may include data from toxicological studies, animal testing, or human clinical trials. If the FDA does not object to the NDI notification within 75 days, the manufacturer may proceed with marketing the product, though the FDA reserves the right to take action if safety concerns arise after the product is available [11].

However, the NDI process is not required for all dietary ingredients. Ingredients that were marketed before October 1994 are exempt, and those that are considered generally recognized as safe (GRAS) also bypass the NDI requirement [10]. The NDI process is, therefore, an essential component of the regulatory framework for new herbal products entering the market as dietary supplements.

8.1.1.3 "Generally Recognized as Safe" Status

GRAS status is another regulatory classification that plays a significant role in the marketing of herbal products, particularly as dietary supplements. The GRAS designation applies to substances added to food, including dietary supplements, that are considered safe based on a long history of common use in food or on the consensus of qualified experts [12]. Ingredients with GRAS status are not subject to the NDI notification process or the same level of premarket scrutiny required for drugs.

For a herbal product or ingredient to be considered GRAS, there must be evidence of its safety under the conditions of intended use, typically demonstrated through scientific research or historical use. Manufacturers are responsible for determining whether their products meet GRAS criteria, and they can voluntarily notify the FDA of their GRAS determination, although this is not mandatory.

GRAS status has been granted to many herbal ingredients commonly used in dietary supplements, including certain botanicals, essential oils, and plant extracts. However, it is essential to note that GRAS status only applies to specific uses of an ingredient in food or supplements, and the safety of the ingredient must be reassessed if its intended use changes, such as when transitioning from food to a medicinal product [12].

8.2 REGULATORY BODIES AND FRAMEWORKS

8.2.1 The Food and Drug Administration

The FDA plays a pivotal role in the regulation of herbal medicines in the United States (Table 8.2). While the FDA's primary mandate involves ensuring the safety and efficacy of pharmaceutical drugs, it also oversees dietary supplements and herbal products under the DSHEA of 1994 [13]. According to DSHEA, herbal products can be marketed without premarket approval, provided they do not make claims to cure, mitigate, or treat disease [4]. This framework limits the FDA's regulatory power over herbal supplements compared to conventional drugs, but the agency still monitors labeling and can take action against misbranded or unsafe products [3].

However, when herbal medicines are used in clinical trials, particularly those aimed at demonstrating therapeutic effects, they fall under the more stringent regulatory category of drugs. In this context, the FDA requires an IND application to be submitted before human trials can begin [14]. The IND application must contain preclinical data on toxicity, dosing, and safety, as well as detailed information on the manufacturing and controls of the herbal product. This ensures that participants in clinical trials are not exposed to unnecessary risks [5].

The FDA also enforces the current Good Manufacturing Practices (cGMP) regulations, which are essential for the quality control of herbal products used in clinical settings. Compliance with these regulations is crucial to ensure the consistency, purity, and potency of the herbal medicine. Noncompliance can lead to significant consequences, including the halting of clinical trials or the rejection of product approval applications.

8.2.2 Other Relevant Agencies (NIH, FTC, etc.)

In addition to the FDA, other regulatory and governmental bodies play significant roles in overseeing the use of herbal medicines in the United States.

- **National Institutes of Health (NIH)**: Through its the National Center for Complementary and Integrative Health (NCCIH), the NIH supports research into the efficacy and safety of herbal medicines. The NCCIH funds clinical trials and provides guidelines for the conduct of research involving herbal products. Its role is critical in advancing the scientific understanding of herbal medicines and ensuring that they are backed by robust evidence [16].
- **Federal Trade Commission**: The FTC is responsible for monitoring advertising and marketing claims made about herbal products. While the FDA regulates product labeling, the FTC ensures that any marketing materials, including online advertisements, are truthful and not misleading. If an herbal product claims to cure a disease without scientific evidence to back it up, the FTC can take enforcement actions, including issuing fines or requiring the company to retract its claims.

8.2.3 Key Legislation and Regulations (DSHEA, FD&C Act)

8.2.3.1 Dietary Supplement Health and Education Act of 1994

The DSHEA is the foundational legislation governing herbal medicines in the United States. Under DSHEA, herbal medicines are categorized as dietary supplements, which means they are not subject to the rigorous premarket approval process required for drugs [17]. However,

TABLE 8.2
FDA's Regulatory Pathways for Herbal Medicines

Regulatory Pathway	Legal Basis	Requirements for Approval
Dietary supplement	DSHEA (Dietary Supplement Health and Education Act)	No premarket approval; compliance with GMP standards required
New drug	FD&C Act (Food, Drug, and Cosmetic Act)	IND application, clinical trials, NDA submission, FDA review
GRAS (generally recognized as safe)	FD&C Act, GRAS Notification Process	Evidence of safety through scientific data or a history of safe use

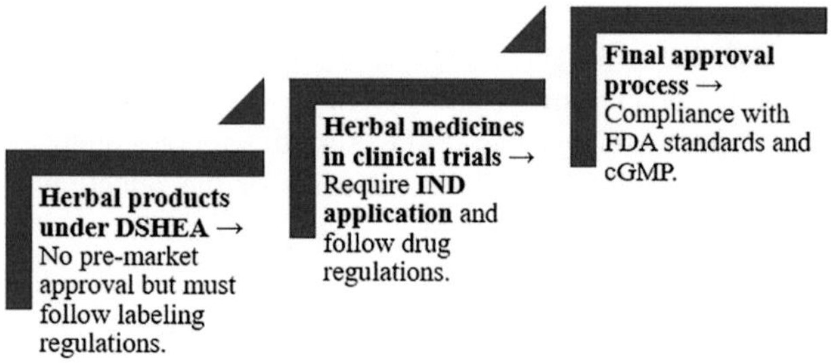

FIGURE 8.2 FDA's regulatory framework for herbal medicines.

manufacturers are required to ensure the safety of their products before marketing them and must provide labeling that is not false or misleading [4]. The DSHEA also places the burden of proving safety concerns on the FDA, meaning the agency can only take action against products after they are on the market if they are found to be unsafe (Figure 8.2).

DSHEA allows for health claims related to the structure or function of the body (e.g., "supports immune health") but prohibits disease claims unless the product has undergone FDA drug approval. This presents a unique challenge for manufacturers of herbal medicines that are intended for therapeutic use in clinical trials, as they must navigate both the dietary supplement regulations and those applicable to investigational drugs.

8.2.3.2 Federal Food, Drug, and Cosmetic Act

The FD&C Act is another crucial piece of legislation governing the safety and efficacy of drugs, including herbal products classified as drugs for clinical trial purposes [18]. Herbal medicines that claim therapeutic benefits must comply with the FD&C Act, requiring them to meet the same standards for efficacy and safety as conventional pharmaceuticals [19]. This involves extensive preclinical testing, submission of an IND application, and adherence to cGMP during production.

Under the FD&C Act, the FDA has the authority to regulate the marketing and distribution of any product considered a drug, including those derived from herbs. This law ensures that any herbal medicines entering clinical trials or the marketplace have been rigorously evaluated for safety and effectiveness.

8.3 PRECLINICAL REQUIREMENTS

8.3.1 Good Laboratory Practices

Before herbal medicines can enter clinical trials in the United States, they must undergo rigorous preclinical testing to assess their safety and biological activity. All preclinical studies intended to support IND applications must adhere to Good Laboratory Practices (GLP), a set of guidelines designed to ensure the quality, integrity, and reliability of nonclinical laboratory studies [15]. GLP regulations, outlined in 21 CFR Part 58, apply to laboratories conducting studies on the toxicology, pharmacology, and biological effects of herbal products.

GLP compliance is critical because it ensures that data from preclinical studies are robust, reproducible, and acceptable to regulatory authorities such as the FDA. Under GLP, laboratories must follow strict guidelines regarding study planning, conduct, and reporting. These include the maintenance of detailed records, proper handling of test materials, qualification of personnel, and validation of instruments used in the research process [20]. For herbal products, GLP-compliant studies are essential in providing the necessary evidence to establish the safety profile before advancing to human clinical trials.

GLP also plays a role in botanical characterization and standardization (discussed later), ensuring the consistent quality of herbal products used in preclinical studies. The enforcement of GLP standards by the FDA provides confidence that the study results are accurate and reliable, reducing the risk of harm to humans when clinical trials begin.

8.3.2 Toxicology and Pharmacology Studies

Toxicology studies are a cornerstone of preclinical testing for herbal medicines. These studies evaluate the potential toxic effects of herbal ingredients and determine safe dosage levels for human use. Both *acute toxicity studies* (to assess short-term exposure) and *chronic toxicity studies* (to assess long-term exposure) are performed using animal models [21]. The objective is to identify any adverse effects caused by the herbal medicine, including organ toxicity, carcinogenicity, and reproductive toxicity.

For herbal medicines, toxicology studies can be complex because they often contain multiple active compounds, and interactions between these compounds may lead to unpredictable outcomes. For this reason, standardized extracts and well-characterized botanical materials are crucial for ensuring the reproducibility and reliability of toxicology data [22]. The results of these studies are used to determine the starting doses for human clinical trials and to establish the maximum tolerated dose (MTD) in humans.

In addition to toxicology studies, preclinical testing also involves *pharmacology studies* to assess the biological effects and mechanisms of action of herbal medicines. These studies examine how the active compounds in the herbal medicine interact with biological systems, such as receptors, enzymes, and signaling pathways. The goal is to provide insight into the therapeutic potential of the herbal product and to support its intended use in clinical trials [23].

Both toxicology and pharmacology data are crucial for compiling an IND application, as they demonstrate to the FDA that the herbal product has a favorable risk-benefit profile for use in human subjects.

8.3.3 Botanical Characterization and Standardization

One of the unique challenges in conducting preclinical studies on herbal medicines is the need for thorough *botanical characterization* and *standardization* of the product.

Unlike synthetic drugs, herbal medicines are often derived from complex mixtures of plant compounds, and the composition can vary depending on factors such as the plant species, growing conditions, harvesting methods, and processing techniques.

Botanical characterization involves identifying and authenticating the plant species used in the herbal product. This is typically done using morphological characteristics, genetic markers, and chemical profiles. Accurate botanical identification is essential to ensure the reproducibility of preclinical studies, as variations in plant species or chemotypes can lead to inconsistent results.

Standardization refers to the process of ensuring that each batch of the herbal product contains consistent levels of active or marker compounds. Standardization is important because it helps guarantee that the product used in preclinical studies has a consistent pharmacological and toxicological profile, which is critical for safety and efficacy assessments [22]. Without proper standardization, variations in the composition of the herbal medicine could result in misleading preclinical data, potentially leading to safety concerns when the product is tested in humans.

Standardization techniques can include quantitative analysis of key compounds using methods such as high-performance liquid chromatography (HPLC) or mass spectrometry (MS) [24]. By ensuring batch-to-batch consistency, standardization supports the scientific validity of preclinical studies and helps meet FDA requirements for botanical drug development.

8.4 INVESTIGATIONAL NEW DRUG APPLICATION

Once preclinical studies demonstrate that an herbal medicine is reasonably safe for use in humans, the next step is to submit an IND application to the FDA [5]. The IND application allows the sponsor (typically the manufacturer or research organization) to conduct clinical trials to investigate the safety and efficacy of the herbal medicine in human subjects. The IND process is an essential regulatory requirement for initiating human trials and eventually seeking approval to market the product as a drug in the United States [5].

8.4.1 IND Application Process

The IND application process begins with the submission of a comprehensive dossier that outlines the details of the proposed clinical trials and provides supporting evidence from preclinical studies. The goal of the IND application is to convince the FDA that the herbal product is safe to administer to humans and that the clinical trials are scientifically sound and ethically designed [5].

There are three types of INDs:

1. **Commercial IND**—submitted by a company intending to market the product as a drug.
2. **Research or Investigator IND**—submitted by a physician or researcher who intends to study the drug for an unapproved indication or in a different population.
3. **Emergency Use IND**—allows for the immediate use of an investigational drug in an emergency situation without an existing IND.

The FDA requires the submission of an IND application at least 30 days before the planned start of any clinical trials. During this period, the FDA evaluates the application to determine whether it is acceptable for the proposed trials to proceed. If the FDA does not issue a clinical hold within the 30-day period, the clinical trials may begin [5].

8.4.1.1 Required Documentation

An IND application must include several critical documents that provide the FDA with a complete understanding of the investigational herbal medicine and the proposed clinical trials. These documents include the following:

1. **Preclinical Study Reports**: Data from the toxicology and pharmacology studies conducted in animals, including safety profiles, pharmacokinetic data, and any adverse effects observed. This information is used to determine the starting doses for clinical trials and to identify any potential safety concerns.
2. **Clinical Protocols**: A detailed plan for the clinical trials, including the study design, methodology, inclusion and exclusion criteria for participants, dosing regimens, and endpoints. The protocol must outline how the safety and efficacy of the herbal medicine will be evaluated in human subjects.
3. **Investigator Brochure**: A document that provides clinical investigators with comprehensive information about the herbal medicine, including preclinical and clinical data, safety precautions, and instructions for administering the product. The investigator brochure helps ensure that clinical trial investigators have sufficient knowledge to conduct the study safely and effectively.
4. **Chemistry, Manufacturing, and Controls (CMC) Information**: Detailed information about the production and quality control of the herbal medicine, including the source of raw materials, methods of extraction or synthesis, and standardization of the product. CMC data are particularly important for herbal medicines, as batch-to-batch consistency must be demonstrated to ensure the reliability of clinical trial results [25].
5. **Informed Consent Documents**: Documents that explain the risks and benefits of participating in the clinical trial and that must be provided to study participants. Informed consent is a fundamental ethical requirement, ensuring that participants understand the study and agree to participate voluntarily.

6. **Other Supporting Documents**: These may include any relevant literature reviews, prior human experience with the herbal product, or special considerations for vulnerable populations such as children or the elderly.

8.4.2 FDA Review and Approval

Once the IND application is submitted, the FDA begins its review process. The FDA's primary concern during the review is the safety of the participants who will be enrolled in the clinical trials. The agency evaluates the preclinical data, clinical protocols, and CMC information to ensure that the proposed trials minimize risks to participants and that the potential benefits outweigh those risks [5].

The FDA may also consult with advisory committees or experts in herbal medicines to assist with the review, especially for complex or novel herbal products. If the FDA has any concerns about the safety of the investigational drug or the design of the clinical trials, it can issue a *clinical hold*, temporarily halting the trials until the issues are resolved. The FDA typically works closely with sponsors to address any deficiencies in the application and to modify the clinical trial design if necessary.

If the FDA does not object within 30 days of receiving the IND application, the sponsor is allowed to proceed with the clinical trials. However, the FDA continues to monitor the trials throughout the investigational process, reviewing safety reports and ensuring that the studies comply with ethical and regulatory standards.

Once clinical trials are successfully completed and demonstrate that the herbal medicine is safe and effective, the sponsor can proceed with submitting an NDA for FDA approval to market the product as a drug in the United States.

8.5 CLINICAL TRIAL PHASES

Once the IND application is approved by the FDA, clinical trials can begin. Clinical trials are divided into four distinct phases, each designed to answer specific questions about the investigational herbal medicine (Figure 8.3). These phases focus on the safety, efficacy, dosage, and side effects of the drug, as well as long-term surveillance post-market approval. The structure of clinical trials is vital for obtaining sufficient data to support the safety and efficacy claims for the herbal product, ultimately leading to its approval as a therapeutic drug (Table 8.3).

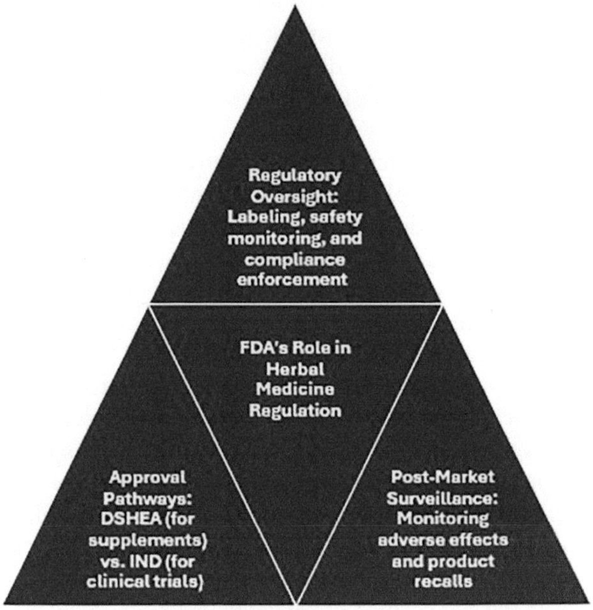

FIGURE 8.3 FDA's role in herbal medicine regulation.

8.5.1 Phase I: Safety and Dosage

Phase I clinical trials are the first tests of the investigational herbal medicine in human subjects. The primary objective of this phase is to assess the *safety* of the drug and determine its *appropriate dosage.*

- **Participants**: Typically, 20 to 100 healthy volunteers or patients with the target condition participate in this phase.
- **Duration**: Phase I trials usually last several months.
- **Focus**: The trial is designed to determine the drug's pharmacokinetics, such as absorption, distribution, metabolism, and excretion (ADME), along with its pharmacodynamics, i.e., how the body responds to the drug [26].

TABLE 8.3
Phases of Clinical Trials for Herbal Medicines

Phase	Objective	Study Design	Regulatory Submission Required
Phase I	Evaluate safety, dosage, and side effects	Small groups of healthy volunteers	IND application
Phase II	Assess efficacy and safety in patients with the condition	Larger groups (100–300)	Annual IND updates
Phase III	Confirm efficacy, monitor side effects, compare with current treatments	Larger groups (1,000+)	NDA submission
Phase IV	Post-marketing surveillance	Real-world patient data	Post-market studies and safety monitoring

- **Outcome**: The key result of phase I is establishing a safe dosage range, identifying any immediate adverse effects, and gathering preliminary information about the drug's interaction with human biology [27].

For herbal medicines, this phase often poses unique challenges, particularly with products composed of multiple plant compounds that could interact unpredictably with human systems. Standardizing the components to ensure batch consistency is essential to accurately assess safety [28].

8.5.2 Phase II: Efficacy and Side Effects

After the drug has been deemed safe in phase I, it moves on to *phase II clinical trials,* which focus on the *efficacy* of the drug and identifying potential *side effects.*

- **Participants**: Typically, 100 to 300 patients with the target disease or condition are recruited. These individuals are more representative of the population for which the drug is intended.
- **Duration**: Phase II trials generally last several months to two years.
- **Focus**: The goal of phase II is to evaluate the drug's effectiveness in treating the targeted condition and to gather additional safety data [29]. This phase also explores short-term side effects and risks.
- **Outcome**: The results help determine whether the drug has a measurable therapeutic effect and justify the need for larger, more definitive studies. Phase II trials often use *randomized controlled trial (RCT)* designs to compare the herbal medicine with a placebo or standard treatment [30].

In the context of herbal medicines, the variability in raw materials can affect efficacy results, which emphasizes the importance of standardized dosing and well-characterized botanical sources.

8.5.3 Phase III: Confirmation and Monitoring

Phase III clinical trials are the most critical phase of drug development. They provide *confirmatory evidence* regarding the drug's efficacy and *monitor adverse reactions* in a larger, more diverse population.

- **Participants**: This phase typically includes 1,000 to 3,000 patients with the targeted condition, making it the largest phase in terms of sample size.
- **Duration**: Phase III trials may last one to four years.
- **Focus**: The trials aim to confirm the effectiveness of the drug across a broader population, assess long-term side effects, and further refine the optimal dosage. These trials are also pivotal for identifying rare side effects that might not appear in smaller trials [29].
- **Outcome**: A successful phase III trial is typically the basis for filing a NDA with the FDA for market approval. The results provide the most substantial evidence that the herbal medicine is both safe and effective for public use.

For herbal medicines, demonstrating efficacy can be more complex due to the multicomponent nature of these products. However, stringent quality controls and the use of GMP ensure consistency and reliability of the botanical product used in the trials [26].

8.5.4 Phase IV: Post-Marketing Surveillance

Once the herbal medicine is approved and marketed, it enters *phase IV,* also known as *post-marketing surveillance.* This phase is critical for detecting any *long-term or rare side effects* and for *monitoring safety* in a broader patient population.

- **Participants**: Phase IV trials involve the general population, including individuals who may not have been included in earlier trials, such as pregnant women, the elderly, or patients with comorbidities.
- **Duration**: These trials are ongoing and may last for several years or even decades.
- **Focus**: The focus is on monitoring adverse effects that may arise from widespread use, long-term safety, and effectiveness. Phase IV studies may also explore additional uses for the drug or test its performance in different populations.
- **Outcome**: Data gathered during phase IV can lead to modifications in the drug's labeling, warnings, or usage recommendations. In some cases, the FDA may withdraw approval if significant safety concerns arise [29].

In the case of herbal medicines, phase IV is especially important for identifying long-term effects that may not have been evident in earlier trials due to the complex interaction of herbal compounds and prolonged use [30].

8.6 REGULATORY GUIDELINES FOR CLINICAL TRIALS

For clinical trials involving herbal medicines in the United States, researchers must adhere to a set of regulatory guidelines designed to ensure the safety, efficacy, and ethical treatment of participants. These guidelines are grounded in well established practices that are globally recognized. Key among them are the *Good Clinical Practice (GCP) guidelines*, the role of *Institutional Review Boards (IRBs),* the necessity for *informed consent,* and the requirement for timely *adverse event reporting.* Compliance with these standards not only ensures that clinical trials meet regulatory expectations but also upholds ethical principles in biomedical research [31].

8.6.1 Good Clinical Practice Guidelines

Good Clinical Practice (GCP) is an international ethical and scientific quality standard for designing, conducting, recording, and reporting clinical trials involving human participants. For herbal medicines, adherence to GCP is crucial to ensure that the results from clinical trials are credible and that the rights, integrity, and confidentiality of trial subjects are protected [32].

- **Scope**: GCP covers all aspects of clinical trials, from the planning and execution of the study to the final analysis and reporting of the results. These guidelines apply to all phases of clinical development.
- **Key Principles**: GCP requires trials to be scientifically sound, with well-defined protocols that prioritize patient safety and ethical considerations. Herbal medicine trials must also ensure standardization of the botanical materials used to maintain consistency and quality throughout the study [26].
- **Implementation**: Researchers are required to conduct trials under GCP guidelines to ensure that the results meet regulatory standards for approval by the FDA. The FDA may conduct inspections of clinical trial sites to verify compliance with GCP [29].

Adherence to GCP is particularly important in herbal medicine trials due to the complex nature of plant-derived products, which often contain multiple bioactive compounds.

8.6.2 Institutional Review Boards

The *IRB* is an essential body in the clinical trial approval and oversight process. It is responsible for reviewing and monitoring research to ensure the safety and well-being of the participants involved.

- **Role**: The IRB's primary responsibility is to review research protocols to ensure that the risks to participants are minimized, the risks are reasonable in relation to the potential benefits, and that informed consent is properly obtained [30].
- **Composition**: IRBs are made up of a diverse group of individuals, including scientists, ethicists, and community members, to provide comprehensive oversight of clinical trials.
- **Approval Process**: For any clinical trial involving herbal medicines, approval by an IRB is mandatory. Without IRB approval, the trial cannot proceed, and any amendments to the protocol must also be reviewed and approved by the IRB [32].

Given the growing interest in herbal medicines, IRBs play a pivotal role in ensuring that the potential risks of using plant-derived compounds are clearly communicated to participants and ethically managed during the trial.

8.6.3 Informed Consent Requirements

The concept of *informed consent* is a fundamental ethical requirement in clinical research. It ensures that participants understand the risks and benefits of the clinical trial and voluntarily agree to participate.

- **Process**: Informed consent involves providing potential participants with detailed information about the study, including its purpose, duration, required procedures, potential risks, and benefits. Participants must also be informed that they have the right to withdraw from the study at any point without penalty.
- **Herbal Medicines**: In trials involving herbal medicines, the informed consent process may include additional elements. Participants must be made aware that herbal products may contain complex mixtures of bioactive compounds, some of which may not be well understood or studied [26].
- **Documentation**: Consent must be documented in writing, and participants should receive a copy of the consent form for their records. For herbal medicine trials, this documentation is particularly critical as it may need to address unique risks associated with plant-derived compounds [30].

Informed consent not only protects the participants' rights but also ensures that researchers are transparent about the trial's potential risks, which is particularly relevant for herbal products with a less-established safety profile (Table 8.4)

8.6.4 Adverse Event Reporting

An essential aspect of any clinical trial is the *reporting of adverse events (AEs)*. This is especially important for trials involving herbal medicines, which may have unpredictable effects due to the complexity of their constituents [33].

- **Definition**: AEs refer to any *unintended and harmful reactions* experienced by participants during the course of a clinical trial, whether they are related to the investigational drug or not [29].
- **Reporting Requirements**: Under FDA regulations, all AEs must be documented and reported promptly. Serious adverse events (SAEs)—those that result in death, life-threatening conditions, hospitalization, or disability—must be reported to the FDA within a specific timeframe [6].
- **Herbal Medicines**: Due to the multicomponent nature of herbal medicines, it can be challenging

TABLE 8.4
Common Challenges in Clinical Trials for Herbal Medicines

Challenge	Description	Mitigation Strategies
Standardization	Difficulty in ensuring consistent active ingredients	Use of standardized extracts and batch testing
Dosage determination	Variability in effective dose due to differences in formulation	Conduct dose-finding studies in early clinical phases
Placebo effect	Strong placebo response in trials with herbal treatments	Blinding and use of placebo controls
Herb–drug interactions	Potential for interactions with conventional medicines	Preclinical studies and thorough patient screening

to attribute AEs to a specific compound. However, proper botanical characterization and the use of standardized products can help mitigate some of these challenges [28].

- **Post-Trial Reporting**: Even after the clinical trial concludes, post-marketing surveillance (phase IV) continues to monitor for any AEs that may arise from the long-term use of the herbal medicine.

Adverse event reporting is critical for ensuring the safety of participants during the clinical trial process, especially when dealing with herbal products that may have complex, not fully understood effects.

8.7 QUALITY CONTROL AND MANUFACTURING

For herbal medicines to be accepted for clinical trials and eventual commercialization in the United States, they must meet rigorous quality control standards. The *GMP* established by the FDA provide the regulatory framework for ensuring the consistent quality of products. In addition, maintaining *batch consistency,* ensuring *quality assurance,* and complying with *labeling and marketing regulations* are essential elements of the quality control process for herbal medicines.

8.7.1 Good Manufacturing Practices

GMP are regulations enforced by the FDA to ensure that products are produced and controlled according to quality standards appropriate for their intended use. For herbal medicines, adherence to GMP is vital because of the inherent variability in plant materials, which may affect the potency, purity, and safety of the final product.

- **Overview**: GMP guidelines for herbal medicines cover the entire production process, from the sourcing of raw materials to the packaging and labeling of the final product. The regulations are designed to prevent contamination, mix-ups, and errors that could compromise product quality.
- **Key Requirements**: GMP for herbal medicines includes standards for facility cleanliness, equipment maintenance, proper handling of raw materials, and trained personnel. It also requires detailed documentation of manufacturing processes, batch records, and quality control measures [34].
- **Herbal Specific Considerations**: Herbal medicines pose unique challenges, such as variability in active ingredients depending on factors like plant origin, harvesting time, and processing methods. GMP regulations, therefore, emphasize the importance of rigorous botanical characterization and quality control throughout the supply chain [28].

The implementation of GMP ensures that herbal medicines are produced consistently and meet the necessary standards for safety and efficacy in clinical trials.

8.7.2 Batch Consistency and Quality Assurance

Achieving *batch consistency* is crucial in herbal medicine production, given the natural variability of plant materials. Batch-to-batch variations in chemical composition can affect the therapeutic efficacy and safety of the product, making quality assurance a critical part of the manufacturing process.

- **Importance of Batch Consistency**: To ensure that each batch of an herbal product has the same composition and potency, manufacturers must standardize the raw materials and manufacturing processes. This includes selecting plants of the same species, grown under controlled conditions, and using consistent harvesting and extraction methods [26].
- **Standardization**: Herbal products are typically standardized to a specific amount of a bioactive compound or group of compounds known to contribute to the product's therapeutic effect. This ensures that each batch delivers the same expected results [35].
- **Quality Assurance Processes**: Quality assurance involves continuous monitoring of the production process to detect and correct deviations. This includes testing raw materials for purity and potency, verifying the identity and concentration of active ingredients, and conducting final product testing for contaminants such as heavy metals, pesticides, or microbial contamination [30].

Maintaining batch consistency ensures that herbal medicines used in clinical trials are of high quality and meet regulatory expectations, which is critical for obtaining FDA approval.

8.7.3 Labeling and Marketing Compliance

Once herbal medicines pass clinical trials and enter the commercial market, they must comply with *labeling and marketing regulations* set by the FDA and the FTC. Proper labeling ensures that consumers are well informed about the product's intended use, potential risks, and benefits.

- **FDA Labeling Requirements**: Herbal medicines sold as dietary supplements must comply with the *DSHEA,* which requires products to be labeled accurately and not make false or misleading claims. Labels must include the name and quantity of each ingredient, a description of the product's intended use, and any potential side effects or contraindications [6].
- **Marketing Claims**: Herbal medicine manufacturers are prohibited from making unsubstantiated health claims. For instance, a product cannot be marketed as curing, treating, or preventing any disease unless such claims have been substantiated through clinical trials and FDA approval as a drug.
- **GRAS and NDI Compliance**: If the herbal product includes ingredients that are classified as *GRAS* or meet the FDA's NDI notification requirements, this must be clearly communicated on the label [12].

Complying with labeling and marketing regulations is crucial for ensuring that herbal medicines are both legally marketed and ethically represented to consumers, thus protecting public health and maintaining trust in the product.

8.8 CASE STUDIES AND EXAMPLES

To further illustrate the regulatory processes involved in the clinical trials and approval of herbal medicines in the United States, this section highlights *successful herbal medicine approvals,* discusses *regulatory challenges and their resolutions,* and offers insights into *best practices and lessons learned.* These real-world examples provide a practical perspective on navigating the regulatory landscape for herbal medicines.

8.8.1 Successful Herbal Medicine Approvals

While the approval of herbal medicines as drugs by the FDA is rare, several products have successfully navigated the regulatory process. These examples demonstrate how rigorous scientific evaluation, adherence to regulatory guidelines, and effective collaboration with regulatory agencies can lead to success.

- **Veregen® (Sinecatechins)**: Veregen, a green tea extract (sinecatechins), was approved by the FDA in 2006 as a botanical drug for the treatment of external genital and perianal warts. It was one of the first herbal medicines to receive drug approval in the United States under the *botanical drug development pathway.* The developers conducted well-designed clinical trials to prove the efficacy and safety of the product, successfully meeting FDA requirements [36].
- **Fulyzaq® (Crofelemer)**: Derived from the sap of the *Croton lechleri* tree, Fulyzaq was approved by the FDA in 2012 for the symptomatic relief of diarrhea in HIV/AIDS patients on antiretroviral therapy. Its approval came after significant research into the botanical source, standardized extraction processes, and multiple clinical trials that established its efficacy and safety [37].

These examples underscore the importance of conducting comprehensive clinical trials and meeting the FDA's stringent regulatory standards for herbal medicines.

8.8.2 Regulatory Challenges and Resolutions

The development and approval of herbal medicines face numerous regulatory challenges, given the complexity of botanical products and the differences between them and synthetic drugs. The following are some common regulatory challenges and how they have been addressed in various cases:

- **Challenge: Variability in Active Ingredients**: Herbal medicines often consist of complex mixtures of active compounds, making it difficult to ensure consistent product quality. This variability can complicate the clinical trial process.
  - **Resolution**: To overcome this challenge, manufacturers have adopted stringent *standardization* techniques, in which the concentration of bioactive compounds is consistently measured and maintained. For example, in the case of *Veregen*, the developers standardized the polyphenol content in green tea extract to ensure batch consistency throughout the clinical trials [38].
- **Challenge: Lack of Preclinical Data**: Herbal medicines may lack comprehensive preclinical data, such as toxicology studies, making it difficult for developers to initiate clinical trials.
  - **Resolution**: Developers can engage with the FDA through *pre-IND meetings* to gain clarity on the specific preclinical data required. This approach was successfully employed by the developers of *Fulyzaq*, who worked closely with the FDA to design preclinical studies that met regulatory expectations [37].
- **Challenge: Regulatory Complexity**: Herbal medicines fall under various regulatory categories

(e.g., dietary supplements, botanical drugs), each with different requirements. Navigating this complexity can be challenging for developers.

- **Resolution**: Successful developers often rely on regulatory consultants and legal experts familiar with FDA guidelines. By seeking expert advice, developers can better understand which regulatory pathway is most appropriate for their product and how to meet the relevant requirements. *Fulyzaq*'s developers initially faced challenges in determining the right regulatory pathway but ultimately succeeded through collaboration with the FDA experts [39].

8.8.3 Best Practices and Lessons Learned

From the case studies and challenges faced by herbal medicine developers, several *best practices* and *lessons learned* can be derived. These can serve as a guide for future developers aiming to bring herbal products to market in the United States.

1. **Early Engagement with Regulatory Authorities**: Herbal medicine developers should initiate dialogue with the FDA early in the development process, particularly through pre-IND meetings. This can help clarify regulatory expectations and identify any potential obstacles.
   - **Lesson**: Early regulatory engagement facilitates a smoother approval process, as seen in the development of *Veregen*, where proactive discussions with the FDA allowed for a clearer understanding of clinical trial requirements [36].
2. **Thorough Preclinical and Clinical Documentation**: Herbal medicines must adhere to the same rigorous scientific standards as conventional drugs. Comprehensive preclinical studies and well-designed clinical trials are essential.
   - **Lesson**: Developers of *Fulyzaq* conducted extensive preclinical studies to address safety concerns, which later contributed to the successful approval of the drug [37].
3. **Standardization and Quality Control**: Ensuring consistent product quality through the standardization of bioactive compounds is critical, especially given the variability of botanical ingredients.
   - **Lesson**: Veregen's developers prioritized standardizing the active polyphenol content in green tea extract, which was key to achieving batch consistency and gaining FDA approval [38].
4. **Adherence to GMP**: Manufacturing herbal medicines under GMP ensures that products are of high quality, safe, and effective.
   - **Lesson**: Both Veregen and Fulyzaq followed stringent GMP standards throughout the manufacturing process, which was crucial for FDA approval [39].
5. **Flexibility and Adaptation to Regulatory Feedback**: Developers must remain flexible and responsive to regulatory feedback throughout the development process.
   - **Lesson**: The development of Fulyzaq involved multiple iterations of clinical trial designs and product formulations based on feedback from the FDA, illustrating the importance of adaptability [37].

By following these best practices and learning from the challenges faced by previous developers, future herbal medicine manufacturers can improve their chances of successful regulatory approval.

8.9 CHALLENGES AND OPPORTUNITIES

The regulation of herbal medicines in the United States presents a complex but evolving landscape, with both significant challenges and potential opportunities. As the demand for herbal products continues to grow, developers must navigate intricate regulatory frameworks while simultaneously driving innovation. This section explores the *challenges* of navigating the regulatory landscape, highlights *opportunities for innovation*, and examines potential *future directions* in the regulation of herbal medicines.

8.9.1 Navigating the Regulatory Landscape

Developers of herbal medicines face several regulatory hurdles that can impede their ability to bring products to market in the United States. Understanding and overcoming these challenges is essential to ensuring the safety, efficacy, and quality of herbal products.

- **Complex Regulatory Classification**: Herbal medicines can be classified as dietary supplements, botanical drugs, or over-the-counter (OTC) products, each with distinct regulatory pathways under the FDA and other agencies [40]. Determining the correct classification can be difficult and has significant implications for preclinical, clinical, and manufacturing processes [38].
  - **Solution**: Developers should engage early with the FDA or hire regulatory consultants to ensure the product is placed in the correct category. Incorrect classification can lead to regulatory delays and compliance issues [39].
- **Standardization of Botanical Ingredients**: Due to the inherent variability of plant-based materials, ensuring batch-to-batch consistency is a major challenge. Lack of standardization can impact both the efficacy and safety of the final product [41].
  - **Solution**: Advances in analytical techniques such as HPLC and MS have enabled better quantification of bioactive compounds, aiding in the standardizing of herbal products [42].

- **Preclinical and Clinical Data Requirements**: Meeting the FDA's stringent requirements for safety and efficacy through well-documented preclinical and clinical studies can be daunting, particularly for developers with limited resources.
    - **Solution**: Collaborative research between academia, industry, and government agencies can help ease the financial burden of conducting large-scale trials. Additionally, developers can take advantage of programs such as the FDA's Botanical Review Team, which provides guidance on clinical trial design and regulatory submissions [43].

8.9.2 Opportunities for Innovation

Despite these challenges, the field of herbal medicine is ripe with opportunities for innovation. Several areas have the potential to not only overcome regulatory hurdles but also to improve the overall quality and accessibility of herbal products.

- **Integration of Traditional Knowledge with Modern Science**: Herbal medicines have long been used in traditional medicine systems, such as *Traditional Chinese Medicine (TCM)* and *Ayurveda.* By integrating traditional knowledge with modern scientific techniques, there is the potential to develop new herbal medicines that are both culturally relevant and scientifically validated [44]. Collaborations between researchers, traditional healers, and biopharmaceutical companies can foster innovation in this field.
- **Personalized Herbal Medicine**: With the growing trend of personalized medicine, there is an opportunity to apply this concept to herbal treatments. Advances in *genomics* and *metabolomics* may allow for the development of individualized herbal treatment plans based on a person's genetic profile or metabolic state [45]. Personalized herbal medicine could be tailored to optimize efficacy while minimizing adverse effects, providing a significant advantage in clinical outcomes.
- **Emerging Technologies in Quality Control**: New technologies, such as *blockchain* and *artificial intelligence (AI),* offer innovative solutions for improving the traceability, safety, and quality control of herbal products. Blockchain can be used to create transparent supply chains, ensuring that herbal products are sourced, manufactured, and distributed according to regulatory standards. AI can assist in predicting potential adverse effects, optimizing clinical trial design, and ensuring regulatory compliance [46].

8.9.3 Future Directions in Herbal Medicine Regulation

As herbal medicine continues to grow in popularity, there is an increasing recognition of the need for a more tailored regulatory framework. Several trends and future directions are likely to shape the regulation of herbal medicines in the coming years:

- **Harmonization of International Standards**: One of the major challenges for the herbal medicine industry is the lack of harmonized regulatory standards across different countries. The development of globally recognized guidelines, such as those proposed by the World Health Organization (WHO) and the International Council for Harmonisation of Technical Requirements for Pharmaceuticals for Human Use (ICH), could help streamline the approval process for herbal medicines worldwide [47]. Harmonization would enable developers to conduct a single set of clinical trials that meets regulatory requirements across multiple jurisdictions, reducing costs and expediting market access.
- **Greater Emphasis on Safety Monitoring**: With the rising popularity of herbal products, there is an increasing need for robust post-marketing surveillance systems to monitor safety. The FDA's MedWatch program currently allows healthcare providers and consumers to report AEs, but there is potential for more proactive monitoring using real-world evidence (RWE) from electronic health records and wearable health devices [48]. These tools can help identify safety concerns earlier and provide valuable data for the continuous monitoring of herbal medicines after they have been brought to the market.
- **Increased Collaboration between Regulatory Agencies and Industry**: Future regulatory trends will likely focus on fostering greater collaboration between regulatory agencies, academia, and the herbal medicine industry. Initiatives like the FDA's Botanical Review Team and the NIH's NCCIH have already made significant strides in this area. Ongoing collaboration will be essential to improve the approval process and ensure that herbal medicines are safe and effective [49].
- **Regulatory Incentives for Herbal Drug Development**: Just as the Orphan Drug Act provides incentives for the development of treatments for rare diseases, there is a growing movement toward establishing incentives for herbal drug development. Such incentives could include tax breaks, grant funding, and priority review by the FDA, which would encourage more companies to invest in herbal drug research [39].

8.10 CONCLUSION

The regulation of herbal medicines in the United States is at a pivotal juncture. While significant challenges remain, there are numerous opportunities for innovation, collaboration, and improvement in the regulatory framework. As science continues to advance and societal demand for herbal products grows, regulatory agencies will likely adapt to better accommodate these complex products. By embracing innovation and adopting a flexible yet rigorous regulatory approach, the future of herbal medicine in clinical practice appears bright.

REFERENCES

[1] Wang, H., & Zhao, L. 2017. "The Historical Role of Herbal Medicines in Global Healthcare Systems." *Journal of Traditional Medicine* 33 (4): 201–209.

[2] Johnson, P. A., & Lee, C. 2021. "The Rise of Herbal Medicine in the United States: Trends, Market Dynamics, and Consumer Preferences." *Journal of Herbal Pharmacology* 22 (3): 200–215.

[3] Dwyer, J. T., Coates, P. M., & Smith, M. J. 2018. "Dietary Supplements: Regulatory Challenges and Research Resources." *Nutrients* 10 (1): 1–24. https://doi.org/10.3390/nu10010041.

[4] "Dietary Supplements | FDA." n.d. Accessed October 17, 2024. https://www.fda.gov/food/dietary-supplements.

[5] "Investigational New Drug (IND) Application | FDA." n.d. Accessed October 17, 2024. https://www.fda.gov/drugs/types-applications/investigational-new-drug-ind-application.

[6] "Guidance for Industry: Questions and Answers Regarding Adverse Event Reporting and Recordkeeping for Dietary Supplements as Required by the Dietary Supplement and Nonprescription Drug Consumer Protection Act | FDA." n.d. Accessed October 17, 2024. https://www.fda.gov/regulatory-information/search-fda-guidance-documents/guidance-industry-questions-and-answers-regarding-adverse-event-reporting-and-recordkeeping-dietary.

[7] Smith, J. A., & Johnson, R. 2021. "The Importance of Regulatory Compliance in the Herbal Product Industry: Safety and Efficacy Considerations." *Journal of Herbal Medicine* 16 (3): 150–157.

[8] "Dietary Supplements: What You Need to Know—Consumer." n.d. Accessed October 17, 2024. https://ods.od.nih.gov/factsheets/WYNTK-Consumer/.

[9] Davis, R. J., & Thompson, M. 2018. "Regulatory Pathways for Herbal Products: From Supplements to Drugs." *Journal of Herbal Medicine Research* 9 (2): 55–63.

[10] "Guidance for Industry: New Dietary Ingredient Notification Procedures and Timeframes—Dietary Supplements | FDA." n.d. Accessed October 17, 2024. https://www.fda.gov/regulatory-information/search-fda-guidance-documents/guidance-industry-new-dietary-ingredient-notification-procedures-and-timeframes-dietary-supplements.

[11] Smith, R. & Green, L. 2020. "New Dietary Ingredient (NDI) Notifications: Understanding FDA Guidelines and Implications for Herbal Products." *Journal of Herbal Medicine* 19 (4): 30–37.

[12] "GRAS Substances (SCOGS) Database | FDA." n.d. Accessed October 17, 2024. https://www.fda.gov/food/generally-recognized-safe-gras/gras-substances-scogs-database.

[13] Miller, K., & Johnson, P. 2019. "The FDA's Role in Regulating Herbal Medicines and Dietary Supplements under DSHEA." *Journal of Regulatory Affairs* 27 (1): 56–64.

[14] Nguyen, T. H., & Martinez, A. 2021. "Navigating the Regulatory Landscape for Herbal Medicines in Clinical Trials: The Role of IND Applications." *Journal of Herbal Medicine* 20 (2): 88–95.

[15] Garcia, M. A., & Patel, S. 2017. "Preclinical Testing Requirements for Herbal Medicines in the U.S." *Journal of Preclinical Pharmacology* 5 (1): 12–19.

[16] "Clinical Trials | NCCIH." n.d. Accessed October 17, 2024. https://www.nccih.nih.gov/research/clinicaltrials.

[17] Jones, T. R., & Baker, S. 2020. "The Dietary Supplement Health and Education Act (DSHEA): Implications for Herbal Medicine Regulation in the United States." *The Journal of Nutrition* 17 (1): 12–20.

[18] Anderson, L. M., & Greenfield, S. 2021. "Understanding the Federal Food, Drug, and Cosmetic Act: Implications for Herbal Products in Clinical Trials." *Journal of Regulatory Science* 8 (3): 45–52.

[19] Miller, A. H., & Thompson, J. 2021. "The Federal Food, Drug, and Cosmetic Act: Implications for Herbal Medicines and Clinical Trials." *Regulatory Affairs Journal* 45 (2): 75–82.

[20] Smith, J., Johnson, T., & Williams, R. 2019. "Good Laboratory Practices (GLP) and Their Impact on Preclinical Research Quality." *Journal of Regulatory Science* 7 (4): 112–125.

[21] Shipkowski, K. A., Betz, J. M., Birnbaum, L. S., Bucher, J. R., Coates, P. M., Hopp, D. C., MacKay, D., et al. 2018. "Naturally Complex: Perspectives and Challenges Associated with Botanical Dietary Supplement Safety Assessment." *Food and Chemical Toxicology* 118: 963–971. https://doi.org/10.1016/j.fct.2018.04.007.

[22] Liu, Y., & Wang, H. 2020. "Standardization of Herbal Extracts and Its Role in Toxicological Studies." *Journal of Herbal Medicine* 14 (4): 95–100.

[23] Parveen, A., Parveen, B., Parveen, R., & Ahmad, S. 2015. "Challenges and Guidelines for Clinical Trial of Herbal Drugs." *Journal of Pharmacy and Bioallied Sciences* 7 (December): 329. https://doi.org/10.4103/0975-7406.168035.

[24] Brown, D., & Patel, S. 2019. "Analytical Methods in Herbal Standardization: HPLC, MS and Beyond." *Journal of Analytical Chemistry* 24 (2): 180–195.

[25] Roe, A. L, McMillan, D. A., & Mahony, C. 2018. "A Tiered Approach for the Evaluation of the Safety of Botanicals Used as Dietary Supplements: An Industry Strategy." *Clinical Pharmacology & Therapeutics* 104 (3): 446–457. https://doi.org/10.1002/cpt.1132.

[26] Lasser, K. L., & Mahapatra, S. 2020. "Pharmacokinetics and Pharmacodynamics in Drug Development: A Review." *Journal of Clinical Pharmacy and Therapeutics* 45 (3): 234–245.

[27] "Step 3: Clinical Research | FDA." n.d.-a. Accessed October 20, 2024. https://www.fda.gov/patients/drug-development-process/step-3-clinical-research.

[28] Smith, J., Thompson, R., & Lee, K. 2020. "Challenges in the Clinical Evaluation of Herbal Medicines: Safety and Standardization Issues." *Journal of Ethnopharmacology* 254: 112–120.

[29] "Step 3: Clinical Research | FDA." n.d.-b. Accessed October 20, 2024. https://www.fda.gov/patients/drug-development-process/step-3-clinical-research.

[30] Shankar, A., & White, J. 2019. "Designing Phase II Trials for Herbal Medicines." *Journal of Clinical Research* 15 (1): 45–52.

[31] Salm, S., Rutz, J., van den Akker, M., Blaheta, R. A., & Bachmeier, B. E. 2023. "Current State of Research on the Clinical Benefits of Herbal Medicines for Non-Life-Threatening Ailments." *Frontiers in Pharmacology* 14.

[32] European Medicines Agency (EMA). 2018. "Guideline Good Clinical Practice E6(R2)." *Committee for Human Medicinal Products* 6 (December 2016): 1–68.

[33] Smith, J., & Patel, R. 2021. "Adverse Event Reporting in Clinical Trials of Herbal Medicines: Challenges and Recommendations." *Journal of Clinical Research* 18 (3): 215–223.

[34] Organization, World Health. 2018. "Guidelines on Good Manufacturing Practices for the Manufacture of Herbal Medicines: In WHO Expert Committee on Specifications for Pharmaceutical Preparations: Fifty-Second Report." *WHO Technical Report Series, No. 1010*, no. 1010, 153–177.

[35] Eisenberg, D. M., & O'Connor, B. 2018. "Standardization of Herbal Products: A Guide to Ensuring Consistent Therapeutic Effects." *Journal of Herbal Medicine* 15 (4): 253–241.

[36] Tang, J., Li, Y., & Lee, H. 2007. "Veregen: A New Botanical Drug for External Genital and Perianal Warts." *Journal of Drugs in Dermatology* 6 (4): 432–436.

[37] Holodniy, M., Kearney, M., & MacKenzie, M. 2013. "Fulyzaq: A Botanical Therapeutic Agent for the Treatment of Diarrhea in Patients with HIV/AIDS." *Journal of Antiviral Therapy*, 18 (2): 109–115.

[38] Farnsworth, N. R., Soejarto, D. D., & Hsu, Y. 2014. "The Importance of Standardization in Herbal Medicines: A Case Study of Veregen (Sinecatechins)." *Journal of Herbal Medicine* 4 (1): 21–27.

[39] Kroll, D. J., & Cordell, G. A. 2015. "Navigating the Regulatory Landscape for Herbal Medicines: The Case of Fulyzaq." *Journal of Dietary Supplements* 12 (2): 137–147.

[40] Baker, S. M., & Nguyen, T. 2020. "Classification and Regulation of Herbal Medicines: Navigating the FDA Pathways." *Journal of Regulatory Science* 6 (2): 75–84.

[41] Thompson, M., & Roberts, J. 2019. "Challenges in Achieving Consistency in Herbal Medicine Production: The Impact of Variability on Efficacy and Safety." *Journal of Herbal Medicine* 15 (1): 45–52.

[42] Fong, H. H. S., & Mitchell, M. 2016. "Innovations in Analytical Techniques for the Standardization of Herbal Products." *Journal of Pharmaceutical Sciences* 105 (8): 2260–2269.

[43] Zhang, H., & Wang, Y. 2013. "Aviating the Regulatory Landscape for Botanical Products: Insights from the FDA's Botanical Review Team." *Journal of Herbal Medicine* 3 (2): 91–97.

[44] Panda, S., Gupta, M., & Sharma, P. 2019. "Bridging Traditional Knowledge and Modern Science: The Future of Herbal Medicine Development." *Journal of Ethnopharmacology*, 234: 67–75.

[45] Chen, Y., Zhang, X., & Lee, S. 2016. "Personalized Herbal Medicine: Integrating Genomics and Metabolomics for Tailored Treatments." *Journal of Personalized Medicine* 6 (2): 20.

[46] Rodriguez, J., Smith, T., & Li, Q. 2020. "Emerging Technologies in Herbal Medicine: The Roles of Blockchain and Artificial Intelligence in Ensuring Safety and Efficacy." *Journal of Herbal Technology* 5 (4): 300–310.

[47] Calixto, J. B. 2019. "Global Harmonization of Herbal Medicine Regulations: The Role of WHO and ICH in Standardizing Approval Processes." *Journal of Ethnopharmacology* 245: 112–118.

[48] Yuan, Y., Xu, Y., & Zhao, Y. 2015. "Harnessing Real-World Evidence for Proactive Monitoring of Adverse Events: Opportunities for the FDA's MedWatch Program." *Journal of Medical Systems* 39 (11): 197.

[49] Wu, Y., Sun, L., & Wang, X. 2014. "Collaborative Efforts between the FDA and NCCIH: Enhancing the Regulatory Pathway for Herbal Medicines." *Journal of Ethnopharmacology* 157: 97–104.

9 Regulatory Aspects of Clinical Trials of Herbal Medicines in Non-European Countries

Uma Ranjan Lal[1], *Megha Soni*[1], *and Snigdha Lal*[2]
[1]Department of Natural Products, National Institute of Pharmaceutical Education and Research, SAS Nagar, Mohali, Punjab, India
[2]Department of Chemistry, Ram Lakhan Singh Yadav College, Bettiah, West Champaran, Bihar, India

9.1 INTRODUCTION

Herbal medicines encompass remedies made from herbs, herbal materials, and plant-based products. Different cultures have developed their own traditional herbal medicine systems, such as Traditional Chinese Medicine, Ayurvedic medicine (Indian subcontinent), Kampo medicine (Japan), Aboriginal medicine (Australia), and Te Rongoā (New Zealand). The recognition of traditional herbal medicine (THM) was highlighted by the 2015 Nobel Prize awarded to Youyou Tu for her discovery of artemisinin from Artemisia annua (Qinghao), which has revolutionized malaria treatment. Herbal medicines have evolved from simple practices to more complex systems; they reflect the interplay of economy, science, and culture over time. Modern medicine and industrialization, with their impact on health and way of living, have also introduced new perspectives on and preferences for traditional herbal remedies [1].

Traditional or folk medicine includes practices and beliefs specific to cultural contexts for treating, diagnosing, and preventing illness. These are termed "complementary" or "alternative medicine" when adopted outside their traditional settings, and their regulations vary widely. For example, in Brazil, it is proposed that herbal drugs can be classified as either "phytotherapeutic medicines" or "traditional phytotherapeutic products" (TPTP), supported by literature from ethnopharmacological studies. This may allow or enhance commercialization without extensive preclinical or clinical safety and efficacy data. Also, the European Union's EC Directive 2004/24 permits herbal products with at least 30 years of medicinal use, including 15 years in the EU, to be classified as "traditional herbal medicine products" (THMP). These products undergo a simplified registration process, although member states may impose additional safety requirements. In contrast, the U.S. Food and Drug Administration (FDA) does not differentiate between herbal and conventional medicines in terms of premarketing safety and efficacy requirements. Consequently, many herbal products are marketed as dietary supplements, which face less stringent regulations. Critics argue that this lenient approach could lead to inadequate safety evaluations, thereby highlighting the need for all medicinal products, regardless of origin, to meet rigorous standards for efficacy and safety [2].

Clinical trials for herbal medicine present unique challenges, unlike those for conventional pharmaceuticals. Randomized clinical trials (RCTs) are the gold standard in conventional medicine, but are difficult for herbal medicines because of sociocultural variables, ethical considerations, and the complex nature of herbal formulations. The variability in plant materials and production methods complicates standardization and reproducibility. Instead of relying solely on RCTs, anecdotal reports, case reports, and case series can be valuable in assessing herbal efficacy. Prospective data collection, particularly with collaboration from herbal practitioners, could provide scientific insights while being more feasible. However, rigorous testing, including chemical profiling and validation, remains essential to ensure quality and efficacy. Integrating conventional testing methodologies with the flexibility of traditional practices is vital for credible evidence in herbal medicine [3].

Clinical trials for herbal medicine, particularly in traditional Chinese medicine (TCM), pose unique challenges due to differences in diagnostic categories and treatment approaches compared with conventional medicine. TCM uses categories like "deficiency of chi and yin," which complicates the adaptation of standard clinical trial designs such as RCTs. Integrating TCM diagnostic criteria with biomedical methods can enhance the external validity of results. Additionally, outcome measures must align with TCM concepts rather than conventional metrics like physical functioning. TCM also requires significant adjustments, such as accounting for individualized treatments; ethical considerations, including publication bias and risk-benefit assessments, are crucial. Establishing standards for dosing,

DOI: 10.1201/9781003522065-9

safety, and efficacy is essential for conducting credible large-scale trials and ensuring accurate risk-benefit determinations in cross-cultural research [4].

Regulatory approaches to traditional medicines vary globally. In China, Japan, and within the European Medicines Agency, traditional medicines are regulated through specific policies and laws, emphasizing safety and efficacy. In contrast, the United States classifies herbs as "dietary supplements," which are not formally regulated as medicinal products. India's approach, governed by the Drugs and Cosmetics Act of 1940 and its 1959 amendment, includes traditional Indian medicines, requiring adherence to the same clinical trial protocols as allopathic drugs. However, traditional and herbal products in India often face less stringent regulation compared with allopathic medicines. Since 1995, herbal drug manufacturing and marketing in India must comply with guidelines set by the Ministry of AYUSH and obtain authorization through licensing. Manufacturers must be certified by drug regulatory authorities and adhere to Good Manufacturing Practices (GMP) and good supply practices for registration and certification. The WHO Research Guidelines for Evaluating the Safety and Efficacy of Herbal Medicines, published in 1993, outline quality assurance requirements, including GMP and post-marketing compliance. Additionally, manufacturers must follow Good Agricultural and Collection Practices (GACP) as per the WHO guidelines from 2003. In developing countries, where over 80% of the population relies on herbal medicine, regulatory standards can be less strict, while developed nations enforce GMP and good clinical practice (GCP). Most Asian countries treat traditional medicine with high regard, akin to modern pharmaceuticals. A significant challenge in herbal medicine is ensuring quality control, as contaminants such as pesticides, heavy metals, and toxins can be present due to environmental or handling issues. Despite variations in regulatory requirements, the registration of herbal medicines typically involves a thorough review of the product's development, pharmacological and toxicological studies, and clinical trials, often guided by the WHO guidelines.

Registration of herbal products requires comprehensive product information, including details such as the list of ingredients, formulation, manufacturing details (including address and license number), brand name, dosage form, intended use, dosage instructions, method of administration, recommended duration of use, potential adverse effects, contraindications, warnings, precautions, significant drug interactions, manufacturing and expiry dates, lot or batch number, and storage conditions. Sophisticated analytical methods such as high-performance liquid chromatography (HPLC), gas chromatography (GC), and GC-mass spectrometry (MS) often aid in generating data for these requirements. The WHO has also established pharmacovigilance practices to evaluate and gather information on the safety and efficacy of herbal products by monitoring adverse drug reactions. Additionally, the European Medicines Agency has set forth scientific guidelines for marketing authorization of herbal medicinal products, focusing on quality, nonclinical, clinical, and safety assessments [5].

9.1.1 Ethical and Regulatory Challenges in Clinical Trials of Herbal Medicines

Herbal medicines face several challenges due to limited scientific evidence supporting their safety and efficacy. Issues such as inconsistent standardization and production practices stem from the complex nature of herbal drugs, which are often composed of whole plants, plant parts, or extracts. Polyherbal formulations, which typically offer enhanced therapeutic potential compared with single herbs, further complicate characterization of bioactive compounds. The efficacy of these products often depends on the combined activity of multiple active ingredients and inert substances. However, predicting the long-term safety and efficacy of such formulations is difficult due to variations in phytochemical concentrations caused by factors such as cultivation season, geographic location, climate, and harvesting methods. These inconsistencies complicate standardization and compromise product quality.

Herbal product trials also face several challenges:

- Inconsistent concentration and composition between formulation batches.
- Variability in the chemical makeup of raw materials.
- Differences in quality testing methodologies and procurement practices.
- Challenges in processing techniques that affect final product composition.
- Difficulties in selecting a suitable active comparator for trials.
- Lack of authentic data when using a generic placebo instead of an active comparator.
- Issues with mimicking the color, smell, and taste of the herbal drug in placebo formulations.
- Variability in treatment outcomes due to the personalized approach in traditional systems like Ayurveda, which tailor treatments based on an individual's constitution (prakriti).
- Patient motivation and adherence to treatment regimens are low in long-term trials due to slow treatment outcomes and the contextual nature of herbal drug use.
- The use of subjective endpoints in trial evaluations may skew results.

Further barriers include financial and logistical constraints faced by small and medium-sized herbal manufacturers, making it difficult to meet regulatory requirements and conduct clinical trials, which typically take 5–10 years and cost billions of dollars. Safety data are more crucial than efficacy data for herbal medicines, necessitating large-scale trials. However, the lack of rigorous human clinical studies makes it hard to draw reliable conclusions about these products' safety and efficacy.

In many countries, herbal products are marketed as dietary supplements, bypassing stringent quality control regulations. Despite this, these products are not free from toxicity risks or potential drug and food interactions. Clinical trials of herbal medicines should monitor improvements in quality of life and symptom relief, similar to modern medical practice. Developing clinically relevant outcomes across diverse populations remains a significant challenge for herbal drug trial investigators [5].

9.1.2 World Health Organization Policy for Clinical Trials

Research on herbal drugs presents several challenges that must be addressed to ensure safety, efficacy, and proper regulation. In 2005, the World Health Organization (WHO) issued operational guidelines to establish the regulatory requirements for clinical trials of herbal products. Research on traditional and complementary medicine has been conducted globally, particularly in the context of the COVID-19 pandemic. The WHO's International Clinical Trials Registry Platform (ICTRP) has listed 4,778 clinical trials related to traditional medicine, including 165 focused specifically on COVID-19. Additionally, the WHO COVID-19 database features 7,516 research articles on traditional medicine's role in managing the disease [6].

Clinical trials typically involve multiple institutions, often spanning different countries, each with its own regulations. This can result in a single trial being registered on multiple platforms, leading to discrepancies in the data across different clinical trial registries. To address this, the WHO ICTRP links global clinical trial registers, providing a single point of access and ensuring clear identification of trials. This platform aims to improve the comprehensiveness, accuracy, and accessibility of clinical trial data, while raising awareness of the importance of trial registration. It also seeks to enhance the capacity for clinical trial registration, promote the use of registered data, and ensure the long-term sustainability of the ICTRP. The ICTRP's global initiative strives to make information on all clinical trials involving humans publicly available, ensuring transparency and fostering collaboration in clinical research, especially in the field of traditional medicine. Through these efforts, the ICTRP hopes to facilitate informed decisions by patients, researchers, and healthcare providers [7].

9.2 REGULATORY ASPECT IN NON-EUROPEAN COUNTRIES

9.2.1 Latin America and Others

9.2.1.1 Canada

A guidance document aims to assist organizations and individuals in navigating the process of obtaining authorization to conduct clinical trials for natural health products (NHPs) in Canada. The legal framework for these clinical trials is outlined in Part 4 of the Natural Health Products Regulations (NHPR), which are overseen by the Natural and Nonprescription Health Products Directorate (NNHPD) of Health Canada. The NNHPD is responsible for reviewing and deciding on applications for NHP clinical trials. Clinical trial data are essential for ensuring the safety, efficacy, and quality of NHPs. The authorization process is designed to protect the safety and ethical treatment of participants and everyone involved in the trial. Detailed protocols and procedures are required to meet the trial's objectives effectively. This guidance should be used in conjunction with the NHPR, published in the Canada Gazette, Part II, on June 18, 2003 [8].

9.2.1.1.1 Regulatory Overview

In Canada, the regulation of NHPs for clinical trials is governed by Part 4 of the NHPR. Section 1(1) of the NHPR defines an NHP as follows:

- A substance listed in Schedule 1 or a combination of substances where all medicinal ingredients come from Schedule 1, a homeopathic medicine, or a traditional medicine used for:
- Diagnosing, treating, mitigating, or preventing diseases, disorders, or abnormal physical conditions and their symptoms in humans.
- Restoring or correcting organic functions in humans.
- Modifying organic functions in humans.

NHPs do not include substances listed in Schedule 2 of the NHPR or those on the Prescription Drug List (Part I or II) under the Food and Drug Regulations. For information on substances used in homeopathic medicines, refer to the Evidence for Homeopathic Medicines Guidance Document. An overview of the NHPR can be found in the Overview of the NHP Regulations Guidance Document.

Part 4 of the NHPR outlines requirements for conducting clinical trials with NHPs, which are similar to those for conventional pharmaceuticals as described in Division 5 of the Food and Drug Regulations, but with additional considerations for NHP-specific aspects.

The main objectives of Part 4 are to:

- Ensure the safety, efficacy, and quality of both the clinical trial protocol and the investigational product(s), including placebo/comparator.
- Protect the safety of clinical trial participants and other involved individuals.
- Adhere to the International Council for Harmonization (ICH) GCP guidelines and federal and provincial laws, with detailed GCP guidelines available in Section 74 of the NHPR and the ICH Guidance Document.

9.2.1.1.2 General Requirements for Clinical Trials

- An NHP cannot be sold or imported for clinical trials without authorization under Part 4 of the NHPR. "Sell" includes any provision of the product, even without monetary exchange. For example, providing an NHP free of charge is still considered selling under Section 2 of the Food and Drugs Act.
- Clinical trials for NHPs (phases I–III) require approval from the NNHPD. Sponsors must notify the NNHPD at least 15 days before the trial starts and report within 15 days if the trial is discontinued.
- A clinical trial application (CTA) must be submitted to the NNHPD. If the NHP does not pose a health risk and is likely to meet trial objectives, the NNHPD will issue a notice of authorization.
- Phase IV trials, which use an NHP within its approved conditions, do not need Health Canada authorization but require Research Ethics Board (REB) approval. NHPs are considered approved if they have a Natural Product Number (NPN), Drug Identification Number (DIN), or Drug Identification Number for Homeopathic Medicine (DIN-HM). Use beyond these conditions is classified as a phase I–III trial.
- If the product license for an NHP in a phase IV trial is suspended or canceled, the sponsor must halt the trial and consult the NNHPD for further instructions.

9.2.1.1.3 Exclusions from NHP Clinical Trials Regulations

- NHPs used in clinical trials do not need a product license because their safety, efficacy, and quality are evaluated during the CTA process.
- A site license is required for the marketing of NHPs but not for clinical trial sites.

9.2.1.1.4 When to Submit a CTA to the NNHPD

(a) **Inadequate Evidence for Safety and Efficacy**
  - Clinical trials may be needed if an NHP previously approved by Health Canada is tested for a condition not covered by its label.
  - Trials are required for NHPs not yet approved by Health Canada, when additional evidence is necessary before marketing.
  - NHPs with no prior human use, such as new isolates or extracts, may require trials to establish their safety and efficacy.
  - Evidence requirements for combination and traditional products are detailed in the Evidence for Safety and Efficacy of Finished NHPs Guidance Document. Traditional products must have at least 50 years of use to be considered.

(b) **Comparative Bioavailability:** Studies assessing the pharmacokinetics of two NHP formulations in healthy volunteers are needed to determine pharmacokinetic properties, typically for single chemical entities.

(c) **N-of-1 Trials:** These trials involve a randomized, multiple-crossover design with a single subject, including treatment and placebo periods, and require blinding of both the subject and investigator. They are suited for chronic, stable conditions with rapid onset and cessation of action. At least two treatment pairs are required for statistical validity.

(d) **Clinical Trials with NHPs and Conventional Pharmaceuticals**
  - Trials using an NHP to address the side effects of a conventional pharmaceutical will be authorized and monitored by the NNHPD as long as the pharmaceutical is used within its approved conditions.
  - Trials comparing an NHP to a conventional pharmaceutical must adhere to the pharmaceutical's approved conditions.
  - Trials investigating an NHP to enhance the efficacy of a conventional pharmaceutical will be authorized by the NNHPD, with possible consultation with the Pharmaceutical Drugs Directorate (PDD).

(e) **NHPs for Non-Self-Care Conditions**
  - Trials involving NHPs for conditions not suitable for self-care are authorized and monitored by the NNHPD. If needed, the NNHPD will forward the application to other relevant directorates.
  - NHPs in clinical trials are not subject to claims for treating or curing conditions listed in Schedule A of the Food and Drugs Act but can be used under medical supervision for such purposes.

When to Submit a CTA to the Pharmaceutical Drugs Directorate

- **NHP/Pharmaceutical Combination:** The PDD will handle authorization and monitoring if a pharmaceutical is used outside its approved conditions or is not approved for sale in Canada, regardless of the NHP's role.
- **NHP with Prescription Ingredient:** The PDD will oversee CTAs for NHPs containing prescription ingredients listed on the Prescription Drug List.

9.2.1.1.5 When a CTA Is Not Required

Observational Studies: These studies do not require Health Canada authorization but must have REB approval. Observational studies monitor outcomes from routine care without manipulating or delivering interventions. If additional samples are collected, the study becomes a clinical trial requiring Health Canada authorization [9].

9.2.1.2 Argentina

In Argentina, the regulation of clinical trials for herbal products falls under the purview of the National Administration of Drugs, Food, and Medical Technology (ANMAT). Herbal products are treated similarly to conventional pharmaceuticals, requiring rigorous assessment for safety, efficacy, and quality before clinical trials can commence. Researchers must submit a detailed protocol for review and approval by ANMAT, ensuring that the study complies with ethical standards and GCPs. Additionally, herbal products used in clinical trials must adhere to specific regulations concerning their classification as either prescription medicines, over-the-counter drugs, or dietary supplements.

In Argentina, herbal regulation was established in 1998 through Resolution 144/98. This regulation introduced guidelines that were partially aligned with, yet distinct from, those for conventional pharmaceuticals. Herbal medicines in Argentina are regulated under different categories: prescription medicines, over-the-counter medicines, and dietary supplements. The law permits medical claims to be made about herbal medicines [10].

9.2.1.3 Brazil

Under Brazilian Law 986/1969, medicinal plants registered as food cannot carry therapeutic claims. However, they can be labeled as "special food" if they present "functional" or "health" claims. According to ANVISA, a functional claim suggests the food supports normal bodily functions, such as growth and maintenance, while a health claim implies a connection between the food and the improvement of a health condition. ANVISA's guideline 18/1999 outlines the basic requirements for functional or health claims on food labels, and guideline 19/1999 details the registration process, which must be backed by robust scientific evidence. Health claims, such as those suggesting a cure or disease prevention, are not permitted. Medicinal plants can also be classified as "new food," which includes foods without a consumption history in Brazil or those used in significantly higher quantities. These products may feature functional health claims but must be supported by strong technical and scientific data, as per ANVISA guideline 16/1999 40.

The National System of Ethics in Clinical Trials (SNEPSH) and the regulations governing clinical trials in Brazil were established by Law #14,874/2024, signed into law by President Luiz Inácio Lula da Silva on May 28, 2024. The introduction of Bill #200 by then-Senators Ana Amélia (PP/RS), Waldemir Moka (MDB/MS), and Walter Pinheiro (PT/BA) in 2017 marked the beginning of a protracted legislative process that came to an end in this law. Bill #6,007/2023 was approved in the House of Representatives following its passage in the Senate. The bill was approved by the Senate with additional amendments on December 13, 2023, which was more than six years later. Nine chapters comprise Law #14,874/2024, with Chapter I providing an overview of the parameters and recommendations for carrying out clinical trials (Articles 1 to 4). The law comprises 65 articles [11].

The regulation of herbal medicines in Brazil has largely remained consistent since the implementation of RDC 48/04. To register herbal medicines, companies must provide proof of safety and efficacy through one of four methods: conducting preclinical and clinical trials as required for other medicines, utilizing data from the "list of simplified registration of herbal medicines," referencing validated literature, or demonstrating traditional use of the product. Validation of safety and efficacy can be achieved by scoring data from a list of technical-scientific literature compiled in IN 05/10, which includes 35 reference books classified into three groups (A, B, and C). These references must provide detailed information about the herbal derivative used, including the plant part, extraction method, recommended dosage, and therapeutic indications.

IN 05/08 provides a "simplified registration species list" that includes 36 plant species. For plants on this list, companies need not submit additional safety and efficacy data if they follow the specified parameters, such as plant organ, chemical markers, and therapeutic indications. Safety and efficacy can also be supported by ethnopharmacological studies, traditional use documentation, or technical-scientific publications demonstrating at least 20 years of proven safety and efficacy. In cases of traditional use, companies must adhere to traditional usage patterns to maintain the chemical profile's similarity with the registered medicinal product. This approach is particularly important when the medicinal plants are used in infusions and decoctions, ensuring that the traditional use aligns with the registered product's chemical profile and efficacy.

Herbal medicine regulation in Brazil has undergone changes through five key rules: Ministerial Directives 22 (1967) and 06 (1995), and RDCs 17 (2000), 48 (2004), and 14 (2010). The most recent, RDC 14/10, introduces updates to improve safety and efficacy standards. These include new quality control measures at each pharmaceutical production stage, the possibility of registering medicines containing herbal drugs as active ingredients if supported by preclinical and clinical trials, and mandatory aflatoxin testing in specific cases as recommended by official monographs or literature. Additionally, it includes provisions for the registration of multicellular fungi and algae in the absence of specific regulations. These changes reflect a more rigorous approach to the registration and control of herbal medicines in Brazil, ensuring higher standards of quality and safety for consumers.

Preclinical and clinical trials are necessary for a particular product only in cases where the desired level of safety and efficacy cannot be achieved through any of the three previously mentioned methods. A guide for conducting preclinical toxicological trials (RE 90/04) is available for herbal medicine products (Brasil, 2004b). The minimal acceptable standards for carrying out acute, subchronic, and chronic toxicological investigations are outlined in this guide. Topical products (skin sensitivity; skin and eye irritation) require further toxicology studies and reproductive toxicity testing. Standardized samples of herbal medicine products or the herbal derivatives that they are derived from will be used in the studies (Brasil, 2004b). ANVISA also released RDC 39 in 2008 as a guide for conducting clinical trials, to supplement the particular requirements for the registration of medicinal products (Brasil, 2008a).

The WHO (2008) states that all phases of clinical trials must adhere to GCP guidelines in order to guarantee that ethics and quality control requirements are met. According to ANVISA's guide on GCP, study proposals must be approved by ANVISA before beginning any research and then be followed by the issuance of a Special Announcement (CE) (Brasil, 2008a) [11].

9.2.1.4 Mexico

Mexico has emerged as a key player in global clinical trials, offering a conducive environment for medical research. The country's appeal lies in its robust healthcare infrastructure, diverse patient population, and favorable regulatory framework. These factors make Mexico an attractive destination for conducting clinical trials. The Federal Commission for the Protection against Sanitary Risks (COFEPRIS) serves as the regulatory authority overseeing the review, evaluation, and approval of all research protocols involving human subjects and/or their biological samples. COFEPRIS is responsible for ensuring that both registered and unregistered investigational products (IPs) used in clinical trials meet safety and efficacy standards. This supportive regulatory environment, combined with the country's healthcare strengths, positions Mexico as a significant hub for clinical research on a global scale [9].

Mexico, while not a major player in global clinical research, has significant potential due to its strategic importance, established research infrastructure, and a suitable study population. Historically, global pharmaceutical companies have been deterred from conducting pivotal drug-approval studies in Mexico due to regulatory and quality concerns. However, there is a growing interest in addressing these challenges to attract major pharmaceutical companies.

The country's medical infrastructure is anchored by the Instituto Mexicano del Seguro Social (IMSS), a vast social security organization that provides healthcare to over 50 million working Mexicans. IMSS operates an extensive network of hospitals and clinics and employs full-time physicians. In addition to IMSS, there are several federal government-financed public hospitals that offer medical care to the indigent population. Mexico is also home to 10 national health institutes dedicated to medical education, research, and healthcare excellence.

Mexico's academic landscape includes two major public teaching institutions, the Universidad Nacional Autónoma de México (UNAM) and the Instituto Politécnico Nacional (IPN). These institutions boast many talented clinical scientists; however, their involvement in clinical research is relatively limited. Furthermore, prestigious private teaching institutions in Mexico also show limited participation in clinical research activities.

As a result, over 70% of clinical trials in Mexico are conducted in government-supported hospitals and public institutions. Placebo-controlled trials, which are essential for determining drug efficacy and safety, are commonly conducted. Most medical research in the country is funded by grants from the Consejo Nacional de Ciencia y Tecnología (Conacyt), with a smaller portion supported by international pharmaceutical companies. As Mexico continues to address its regulatory and quality issues, it is poised to become a more significant player in global clinical trials [12].

9.2.2 Africa

Clinical trials of herbal medicines in Africa reveal strong interest but limited capacity, particularly in achieving publishable results. Many trials have been conducted, yet few have been published, highlighting the need for better training and resources. Challenges include social, financial, and logistical barriers, with over half potentially solvable through targeted training in areas such as protocol design, ethics, statistical analysis, and trial reporting. The MUTHI project is addressing these issues through workshops, online resources, and mentoring programs. Improved training could enhance the quality and impact of herbal medicine research across Africa, especially for prevalent conditions such as HIV/AIDS and malaria [13].

In Africa, fewer than 1% of clinical trials are carried out. Of the 26 oncology clinical trials that took place in Africa in 2019, only six involved participants who were of African descent. Clinical trials for cancer are being conducted across the continent, but there are a number of barriers. A major obstacle is the time to trial activation (TTA), which includes all of the legal and administrative procedures that must be completed before a study can start. This is an important consideration that frequently affects the choice of study location. The Institutional Review Board (IRB), the Pharmacy and Poisons Board, the National Commission for Science, Technology, and Innovation, and the Ministry of Health must all grant TTA approval in turn [14].

9.2.2.1 South Africa

In South Africa, herbal medicines, classified as "Complementary Medicines" (CAM), are regulated under the Medicines and Related Substances Act, 1965 (Act 101 of 1965). This act consists of 40 sections addressing various aspects such as registration procedures, labeling requirements, prohibitions, penalties, the establishment of regulatory bodies, and the powers of the Medicines Control Council (MCC). According to this act, "complementary medicine" refers to any substance or mixture of substances that

- originates from plants, minerals, or animals;
- is used or is intended for use in assisting the body's natural healing process to alleviate, modify, mitigate, or prevent illness or symptoms, or to address an abnormal physical or mental state;
- is used in accordance with the practices regulated by the Allied Health Professions Act, 1982 (Act 63 of 1982).

Previously, there were no specific regulations governing the manufacture, distribution, or marketing of CAMs in South Africa. The country lacked a regulatory body to monitor these medicines. However, South Africa has since established a recognized regulatory authority—the MCC—which enforces the standards set out in the Medicines and Related Substances Act.

The MCC operates through committees comprised of external experts, many of whom are academics from medical and pharmacy institutions. These experts primarily assess the data provided by pharmaceutical companies for product registration [15].

9.2.2.1.1 Recent Developments in CAM Regulation

Previously, there were no registration requirements for CAM products. Labeling was unregulated, and stability data were not needed. However, new regulations introduced by the MCC now require CAM products to be registered and to include stability and toxicity data.

The guidelines aim to provide clear standards for GCP in locally conducted research, ensuring that clinical trials in South Africa adhere to both local regulations and global scientific and ethical standards. They prioritize participant safety and well-being while ensuring credible trial data. They cover the design, planning, management, conduct, and regulation of human clinical trials, including those involving complementary and traditional medicines, medical devices, and other interventions. Ethical principles are referenced from the Department of Health's Ethics in Health Research guidelines (2015).

Clinical trials are essential for evaluating new treatments, requiring scientifically sound and ethical design that balances healthcare advances with participant safety. While ethical principles for clinical trials align with broader health research, special considerations are needed for their complex and invasive designs. The updated SA GCP 2020 guidelines improve trial design, conduct, and reporting while aligning with international (ICH GCP 2016) and local (DoH 2015) standards. The guidelines emphasize participant protection and data validity, consolidating and revising content from SA GCP 2006 to provide clearer, user-friendly guidance tailored to South Africa's context for conducting ethical clinical trials [16].

Clinical trials must be conducted according to a pre-approved protocol, typically developed by a team of experts. This protocol should justify the study and outline the key research questions. It must clearly define the study population, including inclusion and exclusion criteria, and describe the participant recruitment and group-assignment process. The protocol should also specify clinical and biological parameters for monitoring toxicity and efficacy and include a plan for managing emergencies and adverse events.

For each participant, informed consent, medical history, treatment details, and summaries of all tests and examinations should be recorded. Efficacy is measured by specific criteria, such as improvements in clinical symptoms or changes in biological markers, while safety is monitored by observing symptoms and focusing on organs at risk, such as the liver, kidneys, heart, and blood. Strong statistical knowledge is crucial for trial design and data analysis. Although well-conducted RCTs provide reliable evidence of efficacy, placebo-controlled trials are sometimes unfeasible due to ethical or technical challenges.

9.2.2.1.2 Ethics

International ethical guidelines for biomedical research involving human subjects must be adhered to in all clinical trials of traditional medicines. Each trial should be reviewed by an ethics committee based on institutional guidelines. When applicable, patients in trials using a placebo or unproven treatment may receive rescue treatment, which can also serve as a secondary outcome measure. Ethical restrictions on clinical trials, including the use of placebos, vary by country and may be illegal for certain conditions such as cancer. Trials must comply with local laws, and national expert committees are responsible for evaluating authorization applications [17].

9.2.2.1.3 Safety Assessment

The safety assessment must comprehensively address all aspects of the product's safety. The general principle is that if a product has been traditionally used for hundreds of years without reported adverse effects, no strict regulatory action is necessary unless new data prompt a reassessment of the risk/benefit ratio. Such new data are typically obtained through toxicological testing and clinical trials. In the absence of extensive toxicological studies, safety evaluation should rely on documentation showing long-term safe usage. However, chronic toxicity may still go undetected, even with long-term use. Known toxic risks, directions for use, and potential for misuse, abuse, or dependence must be documented. If traditional prolonged use cannot be proven or if safety concerns arise, toxicity data should be provided.

9.2.2.1.4 Efficacy Assessment

The efficacy assessment should encompass all critical aspects, beginning with a review of existing literature. The product's pharmacological and clinical effects should be detailed, especially when the active ingredients are identified. For less severe conditions or preventive measures, traditional use can be considered evidence of efficacy. Reports from doctors, traditional medicine practitioners, or patients may also be factored in. If traditional use is not established, relevant pharmacological or clinical data must be generated through clinical trials.

9.2.2.1.5 *Regulatory Procedure*

(a) **Development:** The process of developing regulations for granting clinical trial authorizations for traditional medicine follows the same stages as those used for regulating traditional medicine practitioners. These stages include advocacy, situational analysis, drafting the initial regulations, holding a consensus workshop, and final adoption by the relevant national authorities.

(b) **Main Elements to be Regulated**

Clinical Trial Authorization: Clinical trials are a mandatory step for obtaining marketing authorization for category 3 and 4 medicines. Any clinical trial requires official authorization, and the following are key regulatory elements:

Article 1: No clinical trial can be conducted in a country without written authorization from the competent authorities.

Article 2: Clinical trial authorization grants a person or legal entity the right to conduct a specific trial.

Article 3: Authorization is subject to an application fee, the amount of which is determined by the competent national authority.

Article 4: The sponsor, whether an individual, organization, group, or other legal or research entity, initiates, organizes, and finances the clinical trial. If multiple parties are involved, they must appoint one entity as the sponsor, responsible for fulfilling the relevant obligations.

Article 5: The principal investigator is the person responsible for directing and overseeing the clinical trial at the designated site.

Article 6: The CTA, submitted to the national authority, must include the following:

- Completed and signed application form by the sponsor
- Clinical trial protocol
- Investigator's brochure (CV)
- Participant information leaflet and informed consent form
- GMP certificates for the medicines
- Signed investigator statements
- Ethics committee approval (for international studies, all participating countries must approve)
- Proof of application fee payment
- Indemnity insurance
- Confirmed study budget

Article 7: The national authority has 90 days to make a decision, which can be an approval, deferment, or rejection. If no decision is communicated within this period, the authorization is considered granted.

Article 8: The sponsor and principal investigator must ensure that the trial follows GCPs as defined by national or internationally accepted guidelines.

Article 9: The principal investigator must report any serious adverse events to the national authority, IRB, and Data Safety Monitoring Board according to established procedures. For multi-center studies, the sponsor must communicate adverse events to all relevant national authorities.

Article 10: Mid-term and final trial reports should be submitted to the national authority as outlined in the trial protocol.

Article 11: Authorized trials may be inspected by national authorities to verify compliance with the protocol. Inspections may lead to trial suspension or termination if noncompliance is detected, regardless of ongoing legal actions.

9.2.2.1.6 *Clinical Trial Authorization Technical Review Committee*

The committee is responsible for evaluating clinical trial applications and providing recommendations to the national competent authority for informed decision-making. Its establishment and functions must be endorsed by the national authority. The following outlines the key elements for forming the technical review committee:

Article 1: A technical committee for reviewing clinical trial authorization applications is hereby established.

Article 2: The committee consists of the following members:

1. Chairperson, appointed by the national authority
2. Secretary, typically the regulatory officer at the NMRA
3. Members, including epidemiologist, pharmaceutical technology expert, pharmacologist, toxicologist, pharmacist, herbalist/phytotherapist, clinician, ethicist, pharmacognosist, other relevant researchers

Article 3: Committee members are appointed by the national authority for a renewable three-year term.

Article 4: The committee may co-opt additional experts as needed, with national authority approval.

Article 5: Members must disclose any conflicts of interest or be excluded from reviewing a trial in which they have an interest.

Article 6: Meetings require a quorum of two-thirds (2/3) of the members.

Article 7: The committee operates under established procedures, with its costs covered by the national authority's budget.

9.2.2.1.7 *Application Procedure*

The process includes three stages that may be adapted based on the circumstances of member states:

- **Ethics Committee Assessment:** This committee assesses the ethical aspects of health research. Its approval is a key component of the application for authorization.
- **Technical Committee Assessment:** The approved protocol is submitted by the applicant to the national authority, which forwards it to the technical committee for review and decision.
- **Decision by the National Authority:** The final decision regarding the clinical trial authorization is made by the competent national authority [18].

9.2.2.2 Mauritius

In Mauritius, conducting clinical trials for herbal products requires navigating a well-defined regulatory framework to ensure the safety and ethical treatment of participants. Here is a summary of the key regulations and guidelines.

9.2.2.2.1 Pharmacy Board Oversight

- **Regulatory Authority:** The pharmacy board, under the Ministry of Health and Wellness, oversees the approval process for all clinical trials, including those involving herbal products.
- **Application Process:** To initiate a trial, sponsors must submit a comprehensive application to the pharmacy board. This application should detail the herbal product, study design, safety protocols, and the process for obtaining informed consent from participants.

9.2.2.2.2 Governance under the Food and Drugs Act

- **Legal Framework:** The Food and Drugs Act serves as the cornerstone for regulating clinical trials in Mauritius. This legislation ensures that all therapeutic goods, including herbal remedies, comply with established safety, quality, and efficacy standards before being tested on humans.
- **Standards Compliance:** Herbal products involved in trials must meet the strict criteria outlined in the act, ensuring that they are safe and ethically tested [19].

9.2.2.2.3 Ethics Committees

Role of HRECs: Any clinical trial, particularly those involving herbal products, must gain approval from a Human Research Ethics Committee (HREC) or a similar ethics review board. These committees review the trial design to ensure ethical conduct and to protect participants' rights.

9.2.2.2.4 Adherence to Good Clinical Practice

International Standards: Trials conducted in Mauritius must follow GCP guidelines, which are globally recognized standards that ensure the credibility of the research and safeguard participants' welfare.

Informed Consent

- **Participant Rights:** Obtaining informed consent is a critical requirement. Participants must be fully informed about the herbal product, the trial's purpose, potential risks and benefits, and must voluntarily agree to participate.

Clinical Trial Registration

- **Transparency:** To promote transparency, clinical trials in Mauritius should be registered with an appropriate clinical trials registry, making information about the study accessible to the public.

Data Protection

- **Privacy Laws:** Data collected during the trial must be handled in accordance with the Data Protection Act 2017, ensuring that participants' personal information remains confidential and secure [20].

9.2.2.3 Egypt

Since 2006, Egypt has adhered to the International Conference on Harmonization Good Clinical Practice (ICH GCP) guidelines, as initiated by the Ministry of Health (MOH) ministerial decree no. 436/2006. These guidelines have regulated clinical trials, first implemented by the National Organization for Research and Control of Biologicals (NORCB). In 2010, the MOH decree no. 399/2010 further regulated clinical trials for biological products, and decree no. 734/2016 extended these regulations to pharmaceutical products through the National Organization for Drug Control and Research (NODCAR). The Egyptian Drug Authority (EDA) replaced both NORCB and NODCAR in 2019, consolidating their oversight responsibilities.

The General Administration of Clinical Trials (GA of CT) within the EDA now oversees the regulatory process for clinical trials, ensuring adherence to GCP principles. This includes reviewing preclinical and clinical data, conducting scientific committee meetings, and offering technical support. The GA of CT is responsible for evaluating clinical research protocols, monitoring ongoing studies, and ensuring compliance through GCP inspections. The administration also evaluates amendments and manages reports on adverse events, ensuring that clinical trials conducted in Egypt meet international standards for safety and effectiveness.

This guideline provides clear instructions for applicants (sponsors, CROs, PIs) on implementing national GCP regulations in Egypt. It applies to all clinical trials involving human participants, including those using new investigational pharmaceutical or biological products, new indications, dosage forms, medical devices, and herbal medicines, as regulated by the EDA under Clinical Trials Law no. 214/2020 [21].

The EDA oversees clinical trials in accordance with the ICH GCP guidelines, WHO guidelines, and subsequent updates. The following is a summary of the EDA's regulatory oversight of clinical trials:

9.2.2.3.1 Principles of GCP (ICH E6 and WHO Guidelines)

1. **Ethical Conduct**: Trials must adhere to the ethical principles of the Declaration of Helsinki.
2. **Risk-Benefit Evaluation**: Trials should proceed only if the anticipated benefits outweigh the risks.
3. **Subject Rights and Safety**: The rights, safety, and well-being of participants take priority over scientific interests.
4. **Adequate Information**: Nonclinical and clinical data must justify the trial.
5. **Sound Protocols**: Trials must follow scientifically valid, detailed protocols.
6. **Compliance**: IRB/Independent Ethics Committee (IEC), EDA, and Supreme Council approvals are required before initiation.
7. **Medical Oversight**: Qualified physicians or dentists must oversee subject care.
8. **Qualified Staff**: All trial personnel must be suitably trained and experienced.
9. **Informed Consent**: Subjects must provide informed consent before participation.
10. **Accurate Records**: All trial data should be recorded and stored for proper interpretation and verification.
11. **Confidentiality**: Subject confidentiality must be protected.
12. **Product Handling**: IPs must be handled in accordance with GMP and EDA regulations.
13. **Quality Assurance**: Systems must ensure participant protection and trial reliability.

Preclinical Results Evaluation (Before First in Human Trial)

1. **Quality Assurance**: Systems must ensure participant protection and trial reliability.
2. **Submission**: Full preclinical data are submitted with a fee, following successful screening.
3. **Evaluation**: Applicants must respond to raised requirements within 15 working days; the final evaluation takes 60 working days after all clarifications.
4. **Final Decision**: Peer review by the EDA advisory scientific committee determines approval.
5. **Clinical Trial Protocol Submission and Evaluation**
6. **Screening**: The clinical trial protocol is screened for completeness within five working days. Responses to requirements must be made within five working days.
7. **Submission**: Complete data are submitted with the necessary fees.
8. **Evaluation**: Applicants must respond to requirements within 15 working days. The evaluation period is 60 working days.
9. **Final Decision**: Peer review is conducted by the EDA scientific committee; approvals are valid for one year and are renewable.

Amendments to Protocol

- Applicants must notify the EDA of any amendments to approved protocols.
- Urgent amendments for participant safety can be implemented immediately, with notification to EDA within 24 hours.
- Amendments affecting safety must be approved by the EDA before implementation.
- Minor amendments not affecting safety may be implemented after 14 days unless otherwise notified by the EDA.

Public Health Emergencies

- Accelerated procedures may be applied to clinical trials in public health emergencies, including fast-track submissions.

Study Initiation and Reporting

1. **Site Activation**: The EDA must be notified when a study begins.
2. **Progress Reports**: Follow-up reports on progress and adverse events are required every three to six months, depending on study duration.
3. **Safety Reporting**: Fatal or life-threatening adverse events must be reported to the EDA within 24 hours, with case reports submitted within seven days. Non-serious adverse events should be reported within 7 days in phase I/II trials and in progress reports for phase III/IV trials.
4. **IMP Destruction**: Investigational medicinal products (IMP) must be destroyed in accordance with EDA regulations, with inspectors present.

   Certainly! Here's a summary of the key points from the guidelines for good regulatory oversight of clinical trials by the EDA:

Completion of the Study

- **Notification:** The sponsor must notify the EDA within 60 days of the trial's end, submitting a summary of the outcomes and all related data and reports.
- **Final Report:** The final clinical study report should be submitted to the EDA's administration of protocols and studies follow-up for review and evaluation.

Early Termination or Withdrawal

- **Premature Termination:** If the sponsor ends the trial early, the investigator must inform the subjects, ensure their follow-up care, and notify the EDA in writing within 15 working days.
- **Protocol Withdrawal:** An applicant can withdraw their protocol or amendment at any time by submitting a formal letter explaining the reasons to the EDA.

Suspension or Termination by EDA

- **Reasons:** The EDA may suspend or terminate trials for noncompliance with GCP, GMP, protocol, or reporting requirements. The decision will be reviewed by the Supreme Council.

Inspections and Reports

- **Inspection Types:** Inspections may be routine, triggered by specific issues, or conducted as follow-up to check on corrective actions.
- **Reports:** The inspection report, including observations, is prepared within 15 working days. The applicant must address the findings with a corrective action plan within 20 working days.

Technical Support

- **Submission:** Applicants seeking technical support should complete the relevant application form and submit it with the required documents.
- **Review:** Technical support dossiers are reviewed within 60 working days.

Principal Investigator Responsibilities

- **Criteria:** Must meet academic, training, and ethical standards and be free from conflicts of interest.
- **Before Study:** Obtain necessary approvals and informed consent, register the protocol, and ensure the proper selection of subjects and research team.
- **During Study:** Conduct research according to the protocol, inform subjects of any amendments, manage safety, and publish results.

Sponsor/CRO Responsibilities

- **Approvals and Funding:** Obtain the required approvals, fund the research, and monitor performance.
- **Agreements:** Enter into agreements with all parties involved and ensure the safety and insurance of subjects.

Key Design Phase Activities

- **Protocol Development:** Design a clear protocol including risk identification and study design.
- **SOPs:** Develop SOPs for all study-related activities.
- **Support Systems:** Develop tools and systems to support the study.
- **Document Generation:** Create all necessary trial-related documents.

Key Conduction Phase Activities

- **Subject Enrollment:** Ensure proper recruitment, eligibility, and informed consent.
- **Investigational Product:** Manage quality, handling, and documentation.
- **Data Management:** Ensure accurate data collection and management.
- **Safety and Monitoring:** Ensure participant safety and perform regular monitoring.

General Considerations

- **Phases Allowed:** Nationally originated interventions can conduct all phases; foreign interventions can conduct phases III and IV, with exceptions for endemic and rare diseases.
- **Authorities Involved:** Approvals must include the IRB, Supreme Council, General Intelligence Agency (if applicable), and drug control entities [22].

9.2.2.4 Sudan

In Sudan, legal provisions mandate that authorization from the Medicines Regulatory Authority (MRA) is required to conduct clinical trials. Additionally, approval from an ethics committee or IRB is necessary for the trials to proceed. However, the law does not require clinical trials to be registered in international, national, or regional registries.

Here's a summary of the key points regarding the conduct of clinical trials from the provided regulations:

1. **Authorization Requirement**: No clinical trial of a regulated product (medicine) can be conducted without prior written authorization from the relevant authority and approval from the minister.
2. **Application Process**
   - Applications must be submitted to the secretary-general, including the prescribed fees.
   - For trials involving animals, details about the animals and their owners must be included.
   - If conducted in a medical institution, the application must be countersigned by a senior medical officer.
3. **Review and Approval**
   - The secretary-general forwards applications to the board for consideration.
   - If the board approves, it seeks the minister's written approval and issues authorization to conduct the trial.
4. **Conditions for Conduct**
   - Trials are subject to conditions imposed by the board and must be conducted in accordance with these conditions.
   - For drugs or products used by adults, voluntary written consent must be obtained from participants.
   - For minors or legally disabled persons, consent must be obtained from their guardians.
   - For animals, consent must be obtained from their owners.

5. **Pre-trial Information and Insurance**
   - The trial conductors must inform participants or animal owners about the trial's aims, risks, and potential adverse effects.
   - Insurance must be maintained to cover any injury or risk to participants or animals.
   - An indemnity must be signed to protect the government, minister, and the board from liability for any injuries or adverse effects.
6. **Monitoring and Suspension**
   - The board has the authority to stop or suspend a trial if it deems it necessary for the public interest.
   - The board will monitor the trial from start to finish to ensure compliance and assess the trial's effectiveness.
7. **Reporting Requirements**
   - A preliminary report on the trial's scientific and ethical evaluation must be submitted within 30 days after completion.
   - A comprehensive report on serious or adverse effects must be submitted within 90 days after completion.
   - The board will also submit an independent comprehensive report to the minister within 90 days, including factual assessments and recommendations [23].

9.2.3 South East Asia

9.2.3.1 Indonesia

In Indonesia, traditional medicine is called jamu. Clinical trials for herbal products like *jamu* are regulated to ensure the safety, efficacy, and ethical standards of these products [24]. Indonesian herbal medicines are classified into three categories based on their preparation methods, claims, and efficacy evaluations: (1) *Jamu*, defined as a herbal medicine whose safety and efficacy are recognized through traditional use and empirical evidence; (2) *Standardized herbal medicines*, herbal products whose safety and efficacy have been verified through preclinical testing, using standardized extracts of plant materials; and (3) *Phytopharmaca*, standardized herbal medicines that have undergone both preclinical and clinical testing to confirm their safety and efficacy [24].

Here's an overview of the relevant regulations and guidelines:

1. **National Agency of Drug and Food Control (BPOM)**
   - **Regulatory Authority:** The Indonesian National Agency of Drug and Food Control (Badan Pengawas Obat dan Makanan, BPOM) is the primary regulatory body overseeing the registration, approval, and monitoring of clinical trials involving herbal products, including *jamu*.
   - **Product Registration:** Before conducting clinical trials, *jamu* products must be registered with BPOM. The registration process includes submitting detailed documentation on the product's formulation, safety data, and intended therapeutic claims.
2. **Regulation of Herbal Medicines**
   - **Classification:** Herbal products in Indonesia, including *jamu*, are classified based on their composition and intended use. *Jamu* products may be classified as follows:
     - **Traditional Medicines (*Obat Tradisional*, OT):** Products that adhere to traditional formulations and have a history of safe use.
     - **Standardized Herbal Medicines (*Obat Herbal Terstandar*, OHT):** Products with standardized active compounds and proven safety and efficacy through preclinical studies.
     - **Phytopharmaceuticals (*Fitofarmaka*):** Herbal products that meet the highest standards of safety, efficacy, and quality, demonstrated through clinical trials.
3. **Ethical Approval**
   Ethics Committees: Clinical trials involving *jamu* must obtain approval from an ethics committee or an IRB. These committees ensure that the trial design is ethical, that participants' rights are protected, and that the benefits of the trial outweigh any risks.
4. **Good Clinical Practice**
   Compliance: Clinical trials in Indonesia must adhere to GCP guidelines, which are internationally recognized standards that ensure the credibility of trial results and the safety of participants.
5. **Preclinical Studies**
   Requirements: Before proceeding to clinical trials, *jamu* products, particularly those classified as OHT or *Fitofarmaka*, must undergo preclinical studies. These studies include toxicity tests, pharmacology, and efficacy studies in animals to ensure the products' safety.
6. **Informed Consent**
   Requirement: Participants in clinical trials must be provided with clear, comprehensive information about the trial, including potential risks and benefits, and must voluntarily agree to participate by signing an informed consent form.
7. **Data Privacy**
   Regulation: Compliance with Indonesia's data protection laws is mandatory to ensure the confidentiality and security of participants' personal information during and after the trial.
8. **Post-trial Obligations**
   Reporting: Upon completion of the trial, results must be reported to BPOM, and any adverse effects must be documented and addressed. These reports are crucial for the final approval and registration of the *jamu* product [24].

9. **Government Support and Regulation**
   Support Programs: The Indonesian government, through various initiatives, supports the development and commercialization of *jamu* products, particularly those that demonstrate potential therapeutic benefits and can contribute to the local economy [25].

9.2.3.2 India

The classification of herbal medicines is determined by factors such as whether they contain processed or unprocessed plant parts and whether they include potentially toxic plants. Safety and efficacy requirements vary depending on how the products are classified and their market availability. Submission requirements for clinical trial data and toxicity studies also differ based on the nature of the herbs and the market presence of the product.

In 2001, the Central Drugs Standard Control Organization (CDSCO) of the Directorate General of Health Services issued GCP guidelines. These guidelines outline the recommended approach for conducting clinical trials on herbal remedies and medicinal plants. The guidelines categorize herbal products based on whether their use and formulation align with traditional medicine literature.

Key points from the guidelines include the following:

- For herbal remedies and medicinal plants intended for evaluation in the allopathic system and potential use in allopathic hospitals, the procedures outlined by the Drugs Controller General of India for allopathic drugs should be followed.
- When a plant extract or compound isolated from a plant is being evaluated for a therapeutic effect not documented in traditional texts, or if the preparation method is different, it should be treated as a new chemical entity (NCE). Accordingly, acute, subacute, and chronic toxicity data must be generated as required by the regulatory authority before clinical evaluation.
- If the plant extract or compound has no historical use and is not mentioned in ancient literature, it should be considered a new drug and must meet all regulatory requirements before clinical evaluation [26].

The guidelines also offer general recommendations for clinical trials involving herbal medicines, including toxicity studies, the importance of standardization, and adherence to GCP principles in all clinical research. Some of these recommendations are:

- Clinical trials should only proceed after herbal preparations have been standardized and markers identified to ensure consistency.
- Plants and herbal remedies should be prepared exactly as described in traditional literature, incorporating GMP standards for consistency.
- For herbal remedies, phase 1 trials should focus on determining the maximum tolerated dose (MTD) and early drug activity measurements.
- If there are concerns about toxicity or if the herbal preparation is intended for use beyond three months, toxicity studies (4–6 weeks in two species) are required for phase 2 trials.
- Phase 3 trials also require 4–6-week toxicity studies in two species.
- Ethical guidelines, including patient information, informed consent, and protection of vulnerable populations, must be strictly followed.
- Clinical trials should be approved by the relevant scientific and ethical committees of the involved institutions.
- Trials should involve a competent Ayurvedic, Siddha, or Unani physician as a co-investigator.

Since the introduction of Drugs & Cosmetics Rule 158B in August 2010, proof of efficacy has become a requirement for licensing patent or proprietary Ayurvedic, Siddha, and Unani (ASU) medicines. This shift has prompted the development of GCP guidelines aimed at ensuring the quality, safety, and therapeutic efficacy of ASU products. Compliance with these guidelines is critical for generating high-quality clinical research data acceptable to regulatory authorities for product registration and market approval, especially for non-classical or non-generic formulations.

The Drugs & Cosmetics Act, 1940, and its associated Rules do not include provisions for clinical trials on traditional medicines. However, between April 2008 and September 2010, the Department of AYUSH granted approval to ten institutes to conduct clinical trials on Ayurvedic and herbal medicines. Details of these trials are provided in the annexure.

Regarding the definition of new AYUSH drugs, including Ayurveda, these are not specified under the Drugs & Cosmetics Act, 1940, and its Rules. In 2013, the government issued "Good Clinical Practice Guidelines for Clinical Trials in Ayurveda, Siddha, and Unani Medicine (GCP-ASU)" to ensure that such studies are conducted scientifically and ethically and that the clinical properties of the ASU medicines being investigated are properly documented. However, these guidelines are not legally binding for conducting clinical trials, and there is currently no proposal to revise them [27].

9.2.3.2.1 New Guidelines for Herbal Medicines

The Ayush Ministry periodically updates regulations for ASU drugs. Previously, there were no specific guidelines for conducting clinical trials for these drugs, but in March 2013, AYUSH released new GCP guidelines specifically for ASU drug trials.

GCP is a set of guidelines that covers the design, conduct, termination, audit, analysis, reporting, and documentation of studies involving human subjects. The primary principle of GCP is that in human research, the interests of science and society must never outweigh the well-being of the study participants. The guidelines aim to ensure that studies are both scientifically and ethically sound, with proper documentation of the clinical properties of the ASU medicine being tested. These guidelines emphasize two key principles: protecting the rights of human subjects and ensuring the authenticity of the clinical trial data for ASU medicines.

The guidelines are based on the 2001 CDSCO document on GCP Guidelines for Clinical Trials on Pharmaceutical Products and are applicable to all ASU medicine research in India, covering all stages of drug development, both before and after product registration in the country. Compliance with these GCP guidelines is mandatory during clinical trials; failure to do so can lead to trial suspension by regulatory authorities. Additionally, the guidelines include provisions for compensating participants in case of adverse events or death during the trial.

From 2017 onward, it has also been mandatory for product labels to include both the manufacturing and expiry dates [28].

9.2.3.3 Nepal

Herbal medicines are often prepared as combination therapies, a practice that dates back to the origins of therapeutic medicine. In Nepal, fixed dose combinations (FDCs) of herbal remedies are widely used. These herbal FDCs are popular because they are believed to improve adherence, reduce side effects, and offer easier access to treatment. However, combination products also carry a higher risk of adverse effects, increase treatment costs, and can lead to ineffective dosages. Despite their extensive use in Nepal, the rationale behind many of these combinations remains unclear and is sometimes questioned. The country lacks comprehensive legislation governing the use of herbal medicines, and many herbal FDCs are not backed by scientific data.

In Nepal, Ayurveda is a predominant traditional medical system that has been practiced for thousands of years. Ethics approval has been granted by the Research Ethics Committee at the Faculty of Medicine and Health Sciences, University of Nottingham, UK (511–2003), as well as by the Ethical Review Board of the Nepal Health Research Council (66/2022). Additionally, a clinical trial license has been obtained from the Department of Drug Administration, Nepal. The study is being overseen by an independent Trial Steering Committee, composed of relevant experts who are responsible for monitoring and providing overall supervision throughout the trial [29].

The WHO has suggested that for herbal medicines with a well-documented history of use, it is ethically acceptable to advance directly to phase 3 clinical trials after completing animal toxicity testing, bypassing phases 1 and 2. Some clinical studies have utilized double-blind, placebo-controlled designs to evaluate various herbal remedies. These include Ginkgo leaf (*Ginkgo biloba*) for dementia, St. John's wort (*Hypericum perforatum*) for depressive disorders, Ginseng (*Panax ginseng*) as a tonic, Feverfew leaf (*Tanacetum parthenium*) for migraine headaches, Garlic bulb (*Allium sativum*) for hypercholesterolemia, Valerian root (*Valeriana officinalis*) as a sedative and sleep aid, Ginger root (*Zingiber officinale*) for nausea and vomiting, Kava rhizome (*Piper methysticum*) to reduce anxiety, Echinacea (*Echinacea purpurea*) for preventing and treating the common cold, and Saw palmetto fruits (*Serenoa repens*) for benign prostatic hyperplasia.

However, these studies have faced criticism for several issues. Concerns include the lack of standardization and quality control in the herbal remedies used, inconsistencies in dosages, insufficient sample sizes, inadequate randomization, and problematic placebos that differ in color, taste, and aroma. These factors can undermine the reliability and validity of the clinical trial results [30].

9.2.3.4 Philippines

In the Philippines, clinical trials are overseen by the Department of Health (DOH) and the FDA, which operates under the DOH. The FDA is tasked with ensuring the safety, efficacy, and quality of food, drugs, cosmetics, and medical devices. While herbal remedies derived from plants have been traditionally used by Philippine indigenous communities for many decades, regulatory mechanisms, as outlined in RA 8423, mandate that herbal products from medicinal plants must demonstrate an empirical record of use and efficacy for at least 50 years before they are recognized. This requirement surpasses the European Union's 2004 Directive on Traditional Herbal Medicinal Products (DTHMP), which stipulates that an herbal medicinal product should be subject to continuous testing for a minimum of 30 years before being exempted from clinical trials and tests for safety and efficacy. However, some experts argue that five decades of traditional use is still insufficient to fully assess the long-term effects of herbal products [31].

9.2.3.5 Ukraine

When developing a clinical trial protocol, it's important to follow key guidelines that ensure the research is conducted ethically, morally, and scientifically. These guidelines include the 2013 Declaration of Helsinki, which outlines the ethical principles for medical research involving humans; the GCP standards described in the ICH E6 guideline; and the EU Directive 2001/20/EC, which focuses on implementing GCP in clinical trials. These regulations were carefully followed during the development of the clinical trial to ensure it met the highest ethical and scientific standards.

This trial will follow internationally recognized GCP standards, as outlined in the ICH E6 R2 Guideline and in line with the 2013 Declaration of Helsinki, while also adhering to local regulations. The investigator must continuously ensure that all team members are properly qualified and fully informed about the trial protocol, any protocol amendments, trial treatments, and their specific responsibilities. Records of team meetings and training should be kept in the trial investigator's site file.

The investigator should also maintain a list of subinvestigators and other qualified individuals assigned significant trial-related tasks. Serious breaches of GCP or significant protocol deviations refer to violations that could substantially impact patient safety, rights, or the reliability of trial data [32].

9.2.3.6 Russia

In Russia, conventional herb-based medicines are subject to rigorous regulation to ensure safety and efficacy. These medicines, which are derived from medicinal plants and contain active plant ingredients, must adhere to strict controls established by government agencies. The federal law "Circulation of Medicines in the Russian Federation," first enacted in September 2010 and updated in 2014 (N-429-F3, 11/22/2014), provides a comprehensive framework governing all aspects of these medicines. This includes regulations for research and development, preclinical and clinical trials, expertise, standardization, and quality control. Additionally, the law covers state registration, manufacturing, storage, transportation, import and export, advertising, sales, and disposal of unused or expired medicines. The guidelines for conducting preclinical trials, proposed by experts in 2012, reinforce the stringent standards. These regulations align with international standards, including those in the European Union and Asian countries, often providing even more rigorous controls. This regulatory approach ensures that plant-originated medicines are thoroughly vetted, emphasizing safety and efficacy in line with global practices [33].

9.2.4 Western Pacific

9.2.4.1 Korea

9.2.4.1.1 Current Status of Herbal Medicine Clinical Trials in South Korea

Clinical trial information in South Korea can be accessed through three primary platforms: the Korea Ministry of Food and Drug Safety (KMFDS) website, the Clinical Research Information Service (CRIS), and the Korea Clinical Trials Information Center (K-CLIC). The KMFDS site provides details on clinical trials approved by the agency, including titles, IPs, and involved institutions. For trials approved after October 26, 2019, more comprehensive information such as participant criteria and outcome measures is available (https://nedrug.mfds.go.kr/searchClinic). CRIS, operated by the Korean Centers for Disease Control (KCDC), is part of the WHO's ICTRP and houses registered clinical trial data (https://cris.nih.go.kr/). K-CLIC offers trial information for both participants and investigators while summarizing the status of ongoing trials (https://www.koreaclinicaltrials.org/). Information specific to traditional Korean medicine (KM) clinical trials can also be found on the National Clearinghouse for Korean Medicine (NCKM) website (http://www.nckm.or.kr/).

Clinical trials in Korea are categorized as either KMFDS-regulated or nonregulated. KMFDS-regulated trial data can be found on the KMFDS website, while nonregulated trials are registered on the CRIS platform if researchers choose to do so. To examine the status of herbal medicine clinical trials, data were collected from the KMFDS and CRIS sites. Typically, the trial titles focus on specific herbal preparations, such as decoctions, "tang," or granules (e.g., Hwangryunhaedok-tang), and rarely use terms like "herbal medicine" or "herb." In some cases, titles omit the herbal medicine name, especially for new drug developments. Given that most herbal medicine trials take place in KM hospitals, the search also targeted trials with "Korean Medicine" mentioned in the institution name.

A single researcher (YC) conducted the searches on April 26, 2020 (KMFDS) and April 23, 2020 (CRIS). The selection process was reviewed by other researchers, and disagreements were resolved by consensus. The inclusion criteria were limited to clinical trials involving herbal medicine, excluding observational and nonherbal studies. Both single- and multi-herbal formulas administered according to East Asian traditional medicine principles were considered, while dietary supplements were excluded, regardless of preparation form or administration duration [34].

9.2.4.1.2 Approval Procedure and Data Requirements for Herbal Medicine Clinical Trials in South Korea

To obtain approval for a clinical trial involving a drug, applicants must submit an application form and relevant data via the KMFDS online platform (https://nedrug.mfds.go.kr). The Clinical Trials Management Division first reviews the documents for completeness and then forwards the application to the Herbal Medicines Division for scientific and technical evaluation. After this review, the application goes back to the Clinical Trials Management Division, which then issues a certificate of investigational new drug (IND) approval. Typically, the process from submission to initial feedback takes around 30 working days. To speed up the process, the IND review and IRB application can be done simultaneously. However, for herbal medicine trials involving treatments used for more than three years or in over 200 cases, conditional IRB approval is required before granting IND approval.

The data requirements for commercial and noncommercial IND applications are largely the same. However, commercial trials prioritize demonstrating safety and efficacy for licensing, which requires more rigorous standards for participant criteria, sample size, outcome measures, and statistical analysis. The following are the key documents and data needed when applying for IND approval. For investigator-initiated trials involving previously approved drugs or natural products, certain data, such as the investigator's brochure, nonclinical data, and prior clinical experience, are exempt. Instead, supporting academic papers demonstrating the scientific validity of the trial should be provided.

- **Study Protocol**: This document outlines the trial's background, purpose, target participants (including criteria for inclusion, exclusion, and dropouts), trial design (e.g., duration, treatment and control groups, blinding, flowchart), termination criteria, investigational product details, dosing plan, trial procedures, data analysis, ethical considerations, and information about the investigator, sponsor, and institution. Protocol guidelines such as the Standard Protocol Items for Clinical Trials with TCM 2018 and the CONSORT Extension for Chinese Herbal Medicine Formulas 2017 can be referenced.
- **Investigator's Brochure**: This provides key information about the investigational product, including dosage, administration method, and safety monitoring, to ensure adherence to the study protocol by investigators and other clinical personnel.
- **Manufacturing and Quality Data**: This includes details on the active pharmaceutical ingredient, dosage, manufacturing process, manufacturer, and quality control results.
- **Nonclinical Data**: Depending on the investigational drug, pharmacological and toxicology data may be required. This can include efficacy tests, safety pharmacology tests, and absorption, distribution, metabolism, and excretion data. Toxicology data, such as single-dose toxicity, repeated-dose toxicity, genotoxicity, reproductive toxicity, and carcinogenicity, must be obtained from GLP-certified institutions. For herbal medicines, nonclinical data requirements may be partially exempted if deemed unnecessary.
- **Previous Clinical Experience**: This should include a rationale and discussion of clinical trials for the drug, especially if it has already undergone clinical testing or been marketed. The submission should include pharmacokinetics, pharmacodynamics, dose response, safety, and efficacy data.
- **Additional Requirements**: Data on the institution, investigator, contract research organization, informed consent, and compensation rules must be submitted. Trials should be conducted in KMFDS-designated institutions, with documentation proving compliance with GMP standards. The development plan and GMP certification for the investigational herbal medicine are also required.
- **Special Data Requirements for Herbal Medicine Trials Used for Over Three Years/200 Cases**: For trials involving herbal medicines, certification that the treatment has been in use for over three years and in more than 200 cases is necessary. This includes an anonymous patient list with medical records and IRB approval. Nonclinical data are not required, but GMP certification is still mandatory. Detailed information on such trials can be found on the NCKM website [35].

9.2.4.2 Japan

Kampo medicines, the primary traditional herbal remedies in Japan, are classified as pharmaceuticals. Rooted in ancient Chinese medicine, they have evolved into a uniquely Japanese practice over time. Ethical Kampo formulations are prescribed by physicians in general medical practice and are covered under the National Health Insurance reimbursement system [36]. In Japan, the Clinical Trials Act (Supplement 7), enacted in April 2017 and effective from April 2018, mandates that clinical research funded by pharmaceutical companies submit research proposals to the national government in advance and follow practices similar to GCP. This law has made conducting post-marketing clinical research on medical products more challenging, both procedurally and financially. Since the implementation of this law, there has been a noticeable decrease in the number of publications and registrations of clinical trials involving Kampo medicines, suggesting that pharmaceutical companies had been actively engaged in research on these traditional medicines up until then.

Currently, clinical trial registration is mandatory before a trial begins. Japan has had three primary clinical trial registration sites since 2005: UMIN-CTR, JAPIC-CTR, and JMA-CCT. In 2018, a new platform, Japan Registry of Clinical Trials (jRCT), was introduced specifically for trials under the Clinical Trials Act. These four registration sites now form part of the Japan Primary Registries Network (JPRN), which integrates with the World Health Organization's ICTRP. There was an increase in registered clinical trials on Kampo medicines starting in 2009, reflecting the spread of registration regulations, but this number has declined since 2017. This decline aligns with a reduction in clinical publications on Kampo medicines and a decrease in RCT reports on Kampo products, both of which are attributed to the Clinical Trials Act.

Before 2018, most registrations were through UMIN-CTR. However, by 2019, the number of registrations via jRCT surpassed those of UMIN-CTR. Trials under the Clinical Trials Act requiring jRCT registration are typically funded by pharmaceutical companies. Consequently, more than half of the Kampo medicine clinical trials registered in 2019 were funded by these companies. It is likely that many trials conducted before 2018, prior to jRCT's establishment, were also funded by pharmaceutical companies. Recently, leading Kampo medicine manufacturers in Japan have focused on collecting basic and clinical data for specific Kampo formulas such as *daikenchuto*, *rikkunshito*, and *yokukansan* [37].

9.2.4.3 China

TCM is largely regarded as a complementary or alternative medical system in most Western countries and is gaining broader acceptance globally. An increasing number of clinical trials on TCM are being conducted internationally, drawing growing interest from scientists around the world in assessing its clinical efficacy through these trials.

To promote the inheritance and innovation of TCM, a guideline has been established to support the creation of an "Evidence System for TCM Registration and Review," which integrates TCM theory, human-use experience, and clinical trials. TCM theory provides the foundation for the rational use of compound preparations, ensuring their alignment with the "theory-diagnosis-formula-medicine" framework. Human-use experience, accumulated through clinical practice, is essential in understanding and verifying the efficacy and safety of these preparations. Clinical trials build on this knowledge to address unresolved issues.

The guideline emphasizes a patient-centered, clinical value-oriented approach to TCM drug development, leveraging TCM's unique strengths. It supports diverse research pathways, including traditional formulas, institutional preparations, and modern scientific formulations. The focus is on collecting human-use experience and generating evidence to support regulatory decisions. The guideline will evolve with the accumulation of practical experience and updates to relevant regulations. In the research and development of new TCM compound preparations based on human-use experience, the clinical benefits, target population, dosage, and efficacy characteristics can often be preliminarily established through this experience, making nonclinical pharmacodynamic studies typically unnecessary. If clinical trials are required, they should be thoughtfully designed in alignment with the formula's characteristics and supported by human-use experience. Both RCTs and real-world study designs, such as pragmatic clinical trials (PCTs), are viable options for conducting these studies [38].

9.2.4.3.1 General Principles

1. **Human-Use Experience**: This guideline bases itself on the clinical practice of fixed TCM formulas or compound preparations supported by TCM theory. This encompasses formula sources, key pharmaceutical data, clinical usage, and related studies to aid in the development or registration of new TCM drugs.
2. **Preparation Process**: For unapproved preparations, the process should align with traditional TCM methods.
3. **Informing Development**: Human-use experience can guide the development of TCM drugs, particularly those based on historical formulas or famous TCM practitioners' creations.
4. **Registration Support**: If human-use experience meets data standards, it can support registration applications and potentially negate the need for nonclinical pharmacodynamic studies. Clinical trials should be designed based on this experience, with RCTs or PCTs as suitable.
5. **Development Paths**: New TCM drugs can follow various development paths based on application category and human-use experience. Specific cases should consult the Center for Drug Evaluation (CDE).

Scope of Application

This guideline applies to the clinical research and development of new TCM compound preparations based on human-use experience, including Class 1.1 TCM compounds and Class 3.2 compounds from ancient formulas.

Information on Human-Use Experience

Source and Evolution of Formula: Includes the formula's origin, TCM theory, ingredient dosages, forms, indications, and changes from the original formulas.

Key Pharmaceutical Information: Encompasses ingredients, preparation methods, dosage forms, and any changes over time.

Clinical Use: Provides details on the use of the TCM compound, including institutions, patient demographics, doses, adverse reactions, and any interruptions in use.

Clinical Practice Data: Derived from hospital information systems and medical records, both structured and unstructured. Previous studies may also contribute data.

Medical Records: Serve as a key source for clinical practice data and need to meet quality standards for analysis and registration. Outpatient and emergency records may lack comprehensive data and should be used carefully. Informatization tools can enhance record quality.

Clinical Study Data: Data from previous studies, whether observational or interventional, are generally more reliable than routine records. Standardization and meta-analysis of these data are recommended for registration purposes.

9.2.4.3.2 Strategies for Clinical Research and Development

Human-use experience can guide research, development, and registration applications for new TCM compound preparations. Historical data from medical records or previous clinical studies should be curated for consistency in analysis. If human-use experience provides sufficient evidence of efficacy and safety, it can directly support marketing registration. Otherwise, further clinical studies are needed. Depending on the quality of the data, exploratory or confirmatory trials such as RCTs or PCTs may be conducted. If no human-use experience is available, conventional research pathways should be followed. The research and development pathways may vary based on submission categories, and applicants should communicate with regulatory agencies for guidance.

9.2.4.3.2.1 Class 1.1 TCM Compound Preparations

Paths ① to ④ focus on Class 1.1 TCM compound preparations:

Path ①: Without human-use experience, follow conventional clinical trials with exploratory and confirmatory RCTs.
Path ②: With weak evidence from previous human-use experience, conduct exploratory studies before confirmatory RCTs.
Path ③: With high-quality human-use experience and positive results, proceed directly to confirmatory RCTs.
Path ④: With high-quality human-use experience and positive results, conduct confirmatory PCTs directly.

9.2.4.3.2.2 Class 3.2 Other TCM Compound Preparations Derived from Ancient Classical Formulas **Paths ⑤ to ⑦ apply to Class 3.2 TCM compound preparations:**

Path ⑤: Use historical human-use experience data to support registration.
Path ⑥: Use prospective studies (interventional or observational) to support registration.
Path ⑦: If historical evidence is insufficient, enhance the evidence through prospective studies.

9.2.4.3.2.3 Curation and Evaluation of Clinical Practice Data of Human-Use Experience Clinical data from human-use experience are often incomplete and must undergo standardized curation to meet analysis and submission standards.

- Data curation

This includes data security, extraction, cleaning, conversion, transfer, storage, and quality control.

- Data quality evaluation

Data quality is evaluated in two steps:

1. Preliminary assessment of source data for analysis, ethical compliance, accessibility, and completeness.
2. Post-curation evaluation for relevance (e.g., variable coverage, accuracy, representativeness) and reliability (e.g., data completeness, accuracy, quality control).

9.2.4.3.2.4 Clinical Study Design Based on Human-Use Experience

- **Study Objectives:** Clearly define the objectives and address scientific questions related to the target population, treatment, and effect measurement. Include primary, secondary, and exploratory objectives.
- **Target Population and Clinical Positioning:** Align the target population and clinical positioning with TCM theory and practice. Set inclusion/exclusion criteria based on study objectives. The criteria can be relaxed for broader indications or be stricter for confirmatory studies. Detail the diagnostic criteria and clarify the functional indication scope.
- **Choice of Controls:** Select active or standard-of-care (SOC) controls, ensuring comparability with TCM. Use established treatments with clear efficacy or well-designed placebo-controlled RCTs.
- **Outcome Variables and Other Study Variables:** Choose primary and secondary endpoints aligned with clinical positioning, such as disease recovery or symptom improvement. Include patient-focused outcomes such as PROs. Collect variables related to treatment, demographics, medical history, and baseline factors impacting outcomes.
- **Data Source and Curation Plan:** Specify the data source and develop a detailed data curation plan, which can be appended to the study protocol.
- **Statistical Analysis Plan:** Detail the statistical analysis plan, including assumptions, models, and sample size estimation considering covariates and data availability after curation. Encourage including all eligible patients, with justification for any selection criteria.
- **Bias Control:** Address potential biases such as selection, confounding, and result-driven biases. Countermeasures should be reflected in the data curation and statistical analysis plans. For exploratory studies, the design should achieve the study's objectives.

9.2.4.3.2.5 Prospective Clinical Study Design If historical data are insufficient, further prospective studies such as RCTs, PCTs, or single-arm trials are needed:

RCTs: Follow international and domestic guidelines.
PCTs: Should mirror real medical practice with broader inclusion criteria, active controls, and a focus on overall effectiveness, including quality of life. Randomization and blinding are preferred but flexible.
Single-arm Trials: Used for rare or critical diseases, emphasizing comparability with external controls and bias control.

9.2.4.3.2.6 Evaluation of Clinical Studies Based on Human-Use Experience

Efficacy Assessment

- Data Adequacy: Ensure information on human-use experience covers essential aspects such as the formula's origin, TCM theory, and clinical data. Evaluate the relevance, reliability, and traceability of the data.
- Protocol Implementation: The study protocol must be well defined, registered, and followed strictly.

Significant changes require ethical review and regulatory communication.
- Report Integrity: Reports should accurately reflect study design, quality control, and analysis, ensuring transparency and robustness.

Safety Assessment

Safety evaluations should detail adverse events, focus on toxic components, and assess the population's tolerance based on nonclinical and human-use data.

9.2.4.3.2.6.1 Benefit-Risk Evaluation Assess whether the study confirms the drug's effectiveness, safe use in the target population, and that the overall clinical benefits outweighing risks [38].

9.2.4.4 Malaysia

With growing interest in herbal medicine research in Malaysia, the guideline has been developed following the "Guideline on Natural Products with Therapeutic Claim: Appendix 7B of the Drug Registration Guidance Document (DRGD), 2022," issued by the National Pharmaceutical Regulatory Agency (NPRA). Its main goal is to provide a clear overview of the requirements and processes for planning and conducting herbal medicine research in Malaysia, covering everything from initial discovery to clinical development.

When investigating herbal medicines in pharmaceutical forms (such as capsules, tablets, and syrups) in Malaysia, adherence to specific regulatory requirements is essential. These include the NPRA's product-related data requirements outlined in the DRGD and the Malaysian Guideline on Clinical Trial Import Licence (CTIL)/Clinical Trial Exemption (CTX). Herbal medicines are regulated under the Poisons Act 1952, the Sale of Drugs Act 1952, and the Control of Drugs and Cosmetics Regulation 1984.

Clinical trials may be sponsored by various entities, including pharmaceutical companies, academic institutions, voluntary groups, and other organizations (e.g., the U.S. NIH). There are two primary types of clinical trials: Investigator-Initiated Research (IIR) and Industry-Sponsored Research (ISR). IIRs are initiated and managed by individual researchers or institutions not affiliated with pharmaceutical companies. In contrast, ISRs are funded by industry organizations that contract with researchers to study interventions for diseases or biomedical conditions. Typically, the sponsor designs the study and owns the protocol.

Before proceeding with clinical trials, safety studies are crucial. Preclinical toxicity studies are conducted to determine the safety profile of herbal products. These studies characterize toxic effects, dose dependence, exposure relationships, and potential reversibility, helping to estimate a safe starting dose for human trials. The recommended duration for repeated-dose toxicity studies should match or exceed the duration of the human clinical trials and be conducted in at least two mammalian species (one rodent and one nonrodent). For longer clinical trials, repeated-dose toxicology studies should also be extended accordingly. For example, six-month rodent and nine-month nonrodent studies typically support trials lasting longer than six months.

Specific toxicity studies should align with the intended duration of the clinical trial. Any relevant safety data from previous human studies can support safety claims and may involve estimating initial safety and tolerability, ADME (absorption, distribution, metabolism, excretion) studies, pharmacokinetics/pharmacodynamics (PK/PD) studies, and documentation such as clinical study reports, published papers, and periodic safety update reports [39].

Calculating the initial dosage for humans is a critical step in ensuring safety during first-in-human (FIH) trials. This calculation should consider all relevant nonclinical data, including pharmacological dose response, toxicological profiles, and pharmacokinetics.

Several key documents must be prepared for submission, tailored to the trial phase, target population, and disease studied. These documents include the following:

Investigator's Brochure (IB): Provides comprehensive information about the investigational product, including preclinical and clinical data.

Clinical Trial Protocol: Details the study design, objectives, methodology, efficacy and safety endpoints, statistical considerations, and overall organization. It should be clear and detailed, serving as a guide for the research process.

Patient Information Sheet (PIS) and Informed Consent Form: These documents ensure that subjects provide informed consent after understanding all aspects of the trial. In Malaysia, these are provided in English and Malay, with translation into other languages such as Chinese and Tamil encouraged.

Clinical Trial Exemption (CTX)/Clinical Trial Import Licence (CTIL): CTX is granted for manufacturing products solely as clinical trial samples, exempting certain regulations under the Control of Drugs and Cosmetics Regulations 1984. CTIL allows the import of products for clinical trials, even if they are not registered.

The approval process for clinical trials involving herbal products or standardized herbal extracts as IPs involves several committees and regulatory bodies, including the National Committee for Research and Development of Herbal Medicine (NRDHM), the Medical Research and Ethics Committee (MREC), and the NPRA. Each body has specific requirements and approval criteria.

This guideline applies to research involving formulated natural products with herbs and considers both local and international regulations. It is specifically tailored to the herbal research environment in Malaysia but can be adapted for other countries as needed. However, these guidelines may not apply to traditional herbal preparations used in specific Traditional and Complementary Medicine (T&CM) practices, which are covered under the "Framework on Traditional and Complementary Medicine Research in Malaysia" guideline.

9.2.4.4.1 Clinical Trial Conduct in Malaysia

Currently, phase 1 clinical trial facilities in Malaysia are limited to Hospital Ampang and Hospital Umum Sarawak (Kuching). The MOH has approved two key guidelines: one detailing the conduct of phase 1 clinical trials in Malaysia (Clinical Research Malaysia [CRM], 2017) and another focused on accrediting facilities for FIH studies (NPRA, 2018b). Sponsors and contract research organizations (CROs) are permitted to conduct FIH studies only at these accredited phase 1 facilities. Trials in other phases can take place at any clinical trial site, including government hospitals, private hospitals, medical teaching hospitals, clinics, or clinical trial centers across the country. These trials are subject to GCP inspections by NPRA inspectors to ensure compliance with the Malaysian Guideline for Good Clinical Practice, ethical standards, and other regulatory requirements (NPRA, 2020) as shown in Table 9.1 [40].

9.2.4.4.2 Timelines

Typically, the assessment of all CTIL/CTX applications will be completed within the following timeframe:

45 working days for phase I trials, as well as for clinical trials involving biological/biotechnological products, cell therapy products, gene therapy products, and herbal products [40].

9.2.4.5 Australia

Regulating clinical trials for herbal medicines in Australia involves specific considerations due to their unique nature. Here's an overview of the key regulations and guidelines:

- **Therapeutic Goods Administration:**
  - **Regulation:** Herbal medicines are regulated under the Therapeutic Goods Act 1989, which mandates adherence to safety, quality, and efficacy standards.
  - **Clinical Trial Approval:** Sponsors must secure approval from the TGA before initiating a clinical trial with herbal medicines. The application should provide detailed information about the herbal product, trial design, safety protocols, and informed consent procedures.
  - **Product Classification:** Herbal medicines are classified as either Listed (lower-risk) or Registered (higher-risk) medicines. This classification impacts the regulatory requirements for conducting trials [41].
- **National Health and Medical Research Council**
  - **Ethical Guidelines:** Clinical trials must adhere to the NHMRC's "National Statement on Ethical Conduct in Human Research," ensuring ethical standards and participant protection.
- **Human Research Ethics Committees**
  - **Approval:** Similar to other clinical trials, those involving herbal medicines require approval from an HREC. The committee reviews the trial's ethical aspects, including risk-benefit analysis and informed consent.
- **Good Clinical Practice**
  - **Compliance:** Trials must follow GCP guidelines, which are internationally recognized standards ensuring the credibility of results and participant safety.
- **Clinical Trials Notification Scheme**
  - **Notification:** Sponsors are required to notify the TGA of their intention to conduct a clinical trial involving herbal medicines. This helps ensure compliance with TGA standards.
- **Informed Consent**
  - **Requirement:** Participants must receive comprehensive information about the trial, including details about the herbal medicine, potential risks and benefits, and must provide voluntary consent.
- **Data Privacy**
  - **Compliance:** Trials must comply with the Australian Privacy Principles (APPs) to protect participants' personal information [41].

9.2.4.6 Mongolia

Traditional Mongolian medicine (TMM) evolved from Mongolian folk medicine and was significantly influenced by Ayurveda and traditional Tibetan medicine.

In Mongolia, clinical trials for traditional herbal medicines are regulated by the MOH. All trials must receive MOH approval and ethical clearance from a relevant ethics committee. Preclinical studies are required to gather safety and efficacy data before human trials. Traditional herbal medicines must adhere to GMP and undergo standardization of active ingredients. Trials are conducted in phases, with detailed study protocols submitted for approval. Adverse events must be reported promptly, and respect for traditional knowledge and cultural practices is emphasized throughout the trial process [42].

9.2.4.7 Taiwan

Policy and decision-makers involved in TCM research in Taiwan are aware of criticisms suggesting that complementary and alternative medicine (CAM) trials are often seen as nonequivalent or inferior to those of conventional medicine. However, conventional medicine also faces its own challenges, including accusations of misconduct in clinical studies and publications. Despite these issues,

TABLE 9.1
Pharmaceutical Data Format for Herbal and Natural Products in Clinical Trials

S No.	Modules	Details Required
1	**Product information**	(a) **Name:** [Product Name]; **Reference Number:** [Reference Number] (b) **Finished Product: Description:** Physical characteristics; (c) **Composition: Active Ingredient(s) and/or Standardized Extract(s):** List of active ingredients/standardized extracts (d) **Other Ingredient(s):** List of active ingredients and/or standardized extracts (e) **Packing/Pack Size:** Brief description.
2	**Standardization of extract**	(a) **Example: Extract Standardized to Contain:** i. X% of compound A (assayed by methods such as HPLC, UV spectrophotometry, etc.) ii. Y% of compound B (assayed by methods such as HPLC, UV spectrophotometry, etc.)
3	**Manufacture of product**	(a) **Name and Address of Manufacturer(s) and/or Repacker(s):** Responsibilities and each production site involved (b) **Certificate of GMP Compliance:** Required for all manufacturers and/or repackers. (c) **Complete Batch Manufacturing Master Formula:** List of ingredients (d) **Manufacturing Process:** Description and principles; flow chart of steps, components used, and in-process controls.
4	**Quality control**	(a) **Control by Manufacturer or External Laboratory:** If external: provide the name and address of the laboratory; tests conducted; and reasons for outsourcing tests (b) **Specifications of Standardized Extracts:** Tests and criteria: Appearance, qualitative and quantitative assays, loss on drying, solubility, microbial limits (total bacterial count, yeast, mold, Salmonella, E. coli), heavy metals (arsenic, mercury, lead, cadmium), and other tests (c) **Certificate of Analysis:** Attach for at least one batch.
5	**Method of identification and analysis of marker compounds**	(a) Description of methods
6	**Finished product quality control**	Tests and Specifications: Appearance, content uniformity, quantitative assay, microbial limits, heavy metals, uniformity of weight, disintegration/dissolution test. (a) **Certificate of Analysis:** Certified by Quality Assurance Manager, including the most recent batch (b) **Validation of Analytical Method:** Reports including specificity, repeatability, reproducibility, linearity, accuracy, detection limit, quantitation limit, and conclusions
7	**Stability of product**	(a) **Storage Condition:** To be included on the label (b) **Proposed Shelf Life:** Provide data if extending shelf life (c) **Stability Studies:** Summary, characteristics, degradation products, and results of at least one batch (d) **Ongoing/Proposed Studies:** Outline of ongoing or proposed studies; **note:** follow ASEAN/ICH Stability Studies Guidelines
8	**Containers/ packaging**	(a) **Immediate Containers/Packaging:** Type, material, capacity, closure, and liner (b) **Other Containers/Packaging** (c) **Dose-Measuring Device/Apparatus/Administration Set (if any):** Description/type, material, capacity (d) **Packaging Inclusions:** Description and composition of desiccants, fillers, etc. (e) **Interaction with Packaging Material:** Indicate if any known interaction
9	**Labeling**	(a) **Refer to Appendix E** (b) **Submit Samples or Drafts of** Labels for immediate and outer packages (c) Original package inserts for comparator products

CAM can still be subjected to rigorous trials; though similar to conventional medicine, challenges related to funding and the practicality of conducting such trials need to be addressed.

Developing new drugs from Chinese Materia Medica (CMM) in the post-genomic era presents significant challenges. Taiwan is at the forefront of TCM research in Asia, with the highest concentration of clinical trial centers meeting the highest standards. The CDE and the DOH Committee on Chinese Medicine and Pharmacy (CCMP) oversee clinical trials at nine specialized herbal medicine centers across the island. The CCMP has also established TCM clinical centers in 10 top teaching hospitals, equipped with the necessary network configurations and standard operating procedures.

To further encourage private sector involvement, the CCMP has received Investigational New Drug (IND) and New Drug Application (NDA) submissions from 14 pharmaceutical companies. After a thorough review, six companies have been approved to conduct clinical trials.

The review process for IND and NDA applications in Taiwan has become more efficient, with review times reduced from 9.9 months (IND) and 2.9 months (NDA) in 1997 to 4.2 months and 1.2 months, respectively, in 2001 [43].

Taiwan and South Korea have similar healthcare systems and National Health Insurance (NHI) frameworks. In Taiwan, the NHI has been reimbursing the majority of residents' medical expenses since 1995, and a year later, traditional medicine (TM) services were also included under NHI coverage. By the end of June 2014, over 99.9% of Taiwan's population was covered by the NHI. The TM services covered by Taiwan's NHI include powdered Chinese herbal preparations, acupuncture, moxibustion, and manipulative therapies. Specifically, 337 types of Chinese herbal formulas and over 500 types of single Chinese herb preparations are covered. In 2011, these TM services represented approximately 3.1% of the total medical expenditures under the Taiwan NHI program [44].

9.2.5 Eastern Mediterranean Countries

9.2.5.1 Saudi Arabia

Before 2009, clinical trials in Saudi Arabia were not overseen by any government body, with local IRBs independently regulating the trials conducted at their respective sites. In 2009, the Saudi Food and Drug Authority (SFDA) established the Clinical Trials Administration. Since 2013, all investigators and sponsors have been required to register early-phase clinical trials (phases I, II, and III) with the Saudi Clinical Trials Registry and obtain SFDA approval before starting the study [45].

In Saudi Arabia, herbal products are classified as traditional products and are permitted if they have been in traditional use for at least 50 consecutive years. The dosage and method of preparation must align with traditional practice.

These products can be categorized based on the type of evidence provided:

1. **Pharmacopoeial Evidence for Traditional Products**
   For products in this category, the medicinal ingredients, quantities, recommended dosage, route of administration, duration of use, dosage form, usage directions, and risk information must match those specified in the Pharmacopoeia. Additionally, the preparation method must follow traditional practice.
2. **Non-pharmacopoeial Evidence for Traditional Products**
   Products in this category require supporting evidence from any two independent sources. These sources can include clinical studies, pharmacopoeias, textbook references, peer-reviewed published articles, or nonclinical studies covering pharmacokinetics, pharmacodynamics, toxicity, reproductive effects, and the potential genotoxicity or carcinogenicity of an ingredient. Alternatively, evidence based on previous marketing experience of the finished product can also be provided to demonstrate safety and efficacy [46].

In Saudi Arabia, the regulation of clinical trials for traditional medicines, including herbal products, is governed by guidelines established by the SFDA. The following outlines the key aspects of these guidelines.

9.2.5.1.1 Regulatory Framework

SFDA Oversight: The SFDA supervises all clinical trials conducted in the Kingdom, including those involving traditional medicines. The Clinical Trials Administration within the SFDA is tasked with the approval, monitoring, and regulation of these trials.

Registration Requirements: It is mandatory for all clinical trials involving traditional medicines to be registered with the Saudi Clinical Trials Registry before they begin. This registration is required for all early-phase trials, including phases I, II, and III.

Ethical Considerations

Informed Consent: Participants must provide informed consent, having a clear understanding of the potential risks, benefits, and objectives of the trial. The consent process must be properly documented.

Ethics Committee Approval: The trial must receive approval from an ethics committee or IRB to ensure that participants' rights and well-being are protected.

Preclinical Studies

Safety and Efficacy Data: Before initiating human trials, preclinical studies must be conducted to gather data on the safety and efficacy of the TM. These studies generally include laboratory and animal testing to evaluate toxicity, pharmacokinetics, and pharmacodynamics.

Documentation and Submission

Study Protocol: A comprehensive study protocol must be submitted to the SFDA for approval, detailing the trial design, objectives, methodology, statistical considerations, and ethical issues.

Adverse Event Reporting: Any adverse events or unexpected outcomes that occur during the trial must be promptly reported to the SFDA.

Traditional Medicine Classification

Herbal and Traditional Products: Traditional medicines, particularly those containing herbal ingredients, are classified based on their historical usage and evidence of safety and efficacy. The

SFDA requires thorough documentation of the traditional preparation methods, dosages, and historical usage.

Evidence Requirements: Depending on their classification, traditional medicines may need to provide evidence from recognized pharmacopoeias, clinical studies, or other credible sources to demonstrate their safety and efficacy [45].

9.2.5.2 Oman

In Oman, the regulation of clinical trials, including those involving traditional medicines, is overseen by the MOH along with other relevant authorities. Although the guidelines for traditional medicines may not be as comprehensive as those for conventional pharmaceuticals [47], the following key elements are generally applicable when conducting clinical trials in the country:

Regulatory Oversight

- **Ministry of Health:** The MOH serves as the primary regulatory authority responsible for overseeing clinical trials in Oman. Approval from the MOH is required for any clinical trial, including those involving traditional medicines, before the trial begins.
- **Pharmacy and Drug Control Department:** This department within the MOH is instrumental in the regulation and approval process for clinical trials involving traditional and herbal medicines.

Ethical Approval

- **Ethics Committee Review:** Clinical trials must be submitted for review and approval by an ethics committee, such as the Research and Ethical Review and Approval Committee (RERAC) under the MOH. This review ensures that the trial meets ethical standards and that the rights and safety of participants are safeguarded.
- **Informed Consent:** Participants must provide informed consent, acknowledging their understanding of the trial's objectives, potential risks, and benefits. The process of obtaining and documenting informed consent is a critical component of all clinical trials [48].

9.2.5.3 Vietnam

Vietnam's MOH plays a central role in overseeing clinical trials within the country. As the authority responsible for approvals, oversight, and inspections, the MOH is also tasked with issuing legal documents, policies, and standards. This includes setting requirements for data and safety monitoring boards and managing the national-level ethics committee.

While Vietnam's capacity for clinical research was initially limited, it experienced significant growth between 2010 and 2015. In 2010, only 25 clinical trials were conducted in the country. However, by 2015, this number had increased to 72 active trials. Half of these were industry-sponsored, while the other half were supported by nongovernmental organizations (NGOs). The majority of these trials (approximately 70%) focused on drug studies, 20% were dedicated to evaluating vaccines, and 10% examined traditional medicines. Among the drug trials, studies on infectious diseases were the most prevalent, followed by research on cancer and cardiology.

Starting in 2011, the MOH began issuing an annual priority list for clinical trials. Building on recommendations from combined assessments, the Ministry mandated in 2015 that all clinical trials conducted in Vietnam be registered on ClinicalTrials.gov. Additionally, a new regulation introduced in 2013 required that any clinical trial seeking MOH approval must be supported by a local People's Committee, which is equivalent to a jurisdiction's governing body.

Recognizing the importance of community involvement, Vietnam's health leadership has been encouraged to expand community engagement in research. NGOs are seen as key players in raising awareness of clinical research and fostering collaboration between communities and researchers. Guiding principles and best practices are available to help establish communities as active partners in biomedical research, ensuring that research efforts are more inclusive and community-focused [49].

9.3 CONCLUSION

Herbal medicines have long been a cornerstone of medical treatment in Asia, Africa, and Latin America. To safeguard this heritage and ensure its integration into contemporary healthcare systems, regulatory bodies in these regions have prioritized the scientific validation of herbal medicines, alongside ethical practices and strong regulatory frameworks. A key challenge in non-European countries is to ensure that the research into herbal medicines meets ethical standards. Herbal product trials often present unique concerns, such as product adulteration, limited data on toxicity, and a lack of standardized dosages. Also, there is little understanding of how herbal remedies may interact with conventional pharmaceuticals, making their safety difficult to assess. This is particularly important in communities where belief in the safety and efficacy of herbal remedies can lead to biases, both in the research process and among participants. To minimize such biases, rigorous study designs—such as RCTs—are essential, as well as clear control groups to compare results. In many non-European countries, local communities hold centuries-old knowledge about the therapeutic properties of plants and herbs. Researchers must respect the intellectual property and cultural heritage of these communities, ensuring that they benefit from any scientific advancements made through research. This collaboration can help ensure that traditional knowledge is not only preserved but also scientifically validated, creating a mutually beneficial relationship.

Despite the value of traditional knowledge, clinical trials are necessary to validate the safety and efficacy of herbal medicines in modern contexts. While traditional use provides anecdotal evidence of efficacy, RCTs remain the gold standard for establishing scientific proof. Herbal medicines are often viewed skeptically by the medical establishment in non-European countries, where modern pharmaceuticals and evidence-based medicine dominate. This makes it more difficult to integrate herbal remedies into mainstream healthcare. However, it is important to emphasize that the benefit-risk assessment for herbal medicines must be case-specific. Not all herbal remedies carry the same risk profile, and the safety and effectiveness of each must be scientifically evaluated.

The development of herbal medicines requires a comprehensive approach, from sustainable cultivation to the production of safe, standardized products. In non-European countries, many herbal remedies are still sourced from wild plants, which can pose sustainability and quality control challenges. Regulatory bodies must enforce GMP to ensure that herbal products are produced under controlled conditions, with standardized dosages and consistent quality. Additionally, clinical trials must follow GCP to maintain research integrity and ensure that claims about a product are reliable and based on sound evidence.

Safety concerns remain one of the biggest obstacles to the widespread use of herbal medicines. Many herbal products have potential long-term risks—such as liver damage, cancer, or other toxic effects—and their interactions with conventional medications are not well understood. Preclinical and clinical evaluations are essential to identify and address these risks. Post-marketing surveillance, in which products are monitored after reaching the market, is also crucial to identify any adverse effects that may emerge once the products are in general use. A "reverse pharmacology" approach, which starts with field data from traditional use and progresses to laboratory studies, can help bridge the gap between cultural knowledge and modern scientific research.

A uniform research policy for herbal medicines is urgently needed across non-European countries. The legal status and regulation of herbal ingredients vary greatly between regions, leading to inconsistent safety standards and quality control. In countries like India, China, and Brazil, where TM is widely practiced, governments have established guidelines for regulating of herbal products. However, these regulations are not always standardized across neighboring countries or regions, creating gaps in oversight. Harmonizing these standards at the global level—such as by adopting guidelines from the WHO—could ensure that herbal medicines meet internationally accepted criteria for safety, efficacy, and quality.

Given the rising demand for herbal remedies, non-European countries must strengthen regulatory frameworks that emphasize quality control, rigorous clinical trials, and post-marketing monitoring. These efforts will help ensure that herbal medicines are both safe and effective for the people who rely on them. By validating traditional knowledge through scientific research and aligning it with international standards, non-European countries can continue to honor their rich medical heritage while enhancing public health outcomes. This balanced approach can pave the way for the global integration of herbal remedies into modern healthcare systems.

REFERENCES

[1] Moreira, D. de L., S. S. Teixeira, M. H. D. Monteiro, A. C. A. X. De-Oliveira, and F. J. R. Paumgartten. (2014). Traditional Use and Safety of Herbal Medicines. *Revista Brasileira de Farmacognosia*, 24(2), 248–257. https://doi.org/10.1016/j.bjp.2014.03.006

[2] Zhang, J., K. Hu, L. Di, P. Wang, Z. Liu, J. Zhang, and P. Yue. (2021). Traditional Herbal Medicine and Nanomedicine: Converging Disciplines to Improve Therapeutic Efficacy and Human Health. *Advanced Drug Delivery Reviews*, 178: 113964. https://doi.org/10.1016/j.addr.2021.113964

[3] Mosihuzzaman, M. (2012). Herbal Medicine in Healthcare: An Overview. *Natural Product Communications*, 7(6): 807–812. https://doi.org/10.1177/1934578X1200700628

[4] Tilburt, J. C., and T. J. Kaptchuk (2008). Herbal Medicine Research and Global Health: An Ethical Analysis. *Bulletin of the World Health Organization*, 86(8): 594–599. https://iris.who.int/handle/10665/270255

[5] Parveen, A., B. Parveen, R. Parveen, and S. Ahmad (2015). Challenges and Guidelines for Clinical Trial of Herbal Drugs. *Journal of Pharmacy & Bioallied Sciences*, 7(4): 329–333. https://doi.org/10.4103/0975-7406.168035

[6] World Health Organization. Clinical Trials. https://www.who.int/health-topics/clinical-trials#tab=tab_2

[7] World Health Organization. (2022). Executive Board 152nd Session Provisional Agenda Item 23. 2EB152/37. https://apps.who.int/gb/ebwha/pdf_files/EB152/B152_37-en.pdf

[8] Clinical Trials for Natural Health Products. (2003). Government of Canada. https://www.canada.ca/en/health-canada/services/drugs-health-products/natural-non-prescription/legislation-guidelines/guidance-documents/clinical-trials.htmlClinical

[9] Research Regulation for Canada and Mexico. (2024). ClinRegs, National Institute of Allergy and Infectious Diseases. https://clinregs.niaid.nih.gov/country/canada/mexico#initiation,_agreements_&_registration

[10] World Health Organization. (2005). *National Policy on Traditional Medicine and Regulation of Herbal Medicines: Report of a WHO Global Survey*. World Health Organization. https://www.who.int/publications/i/item/9241593237

[11] New Law Governs Clinical Trials in Brazil. (2024). Lexology. https://www.lexology.com/library/detail.aspx?g=d9d0620a-fcbd-4717-9d95-f34afb2b73cf

[12] Roberto, C., and A. C. Mario. (2002). Clinical Research in Mexico. *Applied Clinical Trials*, 11(6): 3–5. https://www.researchgate.net/publication/268817409_Clinical_Research_in_Mexico

[13] Willcox, M., N. Siegfried, and Q. Johnson. (2012). Capacity for Clinical Research on Herbal Medicines in Africa. *Journal of Alternative and Complementary Medicine*, 18(6): 622–628. https://doi.org/10.1089/acm.2011.0963

[14] Saleh, M., K. Sharma, A. Shamshudin, I. Obayo, S. Gondi, and N. Karimi. (2024). Regulatory Approval of Clinical Trials: Is It Time to Reinvent the Wheel? *BMJ Global Health*, 9(1): e013727. https://doi.org/10.1136/bmjgh-2023-013727

[15] Department of Health. (2020). *South African Good Clinical Practice: Clinical Trial Guidelines*. Pretoria, South Africa. Third edition: 11–13 [155]. Herbal Medicines: Overview on Regulations in India and South Africa. *World Journal of Pharmaceutical Research*, 6(8): 690–698. https://doi.org/10.20959/wjpr20178-9091

[16] Kumar, V. (2017). Herbal Medicines: Overview on Regulations in India and South Africa. *World Journal of Pharmaceutical Research*, 6(8): 690–698. https://doi.org/10.20959/wjpr20178-9091

[17] Guidelines for Registration of Traditional Medicines in the WHO African Region. (2010). AFRO Library Cataloguing-in-Publication Data. 14–15 https://www.afro.who.int/publications/guidelines-registration-traditional-medicines-african-region

[18] Harmonised Framework for Regulation of African Traditional Medicine Practices, Practitioners, and Products in the ECOWAS Region. (2013). West African Health Organisation. 52–57. https://www.wahooas.org/web-ooas/sites/default/files/publications/2186/harmonised-framework-tmpok_0.pdf

[19] Urquiza-Haas, N., and E. Cloatre. (2023). In the Shadow of the Healing Rainbow: Belonging and Identity in the Regulation of Traditional Medicine in Mauritius. *Griffith Law Review*, 32(2): 236–258. https://doi.org/10.1080/10383441.2023.2249708

[20] Legislation CRRC. (2011). Clinical Trials Act 8 of 2011–1 September 2011. https://crrc.govmu.org/crrc/wp-content/uploads/2023/08/CRRC-CTA-Clinical-Trials-Act-2011.pdf

[21] Matar, A., and H. J. Silverman. (2022). Ethical Analysis of Egypt's Law Regulating Clinical Research. *Journal of Empirical Research on Human Research Ethics*, 17(4): 494. https://www.ncbi.nlm.nih.gov/pmc/articles/PMC9398927/

[22] Guidelines. (2024). Guideline for Good Regulatory Oversight of Clinical Trials by the Egyptian Drug Authority. Central Administration of Biological and Innovative Products and Clinical Studies, General Administration of Clinical Trials. *Egyptian Drug Authority*, 16–20. https://www.edaegypt.gov.eg/media/iqefpbx1/guide-line-gcp-v3.pdf

[23] Karar, M. G. E., and N. Kuhnert. (2011). Herbal Drugs from Sudan: Traditional Uses and Phytoconstituents. *Pharmacognosy Reviews*, 11(22): 83–103. https://doi.org/10.4103/phrev.phrev_15_15

[24] McKenzie, B. Clinical Trials Handbook Asia Pacific. Indonesia. https://www.bakermckenzie.com/-/media/files/insight/publications/2019/healthcare/ap/dsc125067_clinical-trials-handbook-indonesia.pdf?la=en

[25] Amarullah, D. R., and H. Siregar. (2024). JAMU: Traditional Indonesian Medicine. *Journal of Health and Medical Research*, (1): 11. https://www.iomcworld.org/articles/jamu-traditional-indonesian-medicine.pdf

[26] Singh, H. P., S. Sharma, S. B. Chauhan, and I. Kaur. (2014). Clinical Trials of Traditional Herbal Medicines in India: Current Status and Challenges. *International Journal of Pharmacognosy*, 1(7): 415–421. http://dx.doi.org/10.13040/IJPSR.0975-8232.IJP.1(7).415-421

[27] General Guidelines for Clinical Evaluation of Ayurvedic Interventions. (2018). Central Council for Research in Ayurvedic Sciences Ministry of AYUSH, Government of India New Delhi, 3: 43–45. https://ccras.nic.in/wp-content/uploads/2024/07/CCRAS_Guideline-of-Clinical_Evaluation.pdf

[28] Kumar, V. (2017). Herbal Medicines: Overview on Regulations in India and South Africa. *World Journal of Pharmaceutical Research*, 6(8): 690–698. http://dx.doi.org/10.20959/wjpr20178-9091

[29] Chattopadhyay, K., M. Dhimal, S. Karki, P. Regmi, B. Bista, T. K. Biswas, . . . & P. Gyanwali. (2022). A Clinical Guideline-Based Management of Type 2 Diabetes by Ayurvedic Practitioners in Nepal: A Feasibility Cluster Randomized Controlled Trial Protocol. *Medicine*, 101(47): e31452. https://doi.org/10.1097/MD.0000000000031452

[30] Mohan, B. G. (2008). *Aspects of Traditional Medicine in Nepal*. Institute of Natural Medicine University of Toyama. 158. https://www.scribd.com/doc/11869570/Aspects-of-Traditional-Medicine-in-Nepal.

[31] Zarsuelo, M., Z. Zordilla, and D. Anacio. (2018). Review of Regulatory Policies on and Benefits of Herbal Medicine in the Philippines. *Acta Medica Philippina*, 52(5): 31. https://doi.org/10.47895/amp.v52i5.334

[32] Popovych, V. I., I. V. Koshel, O. N. Malofiichuk, L. I. Pyletska, O. A. Semeniuk, O. V. Martynnyk, . . . and O. I. Leta. (2021). Multicenter, Randomized, Open-Label, Comparative Study of Therapeutic Efficacy, Safety and Tolerability of BNO 1030 Application in the Technology of Delayed Prescription of Antibiotics in Patients with Severe Acute Tonsillitis. *medRxiv*. http://dx.doi.org/10.1101/2021.03.10.21253255

[33] Sammons, H. M., M. I. Gubarev, L. V. Krepkova, V. V. Bortnikova, F. Corrick, K. M. Job, C. M. T. Sherwin, and E. Y. Enioutina. (2016). Herbal Medicines: Challenges in the Modern World: Part 2. European Union and Russia. *Expert Review of Clinical Pharmacology*, 9(8): 1117–1127. https://doi.org/10.1080/17512433.2016.1189326

[34] Gad, S. C. (Ed.). (2009). *Clinical Trials Handbook*. John Wiley & Sons. https://www.wiley.com/en-au/Clinical+Trials+Handbook-p-9780471213888#description-section

[35] Ichiro Arai. (2021). Clinical Studies of Traditional Japanese Herbal Medicines (Kampo): Need for Evidence by the Modern Scientific Methodology. *Integrative Medicine Research*, 10(3): 100722. https://doi.org/10.1016/j.imr.2021.100722

[36] Jiang, M., J. Yang, C. Zhang, B. Liu, K. Chan, H. Cao, and A. Lu. (2010). Clinical Studies with Traditional Chinese Medicine in the Past Decade and Future Research and Development. *Planta Medica*, 76(17): 2048–2064. https://doi.org/10.1055/s-0030-1250456

[37] Maegawa, H., T. Nakamura, and K. Saito. (2014). Regulation of Traditional Herbal Medicinal Products in Japan. *Journal of Ethnopharmacology*, 158: 511–515. https://doi.org/10.1016/j.jep.2014.07.012

[38] Dai, L., C. W. Cheng, R. Tian, L. L. Zhong, Y. P. Li, A. P. Lyu, A. W. Chan, H. C. Shang, and Z. X. Bian. (2019). Standard Protocol Items for Clinical Trials with Traditional Chinese Medicine 2018: Recommendations, Explanation, and Elaboration (SPIRIT-TCM Extension 2018). *Chinese Journal of Integrative Medicine*, 25(1): 71–79. https://doi.org/10.1007/s11655-018-2999-x

[39] Guideline for Herbal Medicine Research. (2023) *National Committee for Research and Development of Herbal Medicine (NRDHM), Ministry of Health Malaysia*, 1st Edition, pp. 34–43. https://nmrr.gov.my/storage/media/content/content-attachments/06af2210-3bc4-436f-a381-ef11210e9d27/Herbal-Medicine-Research-Guideline-2023.pdf

[40] Malaysian Guideline for Application of Clinical Trial Import Licence and Clinical Trial Exemption. *National Pharmaceutical Regulatory Agency (NPRA).* Ministry of Health Malaysia, 8th Edition, pp. 1–13. https://npra.gov.my/easyarticles/images/users/1140/Malaysian-Guideline-for-Application-of-CTIL-and-CTX-8th-Ed-Final.pdf

[41] Goods Administration, T. (2021). *Australian Clinical Trial Handbook: Guidance on Conducting Clinical Trials in Australia Using "Unapproved" Therapeutic Goods.* Therapeutic Goods Administration. https://www.tga.gov.au/sites/default/files/australian-clinical-trial-handbook.pdf

[42] Jigjidsuren, A., and B. Oyun. (2022). *Supporting the Regulation of Medicines in Mongolia: Experiences, Lessons Learned, and Future Directions—ADB Support in the Health Sector Reform in Mongolia.* ADB East Asia Working Paper Series. https://www.adb.org/publications/supporting-regulation-medicines-mongolia

[43] Kaphle, K., L. S. Wu, N. Y. J. Yang, and J. H. Lin. (2006). Herbal Medicine Research in Taiwan. *Evidence-Based Complementary and Alternative Medicine*, 3(1): 149–155. https://doi.org/10.1093/ecam/nek016

[44] Huang, C. W., I. H. Hwang, Y. S. Lee, S. J. Hwang, S. G. Ko, F. P. Chen, & B. H. Jang. (2018). Utilization Patterns of Traditional Medicine in Taiwan and South Korea by Using National Health Insurance Data in 2011. *PLoS One*, 13(12): e0208569. https://doi.org/10.1371/journal.pone.0208035

[45] Alsultan, A., W. A. Alghamdi, J. Alghamdi, A. F. Alharbi, A. Aljutayli, A. Albassam, O. Almazroo, and S. Alqahtani. (2020). Clinical Pharmacology Applications in Clinical Drug Development and Clinical Care: A Focus on Saudi Arabia. *Saudi Pharmaceutical Journal*, 28(10): 1217–1227. https://doi.org/10.1016/j.jsps.2020.08.012

[46] Sharma, S. (2015). Current Status of Herbal Product: Regulatory Overview. *Journal of Pharmacy & Bioallied Sciences*, 7(4): 293. https://doi.org/10.4103/0975-7406.168030

[47] Al-Riyami, H., A. Al-Maskari, S. K. Nadar, and M. Almaskari. (2023). Knowledge, Attitudes and Practices Regarding Traditional and Complementary Medicine in Oman. *Sultan Qaboos University Medical Journal*, 23(1): 90. https://doi.org/10.18295/squmj.1.2022.007

[48] Tabatabaei-Malazy, O., Z. Shadman, H. S. Ejtahed, R. Atlasi, M. Abdollahi, and B. Larijani (2018). Quality of Reporting of Randomized Controlled Trials of Herbal Medicines Conducted in Metabolic Disorders in Middle East Countries: A Systematic Review. *Complementary Therapies in Medicine*, 38: 61–66. https://doi.org/10.1016/j.ctim.2018.04.004

[49] Kagan, J., D. D. Giang, M. F. Iademarco, V. T. Phung, C. Y. Lau, & N. N. Quang. (2016). Assessing Clinical Research Capacity in Vietnam: A Framework for Strengthening Capability for Clinical Trials in Developing Countries. *Public Health Reports*, 131(3): 396–403. https://doi.org/10.1177/003335491613100305

10 Good Clinical Practices for Clinical Trials of Herbal Medicine

Sachin Karmakar[1], *Tanzeem Nigar*[1], *Anwesha Das*[1], *Arijit Nandi*[1,2], *and Wei-Yu Lin*[3]

[1]Department of Pharmacy, Sanaka Educational Trust Group of Institutions (SETGOI), Malandighi, Durgapur, West Bengal, India.

[2]Institute for Molecular Bioscience, The University of Queensland, St Lucia, Australia.

[3]Department of Medicinal and Applied Chemistry, Kaohsiung Medical University, Kaohsiung, Taiwan.

ABBREVIATIONS

ADRs	Adverse drug reactions
ASHG	American Society of Human Genetics
CAM	Complementary and alternative medicine
CONSORT	Consolidated Standards of Reporting Trials
CYP450	Cytochrome P450
ECs	Ethics Committees
EHR	Electronic health record
FDA	Food and Drug Administration
GACP	Good agricultural and collection practices
GAP	Good agricultural practices
GCP	Good clinical practices
GMP	Good Manufacturing Practices
H_0	Null hypothesis
H_1	Alternative hypothesis
HDIs	Herb–drug interactions
HHS	Health and human services
HMPC	Committee on Herbal Medicinal Products
ICFs	Informed consent forms
ICH	Indian Conference on Harmonization
IEC	Independent Ethics Committee
IRB	Institutional Review Board
IUPAC	International Union of Pure and Applied Chemistry
QC	Quality control
RAB	Regulation Audit Board
RCTs	Randomized clinical trials
REC	Research Ethics Committee
TCM	Traditional Chinese medicine
TOPO	Target-oriented overall process optimization
US	United States
WHO	World Health Organization
WM	Western medicine

10.1 INTRODUCTION

10.1.1 Overview of Good Clinical Practices

Good clinical practices (GCP) could be a handle that consolidates set up scientific and moral quality guidelines for the plan, recording, conduct, and announcing of clinical research about including the interest of human subjects. Compliance with GCP gives open confirmation that the well-being, security, and rights of investigated subjects are ensured and regarded, reliable with the standards articulated within the affirmation of Helsinki and other globally recognized moral rules, and guarantees the judgment of clinically investigated information. Sometime recently therapeutic items can be presented in advertisements or open well-being programs; they must experience an arrangement of examinations planned to assess security and efficacy within the parameters of harmfulness, strength, dose determination, and field conditions [1].

The ethical performance of clinical trials is the guiding philosophy of GCP guidelines. A trial should not be initiated unless a due evaluation of all the risks and disadvantages that may be expected has been balanced against the expected benefits to the individual participant, as well as to society in general. The pivot of the GCP guidelines is the safety and well-being of the trial subjects, for whose protection the GCP guidelines exist. The security and prosperity of the trial subjects are most critical and ought to win over the interest of science and society [1].

Intention is the foundation of arrangements controlling clinical trials. The Indian Conference on Harmonization (ICH) Rule for GCP characterizes educated assent as a process by which a subject or his lawful agent deliberately affirms a subject's willing consent to participate in a particular study after they have been fully apprised of all the

DOI: 10.1201/9781003522065-10

relevant aspects of the trial that may influence their willingness to participate. The concept is centered on the importance of informed consent, where potential participants are properly informed of the study details, risks, and benefits before agreeing to participate. Educated assent is recorded by means of a written, signed, and dated educated assent form. The educated assent handle depends on three standards [1]:

1. Satisfactory data are given, by and large, of what a sensible individual would need to know in arrange to choose.
2. Members comprehend the data.
3. Assent is given intentionally.

An Institutional Review Board (IRB), also called an officially appointed committee to help, screen, and oversee biological and behavioral tests of individuals, can be referred to as an independent ethics committee or ethical review board. In trying to make a decision regarding whether or not an investigation is to be conducted, they will typically perform a few forms of risk-benefit analysis. Avoiding physical or mental injury to human volunteers is the primary concern of IRBs. US IRBs may approve, require modifications before approval to intended investigations, or reject investigations according to Food and Drug Administration (FDA) and Health and Human Services (HHS) direction, more specifically the Office for Human Inquiry with respect to Assurances. Routine oversight work over "logical," "moral," and "administrative" studies using human subjects may be performed by IRBs [1].

The investigator's educational programs vitae or other articulation of instruction, preparation, and involvement may give starting data about the investigator's qualifications to supply therapeutic care and to conduct clinical investigation. Other sources of data about an investigator's qualifications may incorporate therapeutic authorizing sheets, negligence registries, and/or disciplinary bodies that will have data about the investigator's history of therapeutic practices. References from those common places with the investigator's clinical and/or investigative home may give valuable adjunctive data [1].

Noncommercial clinical investigation plays a progressively basic part in worldwide well-being. Numerous accomplices connect in universal consortia that work beneath the constrained time period of a particular subsidizing period. While one organization, mostly based in the North, is responsible for external finance and arranging subsidizing terms, another association (the sponsor) plans and runs the trial with investigation partners and has ultimate responsibility for the rational, ethical, administrative, and legal conduct of the trial. In fact, recourse to the terms of an external funder may be vital in determining the choice of the support if external subsidizing influences are crucial to most noncommercial investigations. In addition, the competition for getting access to the accessible outside reserves is intense, and noncommercial sponsors may not be in a position to negotiate or deny standard conditions set by a funder [1].

Advocates of clinical trials and contract research firms have to devise, check, and analyze their quality control and quality affirmation systems, along with their key standard operating procedures and other quality reports, in an effort to deliver quality products and services that completely fulfill the needs and desires of customers. Together, the quality affirmation and control systems form the key quality frameworks. Quality control and quality affirmation both make up quality administration. Quality control is centered on satisfying quality prerequisites, while quality affirmation is centered on giving certainty that quality prerequisites are satisfied [1].

Studies that keep up distinguished or identifiable examples must keep subjects' secrecy; that information from these tests ought not to be given to anybody other than the subjects and people assigned by the subjects in writing. Be that as it may, the American Society prescribes that agents educate people that they cannot ensure outright secrecy. The ASHG (American Society of Human Genetics) morals committee recommended that analysts "consider a way of coding tests by a third, free party, who would keep the codes blocked off unless there are particular circumstances in which the code should be broken." In other occasions, the way to secure the privacy of distinguished and identifiable tests is to code them [1].

In clinical trials and other planned ponders, information on the status of members at the time of enrollment and at indicated times from there on is utilized to compare the frequencies of essential and auxiliary results among members. Perceptions, estimations, research facility examples, reactions to meet questions, and other documentation of each study participant's status may be collected over a period of time. Regularly, the information is prepared, put away, and analyzed in the data planning center. The ponder comes full circle with distribution of discoveries and deductions based on factual examination of the information and other considerations. Information administration starts amid the plan of the information collection convention and closes only after the ultimate database for measurable examination and information filing is complete [1].

RECs/IRBs (Research Ethics Committee/Institutional Review Board) look at notices from analysts regarding antagonistic occasions, convention deviations/modifications, and, to a lesser degree, convention infringement. After checking the notices and corrections, IRBs and RECs are to either approve, presume, or revoke existing positive results. Convention changes are usually for significant changes, not calculated or governed ones, such as a change to the consideration screen or phone numbers. Earlier endorsement is additionally required for any major alter, except for critical alterations where prompt activity is required due to security concerns. At that point, the announcement ought to be done as before long as conceivable after the alteration is started. The receipt of notices of convention corrections is free of the proceeding audit process [1].

The principle that potential harms to research subjects must be balanced by expected benefits is common to all universal accounts of ethical conduct in clinical research. Both social benefits, or social esteem, and structured benefits to the investigating member are expected benefits that can warrant human inquiry. The benefit factor of the risk-benefit evaluation in first-in-human trials is therefore grounded exclusively in social respect, and no harmonized benefits can be anticipated. In ethics studies, the concept of social esteem is applied in many different ways and is inherently ambiguous. Since social esteem legitimizes including human members, particularly in early human trials, it is usually tricky [1].

In reaction to these issues, all ethics-related points of interest, counting the strategy and sum of installment to the subjects of the trial, are presently required to be comprised within the educated assent and convention and for review and consideration by the IRB or Independent Ethics Committee (IEC). Simplicity and reproducibility are two of the key pillars of medical research. Falling flat to supply an exhaustive report of the strategy and moral components of a trial might, specifically or in a roundabout way, ruin the worldwide endeavors to handle validity issues and irreproducibility, and the endeavors to ensure the investigation of human subjects [1].

It is worth noticing that, even as the moral and administrative issues relating to clinical trials are being made more rigid, ICFs (Informed Consent Forms) continue to be deficient or misleading. In any case, the investigation presents a terrible picture, and the matter is worth seeking after further advance. We would moreover like to recommend that trials must be endorsed by the morals committee only when their ICFs contain all the obligatory components stated in Schedule Y [1].

The expert needs to verify the nature of every investigational item in question and its specific needs at some point after the assessment has been conducted. For instance, the regulation of high-risk–profile experimental products, like monoclonal antibodies that are able to activate an ominously helpless reaction, calls for particular conditions. An observational period of 30 min in the closeness of a master is suggested; current drugs utilized to treat ominously vulnerable reactions ought to be available [2].

10.1.2 Importance of GCP in Herbal Medicine Trials

All of the elemental moral standards of human support in investigation apply similarly to homegrown cures and inquires including the compounds. Assent would be obtained, the test plan would be strong, benefits and harms would be optimal and balanced for the prospective member, and the topic determination would be balanced.

Concerns that especially apply to clinical trials with homegrown items incorporate:

- Item corruption (has it been archived?)
- Intuitive interactions between homegrown cures and other substances (seldom caught on)
- Regenerative and organ harmfulness information (may be negligible) (Table 10.1) [3]
- Earlier dosage finding (likely to be fragmented)

In spite of the fact that numerous herbs appear positive in preclinical and small-scale clinical outcomes, numerous exorbitant herbal clinical considerations have created negative outcomes. The quality of trials on homegrown cures is profoundly variable, and numerous trials have been found to be of far-fetched quality. The generally few randomized, double-blind tests that get consideration in standard restorative journals are regularly bashed on methodological grounds or interpretation [3].

Quality control of herbal drugs features a direct effect on their security and viability. Natural and rural practices are vital in collecting homegrown materials. The World Health Organization (WHO) has established a series of detailed guidelines and documents related to the quality control and safety of medicinal plants and herbal products. The WHO released the "Quality Control Strategies for Therapeutic Plant Materials." a set of recommended test procedures to quantify the content, character, and quality of therapeutic plant materials, for directing national research institutions involved in healthcare quality control. A new standard named "WHO rules for assessing quality of home-produced medicines with respect to residues and contaminants" was formed in 2007 following the WHO's publication of the "Guidelines on Good Agricultural and Collection Practices (GACP) for Medicinal Plants" in 2003. The European Union, China and Japan have created territorial and national rules for great agricultural and collection homes for therapeutic plants, which guarantee that soil and water system water utilized for homegrown fabric development and engendering are within the limits or free from destructive heavy metals, pesticides, herbicides, and toxicologically perilous substances [3].

TABLE 10.1
The Organ Systems and Their Safety Parameters

Organ System	Safety Parameter
Neurological	Absence of neurologic symptoms [3]
Skin	Clinical evidence that there are no adverse reactions [3]
Musculoskeletal	Absence of myalgias or arthritis; normal values of CPK [3]
Gastrointestinal	Clinical evidence of tolerance [3]
Liver	Normal values of SGPT or SGOT, total bilirubin, and alkaline phosphatase [3]
Kidney	Normal values of creatinine or BUN [3]
Endocrine metabolism and system: potassium/sodium, calcium	Normal values of total protein or albumin, glucose, uric acid, cholesterol, lipase, or amylase [3]
Cardiovascular	Normal blood pressure and EKG [3]
Hematopoietic	Normal values of whole blood count [3]
Additional	More in-depth examination of any organ system that the substance is likely to affect specifically [3]

Standardization of herbal definition requires execution of Good Manufacturing Practices (GMP). In expansion, thinking about different parameters such as pharmacodynamics, pharmacokinetics, dose, solidness, shelf life, harmfulness assessment, and chemical profiling of the homegrown details is considered basic. Heavy metal contamination. Good Agricultural Practices (GAP) in homegrown medicine standardization are similarly critical. In indigenous/traditional frameworks of medication, the drugs are fundamentally apportioned as water decoctions or ethanol extracts. In this way, the therapeutic plant parts ought to be true and free from hurtful material such as pesticides, heavy metals, microbial contamination, radioactive defilement, etc. [3, 4]. The restorative plant is subjected to a single dissolvable extraction once or over and over, or water decoction, or as portrayed in antiquated writings. The extract ought to then be checked for demonstrated natural movement in an exploratory animal model. The bioactive extract ought to be standardized on the premise of dynamic guidelines or major compound(s) in conjunction with fingerprints [5].

All herbal medicine clinical trial data ought to be recorded, dealt with, and put away in a way that permits its precise announcing, interpretation, and confirmation. The privacy of records that might distinguish subjects ought to be secured, regarding the security and privacy rules in agreement with the pertinent administrative requirement(s) [5].

In this stage, Adverse Drug Reactions (ADRs) announcing happens amid clinical trials. The WHO (2004) energized and prescribed recording and detailing the suspected herbal medications unfavorable occasions happening amid clinical trials by following national GCP guidelines for trials on pharmaceutical products (WHO, 2004). In any case,

> there are awesome contrasts between Part States within the definition and categorization of herbal solutions. A single therapeutic plant may be characterized as a nourishment, a utilitarian nourishment, a dietary supplement or home grown pharmaceutical in completely different nations, depending on the directions applying to nourishments and drugs in each nation. This makes it troublesome to characterize the concept of herbal medications for the purposes of national medicate direction, conjointly befuddles patients and consumers)
>
> **(WHO, 2004)**

which influences the announcing of suspected herbal medications ADRs amid this stage [5].

It is, for the most part, contended that the moral, legitimate, and administrative prerequisites for inquiry and clinical trials with homegrown solutions on human creatures ought not to vary in standard from those set for standard pharmaceutical clinical trials. It is vague, however, to what degree there exists harmonization between the rules of distinctive specialists and whether the rules vary from those of customary pharmaceutical trials. There is, moreover, exceptionally little data on completed clinical trials on plant drugs, separated from the common impression that the general quality of proof for elective pharmaceutical randomized clinical trials is destitute, and indeed those trials done on conventional Chinese medication required pressing enhancement [5].

Research concerning domestically developed restorative things must develop from the use of customary models and/or randomized, double-blind, placebo-controlled trials—the "gold standard" of clinical research—to open, observational trials, and coherent research. Over time, the progression plan for herbal remedial things has got to be more closely balanced with the progression of unused chemical substances. Pharmacovigilance is additionally gradually getting to be a standard device to screen the security of herbal therapeutic items [5].

Herbal medications may interact with prescription medications in a variety of body parts, including the kidney, liver, digestive tract, and other areas of activity. The foremost mutual essential instruments for the detailed herb–drug interactions (HDIs) are pharmacokinetic and/or pharmacodynamic interactions. In particular, cytochrome P450 (CYP450) protein induction and/or inhibition and impacts on transport proteins (P-glycoprotein) by HDIs significantly influence the circulating levels of medications and alter the clinical result [5]. The use of pharmacovigilance programs will provide more information to physicians to support the prudent combination of common medications with homegrown medication, as herbal remedies are essential in the healthcare system for treating infections. It will be too helpful to other healthcare experts and buyers to make a sound choice on the safe combination of herbal items with customary medications [5].

10.2 REGULATORY FRAMEWORK

10.2.1 Global Guidelines

Investigation and assessment of herbal solutions without a long history of use, or which have not already been investigated, ought to follow the WHO's research rules for assessing the security and adequacy of herbal medicines. For herbal medications with a well-documented history of conventional use, the following strategies for conducting research and assessing security and adequacy may be followed. Common in surveying the security and/or adequacy of herbal medication, whether determined from a single plant or from a characterized blend of plants, the primary step includes the assessment of written reports. The writing review ought to incorporate reference books, audit articles, precise observation of essential sources, and/or database searches. It ought to be kept in mind, in any case, that reference books and survey articles might contain wrong data. All things considered, these sources will cite essential references that can be consulted for in-depth examination.

The search profile utilized ought to be recorded, as ought to details of any references cited, whether or not they are accessible. The writing ought to then be amplified to assemble data on closely related plant species for chemotaxonomic relationship. GCP guidelines for herbal drugs, according to the International Council for Harmonisation of Technical Requirements for Pharmaceuticals for Human Use (ICH) and the Ministry of Ayurveda, Yoga and Naturopathy, Unani, Siddha and Homeopathy (AYUSH), are intended to promote ethical practice, safety, and clinical efficacy of herbal drug trials. The guidelines stress quality control, standardization, and correct documentation in the research process. They need complete specifications for herbal preparations and medicines, such as botanical identity, details of collection and processing, and suitable analytical methods for quality determination. The guidelines also place special emphasis on the use of clear and consistent nomenclature, complete patient health records, and proper labeling of herbal medicines. In addition, they emphasize the necessity of taking into account the special features of herbal products, including multidimensional chemical structure and long a history of use, while being strictly scientific in clinical trials. In accordance with these principles, researchers and producers can develop and authenticate herbal medicinal products so that they will align with international regulations and ensure trial participants' safety [6, 7].

In the event that a few agents distribute comparative security and/or viability information, they ought to be acknowledged as valuable markers. *In vitro* (biochemical or cellular) security information ought to be seen as markers of potential harmfulness, but not as absolute markers. *In vivo* information from creature studies are more demonstrative of harmfulness and may be considered to be security markers. For both security and adequacy, a pharmacological impact observed *in vitro* or in creature models isn't fundamentally pertinent to people. *In vitro* information ordinarily serves to confirm the detailed component of activity in creatures or humans. Such information has to be affirmed by clinical studies. Well-documented reports of pharmacological action in creatures or people may be seen as having a logical method of reasoning [1].

10.2.2 Regional Guidelines

10.2.2.1 European Guidelines

In order to harmonize appraisal of herbal restorative items and to encourage the marketing of totally different member states of the European Union, a common law was upheld within the European Union [8, 9]. The law created a basic system of fundamental definitions, provided space for decisions to be set down, and indicated the general structure of usual requirements and norms on home-style healing products throughout the EU. In 2004, the European Solutions Organisation created a Committee on Herbal Medicinal Products (HMPC). Strengthening the Community Monographs is the sole responsibility of the HMPC, which suggests harmonized standards for the security and adequacy of herbal substances and arrangements thereof. With regard to the existing level of proof, monographs tend toward well-established use or, if the proof isn't adequate, a traditional use. The HMPC is a fine demonstration of how logical assessment of home-grown drugs can be harmonized and sets acknowledged and science-based benchmarks to guarantee public well-being [10].

10.2.2.2 Chinese Guidelines

Determining how to distinguish between the modernization and Westernization of Traditional Chinese Medicine (TCM) is a crucial aspect of its study and development. For Westernization of TCM, we follow regulatory necessities at fundamental stages of the strategy for pharmaceutical enhancement, including cure revelation, definition, research facility enhancement, animal studies, clinical advancement, manufacturing plan approval and quality control (QC), regulatory submission, review, and handling, in spite of the essential differences between Western medicine (WM) and TCM [10]. It is suggested that administrative requirements be modified to take into consideration the fundamental differences between TCM and WM in order to modernize TCM. In other words, we still ought to be able to see whether TCM is genuinely working with balanced regulatory necessities, utilizing Western clinical trials as a standard for comparison [10].

10.2.2.3 Japanese Guidelines

GMP has to be followed in all pharmaceutical production. As per the Ministerial Law on GMP, every manufacturer should have a proper system of QC at every step of the process, ranging from verification of raw materials to final dispatch of products. Under Article 20, Clause 1 of the Pharmaceutical Issues Act Authorization Organization and Article 96 of the Necessity Law, GMP cannot ensure production lines where lethal medicines are actually chopped or powdered. To ensure Kampo product quality, self-recommended GMP was implemented in 1987, and for OTC products it was passed on in 1992. This is due to the fact that all products have to meet GMP. The two rules were combined into "GMP for Kampo Products" in 2007, and it was reexamined in 2012 [10].

10.2.2.4 Indian Guidelines

The Indian government established a master committee in 1993 to create guidelines for the safety and viability of herbal medicines to be included under the Drugs and Beauty Care Products Act and regulations. It was proposed that no modern herbal solutions, other than those authorized by the authorizing specialists, be permitted to be fabricated or showcased, except for those said in and made in compliance with the formulas given within the "definitive" books for Ayurveda, Siddha, and Unani herbal solutions [10].

10.3 PRECLINICAL TESTING AND SAFETY ASSESSMENT

10.3.1 Phytochemical Analysis and Standardization

Standardization is characterized as the best specialized application of consensual shrewdness consisting of forms of choice in making fitting choices for confirmation, coupled with steady choices for keeping up gotten benchmarks. This view incorporates the case of "unconstrained standardization forms" to create de facto guidelines. Plant-derived substances have, as of late, ended up incredibly intriguing owing to their flexible applications [10]. Restorative plants are the wealthiest bio-resource of drugs for conventional frameworks of medication, advanced drugs, nutraceuticals, nourishment supplements, people drugs, pharmaceutical intermediates, and chemical substances for manufactured drugs [10]. The WHO energizes, prescribes, and advances traditional/herbal cures in national well-being care programs since these drugs are effortlessly accessible at low cost, secure, and individuals have confidence in them. Pharmaceutical extraction techniques include using certain solvents to separate medicinally active plant tissue packages from the inactive or inert components. Amid extraction, solvents diffuse into the strong plant fabric and solubilize compounds with comparative extremity. Phytopharmaceutical and auxiliary plant items of restorative significance include alkaloids, glycosides, terpenoids, flavonoids, and lignans [10].

At present, no official guidelines are accessible for herbal arrangements. These manufacturers have their own specifications, most of which are highly preparatory in nature, and are currently carrying out some testing for their specifications. Separation of all the fixtures' residues as described in a requirement is extremely hard. Design for identification of nearness of individual fixing can be promoted by trying different chromatographic and spectrophotometric techniques, along with analysis of physicochemical characteristics. This is the first essential task, namely, to improve such parameters by which the nearness of the full fixing can be distinguished. Wherever conceivable, these strategies can be connected for quantitative estimation of bioactive bunches of compounds like alkaloids, flavonoids, polyphenolic components, or estimation of specific compounds [10].

10.3.2 Toxicology Studies and Safety Profiling

Inside the European Union, customary herbal arrangements that are regarded as "acceptably secure, however not having a recognized level of efficacy" fit into an exceptional category of drugs ("traditional herbal medicine products"), for which necessities of nonclinical and clinical considerations are less intensive [11, 12]. A heading recommendation conveyed by the Brazilian National Health Surveillance (Anvisa) characterizes a comparatively calm category ("traditional phytotherapeutic products") for selection purposes [12]. With regard to domestic developed solutions, both workplaces show up to be tolerant with regard to affirmation of practicality, and consider long-standing society utilization as a demonstration of security and a waiver of thorough toxicological checking. Nonetheless, certain herbal ingredients and products that have been used traditionally in isolation are known to be hepatotoxic and/or carcinogenic. Drug-metabolizing proteins have also been found to be inhibited or induced by herbal agents. Engineering and clinical intuition are flawed since herbal remedies are usually consumed with prescription medications. An appraisal of the security of herbal arrangements for enrollment purposes needs to consolidate, at minimum, *in vitro* and *in vivo* genotoxicity measures, long-term rodent carcinogenicity tests (for drugs proposed to be ceaselessly utilized for > 3 months or sporadically for > 6 months), regenerative and developmental toxicity studies (for drugs utilized by women of childbearing age), and examination of the impacts on drug-metabolizing chemicals [12].

Advancement of compelling rules for security, viability, and quality is regarded as a principal prerequisite in arranging to set up the evidence base for Conventional Pharmaceutical. The International Union of Pure and Applied Chemistry (IUPAC) has distributed an arrangement of conventions on QC, security, viability, documentation, and standardization of herbal pharmaceuticals in which different noteworthy angles and highlights of analytical chemistry and phytochemistry have been portrayed. In the event that these methodologies are completely actualized by the IUPAC, the WHO will investigate traditional medicine from its cynical view to present-day pharmaceuticals [12].

10.4 STUDY DESIGN AND PROTOCOL DEVELOPMENT

10.4.1 Defining Hypotheses and Objectives

The major aim of the study is to evaluate the main result, which might be the effectiveness of the herbal medication in treating a certain ailment. For instance, "to evaluate the efficacy in reducing symptoms of osteoarthritis," and secondary objectives are supplementary outcomes of interest that may give supportive information regarding the core aim or investigate other impacts of the herbal medication. For example, "to assess the safety and tolerability of [Herbal Medicine X] over 12 weeks of use" [12].

Hypotheses: The null hypothesis (H_0) says that there is no impact of herbal therapy when compared to a control, while the alternative hypothesis (H_1) states that there is a substantial effect. These hypotheses should be stated clearly [13, 14]. For example:

H_0: When compared to a placebo, [Herbal Medicine X] has no effect on osteoarthritis symptoms.

H_1: When compared to a placebo, [Herbal Medicine X] dramatically lessens osteoarthritis symptoms.

10.4.2 Study Population Selection and Inclusion/Exclusion Standards

Target Population: The research population needs to be representative of the population that would utilize the herbal remedy in everyday situations. These might be individuals seeking preventive care in the general community or patients dealing with specific diseases or conditions [15].

Inclusion Criteria: These criteria specify which individuals are eligible to take part in the study. They should be precise and determined by factors such as—Age range: for example, 18–65 years old; Gender: certain studies may focus on a particular gender if the herbal medicine is intended for gender-specific conditions; Health status: specific medical conditions, disease severity, or initial symptom scores; Diagnostic criteria: confirmation of the condition using particular diagnostic tests or clinical assessments; and Previous treatment history: whether patients have not received treatment before or if they have not responded to previous therapies [16].

Exclusion Criteria: These standards are applied to identify participants who should not be included in order to avoid factors that might lead to misunderstanding or needless hazards [17], such as the following: Allergies or Sensitivities: People with documented allergies to the herbal remedy or any of its ingredients; Co-morbidities: Conditions that coexist and may have an impact on the study's findings; Current Medication: Taking medications that might interfere with the herbal medicine; and Lactation/Pregnancy: Due to possible hazards, women who are nursing or pregnant may not be allowed.

10.4.3 Determining Dosage and Administration Routes

Dose Determination: The appropriate amount of herbal medicine to use should be determined by preclinical studies, historical use, or previous clinical trials. If possible, a dose-response study can be useful in finding the best dosage that provides a good balance between effectiveness and safety [18, 19]. Starting Dose: Typically determined based on animal studies or historical usage of the herb. Dose Escalation: If the aim of the study is to find the highest tolerated dose, a dose-escalation method can be employed.

Administration Route: The administration route for an herbal medication should be based on how it is most commonly used or effective. Typical paths consist of—Oral: Tinctures, teas, pills, or capsules; Topical: Lotions, ointments, and creams have localized effects; Inhalation: volatile preparations such as essential oils; and Dose Schedule: Pharmacokinetic data or conventional usage should inform the frequency and duration of dose. Take a daily dosage for eight weeks, for instance [20, 21].

10.4.4 Control Groups and Placebos in Herbal Trials

Control groups are necessary for comparing the effects of the herbal medication to placebo, no therapy, or an active comparator [17]. Examples of control group types are—Placebo Control: The most prevalent kind in studies using herbs. To preserve blinding, the placebo should taste, smell, and look exactly like the herbal remedy [22]. Active Comparator: If the ailment has a conventional treatment, the effectiveness and safety of the herbal remedy can be contrasted with the current course of care using an active comparator group. No Treatment Control: This is occasionally employed when using a placebo would be immoral or unfeasible. Blinding: Double-blind trials are the best way to minimize bias because neither the volunteers nor the researchers are aware of who is given the herbal medication or a placebo [22]. Placebo Design: Creating a placebo that closely resembles the herbal medication is difficult yet necessary. This involves coordinating the hue, flavor, texture, and packaging [22]. To ensure that the groups are comparable at the start of the study, members ought to be arbitrarily designated to either the treatment or control bunch. Depending on the nature of the research, randomization might be block-based, stratified, or straightforward. Outcome Measures: The main and secondary outcome measures must be predetermined and pertinent to goals of the research. They might consist of the following: Clinical Endpoints: Such as improved quality of life, decreased symptoms, or outcomes of clinical testing [22]. Biomarkers: Biological signs of the underlying disease or the way a therapy is working. Adverse Events: Keeping an eye out for toxicities or adverse effects is essential, particularly in herbal studies where the safety profile might not be entirely established [22]. The study design and protocol development have been illustrated in Figure 10.1.

FIGURE 10.1 Study design and protocol development under GCP.

10.5 ETHICAL CONSIDERATIONS

10.5.1 Informed Consent Process

The informed consent process is one of the most crucial ethical considerations in clinical trials including herbal remedies—clearly explaining to potential participants the objectives, methods, risks, benefits, options, and voluntary nature of participation in the study [23]; ensuring that participants understand the information and are capable of making an informed choice about participating. Participants or their authorized representatives must give written consent. Participants are free to leave the study at any moment without incurring any fees. Consent may be reevaluated if new information arises that could impact a participant's decision to continue [24].

The following are obstacles to achieving informed consent in studies including herbal medicine: The intellectual property rights of herbal medicines are not adequately protected by regulations or by other procedures. Certain herbalists lack the necessary abilities to complete the informed consent procedure. Cultural issues frequently entail shared responsibility within families or communities when making decisions about illness care. Complying with informed consent guidelines contributes to the advancement of scientific understanding while upholding participant autonomy in herbal medicine trials. The secret to negotiating the ethical challenges of this profession is constant communication among communities, ethics oversight organizations, and researchers [25, 26].

10.5.2 Respect for Cultural Beliefs and Practices

It is important for researchers conducting clinical trials on herbal medicines to exercise cultural sensitivity and respect for the beliefs and practices of the study population. Participating in the community—To ensure that the study is acceptable and appropriate for the community; involve the community in its design and conduct [26]. Work with elders, traditional healers, and community leaders to understand cultural norms, beliefs, and practices around health and medicine. Modifying Research Protocols: Study personnel should be taught in cultural competency and humility to work effectively with varied communities. Research protocols should be adapted to match cultural practices wherever possible, for example, by using traditional healing rituals or using culturally relevant terminology [27]. Translation of Materials: Translate research instructions, ICFs, and other papers into readily understood languages for participants. Use language consistent with cultural norms; avoid using phrases that might be insulting or unclear [28]. Respecting Intellectual Property: Be mindful of the rights to intellectual property held by those who possess traditional knowledge, particularly in relation to herbal remedies [27]. If traditional knowledge is used in the research, take part in equitable benefit-sharing arrangements. Handling Divergent Worldviews: Have an open discussion to comprehend and accept various viewpoints on what makes for legitimate information. Recognize that disparate cultural worldviews might result in misconceptions regarding research methods and supporting data [28]. These guidelines make it possible to conduct research on herbal medicine in a way that respects regional customs and beliefs, improves scientific understanding, and is sensitive to cultural differences. It is essential that communities, scholars, and those with traditional knowledge work together continuously and share information [29, 30].

10.5.3 Confidentiality and Privacy of Participants

Ensuring the privacy and confidentiality of clinical trial participants, particularly those utilizing herbal medicine, is contingent upon adherence to GCP rules. These regulations are based on moral precepts that put trial subjects' safety, rights, and welfare first [31]. (a) Every participant must provide their free and informed permission before beginning a clinical trial, according to GCP regulations. Ensuring that participants are completely informed of how their data will be used is essential to protecting their privacy during the consent process [32], (b) and it stresses how important it is to keep documents that might identify trial participants private. Respecting privacy guidelines in line with relevant legal obligations is part of this. To guarantee that personal data are handled safely and are only accessible by authorized individuals, researchers must have policies in place [33]. (c) Clinical trials have to be carried out according to procedures approved in advance by an IEC or IRB. The purpose of this monitoring is to guarantee that participants' rights and privacy are upheld during the trial [34]. (d) Although anonymizing data is a frequent practice to preserve participant privacy, it is not always possible, particularly in pragmatic clinical trials where the study may require identifying information. Consequently, the necessity for data and the preservation of personal privacy must be balanced [35]. (e) The Declaration of Helsinki and other international ethical norms are in accordance with GCP principles, which further uphold clinical research's dedication to participant anonymity and ethical care [36, 37].

Consent forms are often long and confusing to participants, which might make it harder for them to decide whether to participate. In light of recent data breaches in the healthcare industry, researchers must implement clear procedures that foster participant confidence in order to allay public worries about privacy. It is crucial to take into account the cultural background and the possible impact of traditional beliefs on participants' expectations about privacy and consent while conducting studies utilizing herbal medications.

10.5.4 Role of Ethics Committees and IRBs

IRBs and ethics committees are crucial for monitoring investigations involving human beings and maintaining ethical standards across the whole study process. Although they serve comparable purposes, their objectives and scope differ [38].

10.5.4.1 IRBs

An administrative entity called an IRB was created to safeguard the welfare and rights of research subjects. The following are an IRB's main duties: reviewing research proposals. IRBs assess research procedures to make sure they adhere to legal and ethical requirements. This entails evaluating the study's overall scientific worth as well as the possible hazards and benefits to participants and the effectiveness of informed consent procedures [38, 39]. Approval Authority: Research studies might be approved, disapproved, or require revisions by an IRB. This guarantees that no human subject study may begin without prior permission from the IRB. Ongoing Monitoring: In addition to overseeing authorized projects continuously, IRBs make sure that the research process is conducted ethically by following established rules. Periodic updates from researchers could be necessary for them to keep an eye on compliance. Composition: A varied range of individuals, including scientists, non-scientists, and community representatives, often comprise an IRB in order to provide a comprehensive viewpoint on the ethical consequences of the research [40].

10.5.4.2 Ethics Committees

A wider range of situations, such as healthcare facilities and research settings, are served by ethics committees (ECs). Their principal roles consist of the following: ECs evaluate research proposals and clinical practices to make sure they adhere to ethical guidelines, taking into account both the rights of individuals and the well-being of the community. They examine potential risks to not only participants but also to the larger community and environment [41]. ECs frequently offer advice on moral conundrums that researchers and medical professionals encounter. They assist in creating institutional policies linked to ethics, provide instructional programs, and foster conversations about ethical concerns [42, 43]. ECs are not able to compel compliance, even though they are able to suggest measures. They provide an advising role, guiding stakeholders through difficult moral dilemmas while upholding the decision-makers' independence. ECs, like IRBs, are made up of individuals with diverse backgrounds and areas of expertise, including both science and non-science. This variety aids in pointing out moral issues that researchers would miss on their own [44].

In conclusion, ECs and IRBs are critical to upholding moral norms in medicine and research. IRBs oversee the protection of research subjects and make sure that regulations are followed; ECs, on the other hand, oversee, advise, and educate on ethical matters in a wider range of institutional contexts. Collectively, they support the safeguarding of patients' rights within the healthcare system and the integrity of research.

10.6 STANDARDIZATION AND QUALITY CONTROL OF HERBAL PRODUCTS

10.6.1 Sourcing and Identification of Herbal Materials

About 75% to 85% of the crude material utilized in the botanical trade is obtained from the wild. Some companies are now trying to get at least some of their crude material from sustainable and ethical sources due to increasing demand for high-quality botanical products, while others contribute little to this part of their business. Most businesses are remote and unaware of the conditions under which raw materials are gathered, and typical business practices encourage poor treatment of species that offer little value to cultivators and collectors. In any case, there too exist openings to make change in this segment [42].

Contaminants in herbal solutions are classified into physicochemical contaminants and organic contaminants. An assortment of agrochemical operators and a few natural solvents may be present as buildups in homegrown drugs. Contamination should be avoided and controlled through quality affirmation measures such as GACP for therapeutic plants and GMP for herbal drugs. Chemical and microbiological contaminants can result from the use of human excreta, animal fertilizers, and sewage as fertilizers. As is stated in the WHO rules on GACP for therapeutic plants, it is not permitted to use human waste as fertilizer, and creature fertilizers ought to be altogether composted. Poisonous components and other chemical contaminants, including solvents from items planned for use in households and industrial chemicals, can be concentrated in composted sewage. Hence, care ought to also be exercised with sewage administration in agrarian regions [43].

Botanical texture verification can be an important initial step in the utilization of such materials, both in investigation as well as in commercial goals. Functional and true suppositions in terms of identification of locally-made fixes can easily be drawn from only a very small amount of analysis and comparison of valid and invalid experimentation of intact, cut, or ground plant texture. Microscopy is a valuable method for the identification of both botanical and non-botanical adulterants, such as inorganic minerals, pharmaceuticals, and microbiological contamination. The reliability and ability of microscopy as a method of botanical confirmation have improved because of advances in amplification device development, light and fluorescence alterations, independent ordering, and electron magnifying focal point examination. Organoleptic examination, utilized in combination with advanced microscopy equipment, gives enhanced precision for botanical confirmation [44].

Concerns over the inconsistent makeup of herbal remedies and sporadic incidents of intoxication by adulterants and/or hazardous ingredients have persisted for a long time. Herbal solution QC is intended to ensure the solutions' viability, security, and consistency. It has been demonstrated that chemical fingerprinting is a useful technique for homegrown medicine QC. An intriguing design that illustrates the proximity of multiple chemical markers inside a test may be a chemical unique fingerprint [45].

Herbal-based healthcare has increasingly become fashionable globally over the last two decades. This has produced a boom in the illicit drug industry. For the herbal industry to prosper, appropriate and speedy internationally recognized certification of the plant or plants is essential. The traditional approach is to hire a skilled taxonomist who uses traditional and contemporary methods to determine the correct identification of plants. The process, however, becomes more laborious and cumbersome for large-scale bulk identification verification. Instead, DNA barcoding offers a rapid and reliable kit that guarantees efficient and dependable species identification. Standardized tag–dependent DNA standardized tags are based on standardized tag loci with adequate capacity to carry enough information necessary for the discrimination of closely related plant species, as well as identification of sophisticated confusing species. Some genuine convincing DNA barcodes that were once adopted to be applied to plant recognized verification have been produced locally as follows: matK, rbcL, trnH-psbA, ITS, trnL-F, 5S-rRNA, and 18S-rRNA. Creating advances in DNA barcoding coupled with next-generation sequencing and high-resolution dissolving twist examination have cleared the way for viable species-level assurance recovered from processed domestically developed products [46, 47]. The multilocus methodology's progress and use have opened up so far untapped possibilities for the DNA barcode–based plant recognition proof for the herbal medicine sector. DNA barcoding should function in harmony with other elements of the framework's scientific approach for reliable and effective identification of the drug's natural constituents and thorough QC. We suggest that, in addition to need-based transcriptomics and proteomics, DNA barcoding should be utilized in conjunction with metabolomics to effectively resolve confirmation issues pertaining to the herbal market [47].

10.6.2 Preparation and Standardization of Herbal Extracts

Depending on the nature of the plant, the mode of application, and the intention, herbal medicines are generally made in a number of ways using adjunct ingredients.

The art of preparation of herbal medicines depends, in a sense, upon the processing of plants in order to release their full medicinal impact. Herbs are used outdoors and also indoors, in showing disdain for the reality that, more often than not, they are more practical, with repairing properties utilized indoors. Implantations (hot teas), decoctions (bubbled teas), tinctures (alcohol–water extracts), macerations (cold-soaking), saturation, absorption, internal inhalation of powdered plants (like snuff), steam inhalation, and, in fact, fragrance-based treatment and dry courses of action are the foremost internal courses of action, while washes, compresses, poultices, salves, and emollients are the foremost external courses of action [48]. The growth of a medicinal plant is inseparably associated with certain chemical groups, such as essential oils, alkaloids, acids, steroids, tannins, saponins, and glycosides. Since the majority of these chemicals are soluble in hot or cold solvents such as water, alcohol, or other general solvents, each of these categories of chemicals can have an optimum practical extraction technique that allows the chemicals to be extracted from the plant and incorporated into the desired herbal medicine. For example, blends and decoctions are favored for water-dissolvable chemicals (e.g., anti-inflammatory plant steroids to treat joint pain), while tinctures are favored for alcohol-dissolvable chemicals (anti-bacterial alkaloids) [47].

The purity, concentration, and therapeutic action of plant extracts taken directly from crude plant tissue are variable. By the inclusion of excipients or by mixing herbal drugs or herbal steady courses of action, standardization is the process of transforming the locally produced cure arrangement into a specified substance of a constituent or an assembly of substances with known supportive development. Well-established substances in stable concentrations of high-quality extracts are referred to as standardized extracts, and they are subject to rigorous quality standards throughout the whole process of manufacturing, sourcing, and designing dosage forms. There is no regulatory definition of dietary supplement standardization to date. Thus, the term "standardization" can be applied to describe a variety of tried-and-true activities of botanicals on the body. There are two forms of standardization. (i) Sincere standardization means that a particular phytochemical or combination of constituents is known to possess activity; e.g., products prepared from ginkgo contain 26% flavones and 6% terpenes. Such compounds, not exclusive to the entire herb, are highly concentrated and are usually considered to be phytopharmaceuticals. They are usually infinitely more effective than the whole herb. (ii) This sort of standardization is based on makers guaranteeing the presence of a certain rate of marker compounds; these are not pointers of therapeutic development or quality of the herb [47]. The standardization of herbal drugs is illustrated in Figure 10.2.

10.6.3 QC Measures

QC of the quantity and safety of herbal drugs and cosmetics is important. The state of a substance that is selected due to its character, ethics, composition, or other physicochemical or normal characteristics, or by the production types, can be described as quality. QC can be applied to denote the aspects that consist of maintaining the authenticity and quality of a manufactured good. All root preparations of plants, solutions, or cosmetics are required to fulfill the basic criteria of utility and safety, which are ensured through practical, experience-based testing. Three basic pharmacopoeial definitions upon which QC is founded are (a) character (botanical character); (b) immaculateness (free from contaminants); and (c) substance or degree (quantity of active ingredients in optimal percentage). After all, the following are suggested first steps to be undertaken for an

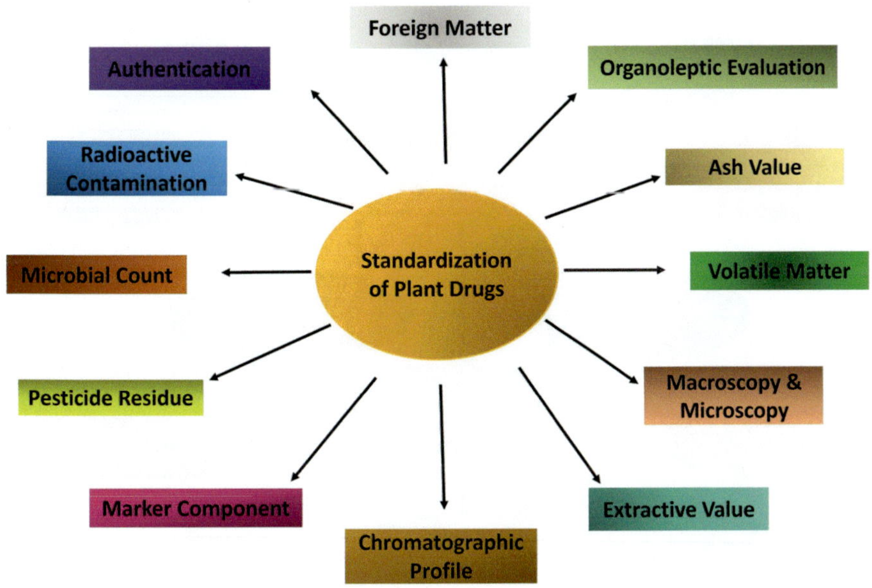

FIGURE 10.2 Standardization of herbal drugs.

age of high-quality locally produced cure: (i) proper botanical recognized identification; (ii) phytochemical screening; and (iii) standardization. Requirements such as source, type of action, physical constants, debasement, impurities, dampness, cinder material, and soluble deposits have to be analyzed in order to determine identification and purity. Most herbal drugs have indefinite active constituents, which are hard to test in the material. Be that as it may, markers can be utilized for control purposes, but the marker may be independent of any helpful development [47].

10.7 DATA MANAGEMENT AND RECORDKEEPING

Fastidious recording of side impacts happening in patients and cautious recordkeeping of all information collected are trademarks of a great study. Furthermore, an adequate number of patients taking a specific herbal cure ought to be watched, and it is essential to prepare eyewitnesses to look for and record side impacts [47].

In case herbal medication is to be really intriguing, i.e., an association, at that point the trade or sharing of information between diverse specialists would appear imperative to the process. All individuals ought to know what the other individuals are considering, what they have found, what conclusions they have come to, and what they are suggesting toward the upkeep of a patient's health—not just at the time of a referral but on a continuing care premise as a patient's well-being and illnesses advance. The need to share information goes past the immediate interdisciplinary group. In the future, other common or pro-ordinary specialists are likely to be included in the patient's care. A patient may move and alter their common specialist and complementary and alternative medicine (CAM) advisers. Or, other professionals may end up included in the patient's care as their condition advances and changes. The concept of an intrigue group must allow for professionals who are not individuals, and the concept of coordinated healthcare must permit treatments that have not yet been connected and clinical settings that do not yet exist. Moreover, the sharing of information needs to be future-focused as well as centered on the prompt intelligence among current professionals. The coordinated exchange of information between clinical and electronic health record (EHR) frameworks is risky, since distinct EHRs used by different professionals are regularly organized in an unexpected way, and reformatting the information from one framework to coordinate the arrangement of another framework is complex and costly [47].

10.8 STATISTICAL ANALYSIS AND INTERPRETATION

There are incalculable scientific publications on homegrown drugs, but tragically many of them do not accurately report their pharmacological, biological, and chemical perspectives, including the constituents and stability of the herbal/extract preparations; thus, their security, adequacy, and consistency may not be demonstrated. For creating a present-day medication from homegrown drug(s), total pharmacological and chemical characterizations of their bioactive metabolites have to be well organized. Reproducibility requires the improvement, evaluation, and standardization of the chemical, organic and pharmacological strategies based on the current state of the art [20].

Another imperative zone of medicare improvement is to consider the basic chemical databases (Figure 10.3) with information bases on target qualities and proteins and make modern chemical substances through computational atomic modeling for pharmacological assessment [47].

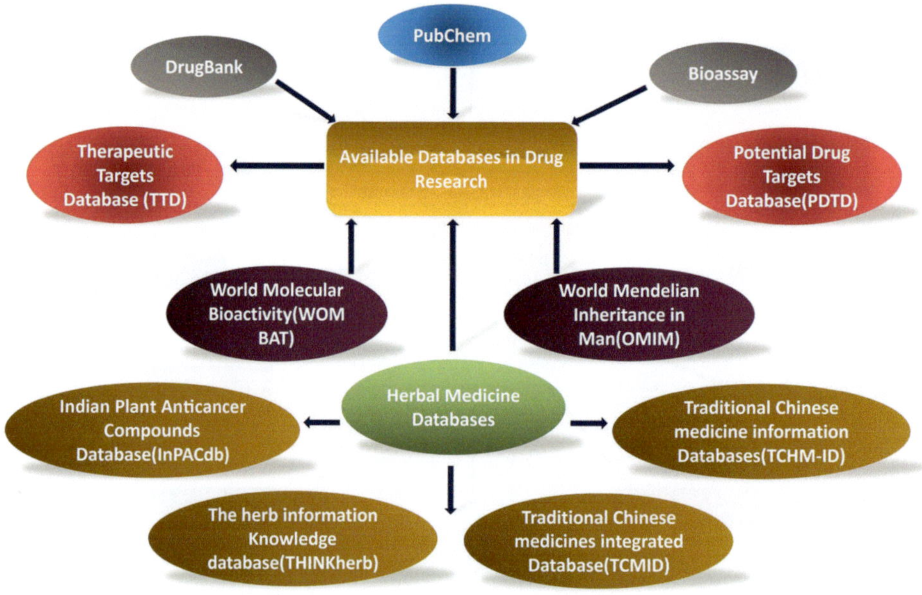

FIGURE 10.3 Available drug databases and herbal drugs databases for research.

One of the strategies that can be used for keeping the quality of pharmaceutical drugs compounded locally at par all the time is target-driven overall process optimization, or TOPO. The four building blocks that constitute the TOPO methodology are target definition, information pretreatment, preparation modeling, and overall process optimization. The Bayesian approach is coordinated into the optimization step [46]. As part of the TOPO tool, several units of the generating framework are optimized in sequence to generate optimal working conditions that are compatible with the quality target. Graphical estimates of the Bayesian back-predictive dispersion and the prior target attainment were employed to monitor the effects of TOPO. The generation process was screened and optimized by balancing the likelihood direction. A seven-step preparation for production that was utilized to produce *Lonicerae japonicae* extract was successfully linked with the proposed TOPO method. It is illustrated that TOPO may keep the generation process in line with the predefined target and diminish the variability of the ultimate items [47].

The Consolidated Standards of Reporting Trials (CONSORT) explanation was first distributed in 1996 and reexamined in 2001. This articulation comprises a 22-item checklist and flow chart to direct creators, peer commentators, editors, and perusers on the basic data required in reports of two-group parallel randomized clinical trials (RCTs). The CONSORT explanation is embraced by leading therapeutic diaries, publication groups, professional social orders, and funding bodies. Since its initiation, several expansions and articulations of the Associate for specific settings have been constructed. The CONSORT checklist application to RCTs of herbal medicinal products is illustrated in this study. Proper documentation of all usage features should be detailed in controlled trial reports of herbal medicines, investigation, outcomes, and translation, as prescribed within the Partner articulation. A few things recommend that reports of CAM RCTs insufficiently portray vital perspectives of their strategy [47].

10.9 CHALLENGES AND CONSIDERATIONS IN HERBAL MEDICINE TRIALS

Two fundamental questions of ethics, which present unique challenges and are peculiar to clinical research investigating raw extracts or sections from herbal medicine, are rational justification and opportunity appraisal. There are impressive challenges for both clinical examiners and morals committee individuals in managing such issues; numerous of them remain uncertain, resulting in a wide variety in moral necessities, legitimization, and choices. In spite of a surprising surge in herbal medication inquiry universally, a number of clinical examiners or indeed morals committee individuals have constrained certainty in managing related moral issues [47].

New studies may be required to further improve the evidence-based herbal therapy. For instance, certain agents might ponder a three-arm study whereby research compares the clinical appropriateness of a standard course of herbal medicine, an individualized plan of herbal medicine, and a dynamic placebo control. Formulating a control, especially a placebo, that is not different from the subject herbal drug is another pragmatic barrier to a randomized controlled solution. Herbal arrangements, especially within the frame of decoction, may have an impossible-to-miss color, particular scent, and one-of-a-kind taste, resulting in a significant challenge for selecting and fabricating a coordinating control. Besides, a legitimization for the use of a placebo, in the event that required, in herbal medicine trials may need extra contemplations with respect to social recognition and community. The use of a fake treatment may be seen as cheating in a few communities, so it may not be ethically satisfactory in certain settings [47].

The daily print scope of clinical trials for herbal remedies was more negative than that of pharmaceutical studies, demonstrating contempt for the generally favorable outcomes and tone of the clinical trials. Typically, this is frequently inverse to most conveyed research on media scope of CAM [48]. They take into account a far wider range of medicines, and the media substance is by and large recounted instead of evidence based. Without a doubt, writers are showing a degree of skepticism uncommon for therapeutic reporting. It is conceivable that once stood up to with genuine proof, writers are more critical or uncertain.

Both shared assumptions and CAM-related nuances can be considered more newsworthy than contrary to the trend of positive proclaiming when faced with demonstration. Journalists using press releases from high-impact, peer-reviewed journals might have covert biases toward mainstream medicine and conventional methods. Additionally, scholars turn to trusted sources inside the biomedical community for comments on clinical trials, both herbal and pharmaceutical, conceivably leading to a biomedical inclination in enumerating trial outcomes [47, 49].

With the time of fast-developing science and technology, a person can receive an inclination toward avoiding the old knowledge and virtues, as well as traditional medications as a whole. A person should keep a watch over the ancient information in pursuing herbal medicines, even though the "post-genomic" age has great opportunities for screening medicinal plant bioactive molecules. The majority of the medicinal properties of plant species were found through the accumulation of information from their wide use in traditional herbal medicine. Thus, traditional medicine knowledge is important to enhancing the rate of success of medicinal discovery from natural medicines. Generally, the success rate of the manufactured course for developing new medicinal specialists may be 1/10,000; however, the success rate with regard to modern restorative moieties based on medicinal plants utilized in conventional therapeutic frameworks can be as high as 1/4 or more [46].

10.10 CASE STUDIES AND EXAMPLES

10.10.1 *Ginkgo biloba*

The anticipation of crippling inveterate infections in elderly people may be a major clinical and public health goal. Dementia, particularly Alzheimer disease, is a predominant chronic disease currently influencing more than 5 million individuals within the United States and is a driving cause of age-related disability and long-term care placement. The herbal item *G. biloba* is endorsed in a few regions of the world for the conservation of memory; however, there are no medicines endorsed for the primary prevention of dementia. Within the United States, worldwide sales of *G. biloba* surpass $249 million annually.

Oxidative stress may quicken the cascade of AD neurotic changes that can lead to dementia and cerebrovascular illness. A major component by which G. biloba is proposed to exert its impact is by the activity of different antioxidants. More recently, an *in vitro* consideration demonstrated that ginkgo extract has an anti-amyloid accumulation impact, proposing another mechanism whereby G biloba may be useful in dementia avoidance [47].

10.10.2 Echinacea

A few hundred scientific considerations on echinacea, including a dozen randomized trials that tried echinacea for anticipating or treating the common cold, had already been distributed by the mid-1990s, when echinacea had become prevalent within the United States. About all of these early trials detailed either statistically significant advantage or patterns toward advantage. In any case, all were manufacturer-sponsored and of fair to poor quality. In this setting, we found it fundamental to conduct our own trial from 1999 to 2000, which yielded negative effects. A few modern trials have been distributed since that point, some with positive outcomes and some with negative outcomes. Orderly surveys and meta-analyses have differed in their consideration criteria, audit strategies, outcomes, and interpretation. We outlined and conducted this trial since the viability of echinacea was still hazy [49].

10.10.3 Curcumin (Turmeric)

Logical considerations uncover that curcumin could be a powerful operator in overseeing and anticipating torment and fiery disorders, which are connected to its antioxidant, anti-inflammatory, pain-relieving, antimicrobial, and neuroprotective impacts. Various records highlight the golden properties of turmeric and curcumin with completely different obvious definitions in treating and avoiding fiery and difficult disorders, including provocative bowel infections, osteoarthritis, skin, and verbal aggravation, as well as neuropathic torment. Extra high-quality inquiry is required in this field to authoritatively affirm the security and viability of herbal items. Because of lacking clinical trials, additional high-quality studies are required to firmly set up the clinical adequacy of this plant. In addition, pulling in researchers' consideration to herbal definitions and examinations in this region, as well as moderate ventures to bolster clinical trials, will be valuable to reaching the conclusive objective of analysts [50].

10.10.4 *Ashwagandha*

A few studies on the adaptogenic effects of *ashwagandha* have been conducted in the last few years. Mice were utilized to assess the anti-stress effects of *Withania somnifera* alcoholic root extract by measuring how well they could swim in water. It seemed that, during stress, *W. somnifera* triggered a system of generalized increased resistance. In a 60-day therapeutic experiment, Chandrasekhar et al. investigated the efficacy and safety of full-spectrum, high-concentration ashwagandha root extract in reducing stress and anxiety in 64 volunteers. Additionally, the study showed a significant decrease in blood cortisol levels, and all stretch evaluation scale values observed in comparison to the fictitious treatment group were also observed to be overweight and pushed. According to a different investigation, pre-treating rats with a fluid suspension of extracted ashwagandha roots was effective in lowering the levels of cortisol and ascorbic acid in animals that were forced to swim. Pre-treating mice with *Withania somnifera* root extract improved their swimming time, according to another researcher who assessed the impact of ethanolic *Withania somnifera* root extract against powerful stress induction in mice. Encouragingly, the whole ashwagandha treatment restored the alterations in blood glucose, lipid, and plasma cortisol levels brought on by stress. Comparative effects of ashwagandha root extract were also observed in adults who were overweight and under pressure [48].

10.11 CONCLUSION

In this situation, where traditional medication and information are on the verge of being terminated, there is a need to act firmly and to thoroughly ensure and protect our legacy. It is presently laid on the shoulders of the administrative bodies to screen controlled and quality streams of herbal items and to encourage their advancement to clinical trial stages.

REFERENCES

[1] World Health Organization. (2005). Handbook for good clinical research practice (GCP): Guidance for implementation.

[2] Méthot, J., Brisson, D., & Gaudet, D. (2012). On-site management of investigational products and drug delivery systems in conformity with Good Clinical Practices (GCPs). Clinical Trials, April; 9(2), 265–271.

[3] Gennari, A., Van Den Berghe, C., Casati, S., Castell, J., Clemedson, C., Coecke, S., . . . & Gribaldo, L. (2004). Strategies to replace *in vivo* acute systemic toxicity testing: The report and recommendations of ECVAM workshop 50. Alternatives to Laboratory Animals, 32(4), 437–459.

[4] Nandi, A., Mandal, N., Das, A., & Dey, Y. N. (2024). Approaches based on network pharmacology and molecular docking to predict the molecular mechanism of plumbago zeylanica's anti-inflammatory action. Journal of Biological Regulators and Homeostatic Agents, 38(3), 2055–2067.

[5] Guideline, I. H. T. (1996). Guideline for good clinical practice E6 (R1). ICH Harmon Tripart Guidel, 4.

[6] Patwardhan, B. (2011). Ayurveda GCP guidelines: Need for freedom from RCT ascendancy in favor of whole system approach. Journal of Ayurveda and Integrative Medicine, 2(1), 1.

[7] Nandi, A., & Das, A. (2021). Recent advances in natural product-based NF-κ B inhibitors as anticancer and antiinflammatory agents. Rheumatology Current. Research, 11(9), 19–21.

[8] Knöss, W. (2015). Harmonization of regulatory requirements in Europe to ensure quality, safety and efficacy of herbal medicinal products. In Evidence-based validation of herbal medicine, January 1 (pp. 201–216). Elsevier.

[9] Nandi, A., Das, A., Dey, Y. N., & Roy, K. K. (2023). The abundant phytocannabinoids in rheumatoid arthritis: Therapeutic targets and molecular processes identified using integrated bioinformatics and network pharmacology. Life, March 5; 13(3), 700.

[10] Carvalho, A. C., Ramalho, L. S., Marques, R. F., & Perfeito, J. P. (2014). Regulation of herbal medicines in Brazil. Journal of Ethnopharmacology, 158 Pt B, 503–506.

[11] Nandi, A., Chattaraj, B., Das, A., Prasad, R., & Dey, Y. N. (2024). Inhibiting brushite crystal growth: Molecular docking exploration of Enhydra fluctuans phytoconstituents and their interaction with human serum albumin. Journal of Biomolecular Structure and Dynamics, 1–10.

[12] Guo, R., Canter, P. H., & Ernst, E. (2007). A systematic review of randomised clinical trials of individualised herbal medicine in any indication. Postgraduate Medical Journal, 83(984), 633–637.

[13] Salm, S., Rutz, J., van den Akker, M., Blaheta, R. A., & Bachmeier, B. E. (2023). Current state of research on the clinical benefits of herbal medicines for non-life-threatening ailments. Frontiers in Pharmacology, 14, 1234701.

[14] Jash, R., Maji, H. S., Chowdhury, A., Maparu, K., Seksaria, S., Gupta, P., . . . & Prasanth, D. N. K. (2025). Oxymatrine ameliorates epithelial mesenchymal transition in IgA nephropathy induced rats. Tissue and Cell, 93, 102671.

[15] Parmar, M. P., Das, A., Vala, D. P., Bhalodiya, S. S., Patel, C. D., Balachandran, S., . . . & Patel, H. M. (2025). QSAR, antimicrobial, and antiproliferative study of (R/S)-2-thioxo-3, 4-dihydropyrimidine-5-carboxanilides. ACS Omega.

[16] Emanuel, E. J., Wendler, D., & Grady, C. (2000). What makes clinical research ethical? Jama, 283(20), 2701–2711.

[17] Das, A., Nandi, A., Kumari, V., & Alvala, M. (2023). FBDD & De Novo drug design. Applied Computer-Aided Drug Design: Models and Methods, 159.

[18] Sorkin, B. C., Kuszak, A. J., Bloss, G., Fukagawa, N. K., Hoffman, F. A., Jafari, M., . . . & Pauli, G. F. (2020). Improving natural product research translation: From source to clinical trial. FASEB Journal: Official Publication of the Federation of American Societies for Experimental Biology, 34(1), 41.

[19] Tilburt, J. C. (2010). The role of worldviews in health disparities education. Journal of General Internal Medicine, 25 Suppl 2(Suppl 2), S178–S181.

[20] Das, A., Nandi, A., Kumari, V., & Chander, S. (2024). Immunoinformatics in drug designing. In Computational biology in drug discovery and repurposing (pp. 127–148). Apple Academic Press.

[21] Costa-Font, J., & Sato, A. (2024). Cultural persistence and the 'herbal medicine paradox': Evidence from European data. Journal of Health Psychology, 0(0).

[22] Narayana, R. R., Gowda, K. P., Ahamed, S. M., & Badami, S. (2010). ICH guidelines with special emphasis on good clinical practice guidelines (GCP). Research journal of Pharmacology and Pharmacodynamics, 2(1), 27–32.

[23] Letsiou, S., Madesis, P., Vasdekis, E., Montemurro, C., Grigoriou, M. E., Skavdis, G., Moussis, V., Koutelidakis, A. E., & Tzakos, A. G. (2024). DNA barcoding as a plant identification method. Applied Sciences, 14(4), 1415.

[24] Kumari, V., Nandi, A., Das, A., & Chander, S. (2024). Bioinformatics for determining the active site of a target proteins. In Computational biology in drug discovery and repurposing (pp. 43–59). Apple Academic Press.

[25] Grady, C. (2015). Institutional review boards: Purpose and challenges. Chest, 148(5), 1148–1155.

[26] Bhattacharya, T., Tiwari, S., Nandi, A., Lin, W. Y., & Arumugam, S. (2024). Development of microRNA as prognostic markers in head and neck cancer. In Diagnostic, prognostic and therapeutic role of MicroRNAs in head and neck cancer (pp. 193–205). Academic Press.

[27] Kurihara, C., Kerpel-Fronius, S., Becker, S., Chan, A., Nagaty, Y., Naseem, S., Schenk, J., Matsuyama, K., & Baroutsou, V. (2024). Declaration of Helsinki: Ethical norm in pursuit of common global goals. Frontiers in Medicine, 11, 1360653.

[28] Mahdi, Y. K., Mogana, R., Kumar, D., Nandi, A., Das, A., Atia, Y. A., . . . & Dey, Y. N. (2023). *In silico* molecular docking and scrutinizing druglike properties of selected phytoconstituents against TAK1, Xanthin Oxidase, and IL-1 β targets in gouty arthritis. Current Trends in Biotechnology and Pharmacy, 17(4), 1496–1505.

[29] Moher, D., Schulz, K. F., Altman, D. G., & CONSORT. (2001). The CONSORT statement: Revised recommendations for improving the quality of reports of parallel group randomized trials. BMC Medical Research Methodology, 1, 2.

[30] Nandi, A., Das, A., Lin, W. Y., & Chakraborty, P. (2023). Antiviral drugs in past and current pandemics. In Current trends in the identification and development of antimicrobial agents (pp. 167–194). Bentham Science Publishers.

[31] Tsfati, Y. (2003). Media skepticism and climate of opinion perception. International Journal of Public Opinion Research, 15(1), 65–82.

[32] Parveen, D., Das, A., Amin, S., Alam, M. M., Akhter, M., Ahmed Khan, M., Ali, R., Anwer, T., Sheikh, K. A., Azam, F., & Shaquiquzzaman, M. (2024). Effectiveness of estrogen and its derivatives over dexamethasone in the treatment of COVID-19. Journal of Biomolecular Structure and Dynamics, March 3; 42(4), 1858–1874.

[33] Koonrungsesomboon, N., Morakote, N., & Karbwang, J. (2021). Ethical considerations and challenges in herbal drug trials with the focus on scientific validity and risk assessment. Phytotherapy Research, May; 35(5), 2396–2402.

[34] Singh, D., Aggarwal, A., Patel, R., Das, A., Sharma, S., Behera, B., Panigrahy, R., Maity, S., Kirane, A. R., Kharkwal, H., & Sankaranarayanan, M. (2023). Design, synthesis, anticancer and antimicrobial studies of 2-phenylthiazolidin-4-one glycinamide conjugates. Chemistry Select, November 13; 8(42), e202301806.

[35] Nandi, A., Auti, P. S., Jagtap, U. A., & Paul, A. T. (2024). Investigating the role of indole and quinazolinone-based hybrid analogues with ketoamide fragment and alkyl extension for potential PL inhibition. Journal of Molecular Structure, April 5; 1301, 137337.

[36] Bhattacharya, T., Nandi, A., & Chander, S. (2024). Gallstone: A factor for acute pancreatitis. In Gallstone formation, diagnosis, treatment and prevention, January 1 (pp. 117–141). Academic Press.

[37] Boon, H., Caulfield, T., & Bubela, T. Herbal remedy clinical trials in the media: A comparison with the coverage of conventional pharmaceuticals.

[38] Chattaraj, B., Nandi, A., Das, A., Baidya, A., Mahata, S., Chowdhury, A., Mitra, S., Roy, S., Chakraborty, S., & Dey, Y. N. (2023). Enhydra fluctuans Lour. aqueous extract inhibited the growth of calcium phosphate crystals: An *in vitro* study. Food Chemistry Advances, October 1; 2, 100287.

[39] Rathod, B., Joshi, S., Regu, S., Manikanta, K. V., Kumar, H., Dubey, S., Chowdhury, A., Shaikh, R. P., Das, A., Patel, S., & Satvase, R. (2024). Design and synthesis of novel quinazolinone-based pyruvate kinase M2 activators as selective inhibitors of oral cancer cells. Journal of Molecular Structure, May 15; 1304, 137595.

[40] Singh, D., Patel, R., Aggarwal, A., Das, A., Sharma, S., Behera, B., Panigrahy, R., Kirane, A. R., Kharkwal, H., Kumar, P., & Bokolia, N. P. (2024). Phenylthiazolidin-4-one piperazine Conjugates: Design, Synthesis, anticancer and antimicrobial studies. Results in Chemistry, January 1; 7, 101237.

[41] Bhattacharya, T., Nandi, A., Das, A., & El-Shazly, M. (2024). Role of liver in gallstone formation. In Gallstone formation, diagnosis, treatment and prevention, January 1 (pp. 51–70). Academic Press.

[42] Pan, S. Y., Zhou, S. F., Gao, S. H., Yu, Z. L., Zhang, S. F., Tang, M. K., Sun, J. N., Ma, D. L., Han, Y. F., Fong, W. F., & Ko, K. M. (2013). New perspectives on how to discover drugs from herbal medicines: CAM' S outstanding contribution to modern therapeutics. Evidence-Based Complementary and Alternative Medicine, 2013(1), 627375.

[43] Madan, A., Garg, M., Satija, G., Sharma, B., Shaquiquzzaman, M., Akhter, M., Iqubal, A., Khan, M. A., Parvez, S., Das, A., & Sheikh, K. A. (2023). SAR based review on diverse heterocyclic compounds with various potential molecular targets in the fight against Covid-19: A medicinal chemist perspective. Current Topics in Medicinal Chemistry, May 1; 23(14), 1319–1339.

[44] Yang, G., Wang, Y., Sun, J., Zhang, K., & Liu, J. (2016). Ginkgo biloba for mild cognitive impairment and Alzheimer's disease: A systematic review and meta-analysis of randomized controlled trials. Current Topics in Medicinal Chemistry, February 1; 16(5), 520–528.

[45] Chattaraj, B., Khanal, P., Nandi, A., Das, A., Sharma, A., Mitra, S., & Dey, Y. N. (2023). Network pharmacology and molecular modelling study of Enhydra fluctuans for the prediction of the molecular mechanisms involved in the amelioration of nephrolithiasis. Journal of Biomolecular Structure and Dynamics, December 29; 41(24), 15400–15410.

[46] Nandi, A., Nigar, T., Das, A., & Dey, Y. N. (2023). Network pharmacology analysis of Plumbago zeylanica to identify the therapeutic targets and molecular mechanisms involved in ameliorating hemorrhoids. Journal of Biomolecular Structure and Dynamics, November 7, 1–5.

[47] Nandi, A., Dey, Y. N., Maity, D., & Das, A. (2024). Vaccine development strategies and impact. In Diagnosis and analysis of COVID-19 using artificial intelligence and machine learning-based techniques, January 1 (pp. 149–180). Academic Press.

[48] Barrett, B., Brown, R., Rakel, D., Mundt, M., Bone, K., Barlow, S., & Ewers, T. (2010). Echinacea for treating the

common cold: A randomized trial. Annals of Internal Medicine, December 21; 153(12), 769–777.
[49] Chattaraj, B., Nandi, A., & Lin, W. Y. (2024). Role of the gallbladder in our metabolism and immune system. In Gallstone formation, diagnosis, treatment and prevention, January 1 (pp. 23–38). Academic Press.
[50] Khan, A., Sharma, S., Das, A., Alam, M., Syed, M. A., & Haque, S. E. (2024). D-Pinitol improves cognitive dysfunction and neuronal damage induced by isoproterenol via modulation of NF-κB/BDNF/GFAP signaling in Swiss albino mice. Iranian Journal of Basic Medical Sciences, 27(3), 326.

11 Regulatory Aspects, Challenges, and Opportunities in Clinical Trials of the Indian System of Medicines

Maya Elizabeth Thomas[1], *Anoop Kumar*[2], *and Devesh Tewari*[3]
[1]School of Pharmaceutical Sciences, Delhi Pharmaceutical Sciences and Research University, New Delhi, India
[2]Department of Pharmacology, Delhi Institute of Pharmaceutical Sciences and Research, Delhi Pharmaceutical Sciences and Research University, New Delhi, India
[3]Department of Pharmacognosy and Phytochemistry, School of Pharmaceutical Sciences, Delhi Pharmaceutical Sciences and Research University, New Delhi, India

11.1 INTRODUCTION

11.1.1 History and Evolution of Indian Systems of Medicine

The Indian Systems of Medicine (ISM) comprises several distinct practices, including Ayurveda, Unani, Siddha, Yoga, Sowa-Rigpa, and Homeopathy. Ayurveda, often considered one of the oldest traditional medicine systems, has a history dating back over 5,000 years, with its origins believed to be divine, passed from the Hindu god Brahma to ancient sages for the well-being of mankind. The knowledge of healing properties was composed in the form of "Shlokas" in the four Vedas: Rig Veda, Yajur Veda, Sama Veda, and Atharva Veda. The Rig Veda is the most prominent, describing 67 plants and 1,028 Shlokas, while the Atharva Veda and Yajur Veda describe 293 and 81 medicinally useful plants, respectively [1]. This ancient knowledge was later compiled and edited into foundational texts like the Charaka Samhita, which describes all the aspects of Ayurvedic medicine, and the Sushruta Samhita, which details the science of surgery. India has one of the world's oldest medical systems, known as Ayurvedic medicine or Ayurveda. *Ayur* in Sanskrit means "life," and "veda" means "science" or "knowledge"; thus, Ayurveda is the science of life that has evolved in India over thousands of years. In modern times, Ayurveda is considered complementary alternative medicine (CAM), herbs and specialized diets [2]. The holistic approach toward patient care and empathetic behavior are the strengths of Ayurveda; it aims at reaching the inner consciousness of the patient to cure the root cause of disease [3]. The Unani system, with its roots in Greece, was introduced to India by Arabs and has since made a firm footing in the country, where it is officially recognized by the government for clinical practice and research [1]. In 1971, a research council named the Central Council for Research in Indian Medicine, Homoeopathy and Yoga (CCRIMH) was established. Subsequently, this council was bifurcated to create four separate councils: the Central Council for Research in Ayurveda & Siddha (CCRAS), the Central Council for Research in Unani Medicine (CCRUM), the Central Council for Research in Homoeopathy (CCRH), and the Central Council for Research in Naturopathy and Yoga (CCRNY) [4].

11.1.2 Importance and Objectives of Clinical Trials for ISM

The primary objective of conducting clinical trials for ISM is to generate scientific evidence to prove the safety, efficacy, and quality of its therapies and products [5]. Historically, the effectiveness of ISM has been based on empirical knowledge and anecdotal evidence passed down through generations [1]. However, in an era of evidence-based medicine, this is no longer sufficient for global acceptance or integration into mainstream healthcare. Clinical trials are emerging as an important activity in India because they are an essential component of the drug discovery and development program. The only robust way to evaluate a new medicine is by conducting properly designed clinical trials [6]. Clinical trials are essential for identifying potential risks associated with traditional herbal medicines, which can only be established once safety and efficacy are proven [4]. They are also a mandatory requirement for obtaining regulatory approval for ISM drugs to be prescribed and commercialized [7]. Furthermore, well-designed clinical trials contribute to building a reproducible and consistent body of evidence, which is essential for the global acceptance and commercialization of ISM products [8].

11.1.3 Scope of the Study

This study focuses on the regulatory aspects, challenges, and opportunities specific to conducting clinical trials for ISM. It will not serve as a general guide to clinical research but will concentrate on the unique issues that arise when applying modern clinical trial methodologies to traditional, holistic systems

DOI: 10.1201/9781003522065-11

[7]. The scope will include a detailed analysis of the Indian regulatory framework as outlined in the New Drugs and Clinical Trials Rules, 2019, and a comparative look at international guidelines from agencies such as the World Health organization (WHO) and FDA [6, 7]. This study will also delve into unique design, ethical, and legal challenges, such as the difficulty of standardizing complex herbal formulations and the complexities of ensuring patient safety in a traditionally rooted system [7]. Finally, it will discuss the future potential for ISM through technological advancements such as Ayur informatics, the harmonization of methodologies, and the role of government initiatives in fostering a positive environment for research [9].

11.2 REGULATORY FRAMEWORK FOR CLINICAL TRIALS IN INDIA

The regulatory framework for clinical trials in India has undergone a significant transformation to ensure participant safety and ethical conduct. This evolution has led to the establishment of a robust system that balances scientific rigor with the promotion of traditional Indian medicine. The current framework is characterized by a mix of general guidelines and specific rules tailored to the country's unique healthcare landscape.

11.2.1 Evolution of Regulatory Framework

India's journey in clinical trial regulation began with the Drugs and Cosmetics Act of 1940, followed by the Drugs and Cosmetics Rules of 1945. Early on, the regulatory environment was less structured, with the first significant step being the inclusion of Schedule Y in 1988, which laid down specific requirements for permission to conduct clinical trials [2]. However, a period of controversy from 2011 to 2013, sparked by public litigation over ethical issues and patient compensation, highlighted the need for stricter regulations. This led to a comprehensive review that culminated in the landmark New Drugs and Clinical Trials Rules, 2019, which superseded Schedule Y and aimed to create a more transparent, predictable, and effective system for clinical research [7, 8]. The details of evolution of regulatory framework are presented in Figure 11.1.

11.2.2 Guidelines for Clinical Trials of ISM (AYUSH, GCP)

For the ISM, which include Ayurveda, Yoga, Unani, Siddha, and Homeopathy, the Ministry of AYUSH has developed a specialized set of guidelines that operate in conjunction with the general Good Clinical Practice (GCP) principles [2]. These guidelines are uniquely tailored to address the complexities inherent in traditional medicine, such as the difficulty in standardizing complex herbal formulations and the importance of documenting traditional knowledge. This dual-pronged approach, combining specific AYUSH protocols with global GCP standards, is a critical step in providing a strong scientific basis for traditional Indian medicines and their integration into mainstream healthcare [8]. The flowchart on the guidelines for clinical trials for ISM is shown in Figure 11.2.

11.2.3 The New Drugs and Clinical Trials Rules (2019)

The New Drugs and Clinical Trials Rules, 2019, represent a major overhaul of the regulatory landscape in India [7, 8]. These rules have been designed to streamline the entire approval process by drastically reducing review timelines. For instance, applications for drugs developed in India are now subject to a 30-day review period, with a provision for "deemed approval" if the regulatory body fails to make a decision within the specified timeframe. Furthermore, the rules have significantly strengthened patient protection by requiring all ethics committees to be registered with the Drugs Controller General of India (DCGI) and by establishing a clear, mandatory framework for providing compensation and medical care to trial participants in the case of any trial-related injury.

11.2.4 Key Regulatory Bodies (Ministry of AYUSH, CDSCO, DCGI, CCRAS)

The regulation of clinical trials in India is a collaborative effort involving several key government bodies. The Central Drugs Standard Control Organisation (CDSCO), headed by the DCGI, serves as the primary authority for granting permission to conduct clinical trials and ensuring compliance with regulations. Parallel to this, the Ministry of AYUSH has the specific mandate of promoting and regulating the ISM [2]. Additionally, the CCRAS, an autonomous body under the Ministry of AYUSH, plays a crucial role in developing and overseeing clinical research on Ayurvedic drugs and treatments, thereby contributing to the evidence-based validation of this traditional system of medicine [10]. An overview of the regulatory bodies in India is presented in Table 11.1.

11.3 INTERNATIONAL REGULATORY FRAMEWORK

11.3.1 WHO Strategy and Guidelines on Traditional Medicine

The WHO plays a crucial role in shaping the global approach to traditional medicine (TM), which it defines as "the sum total of the knowledge, skills, and practices based on the theories, beliefs, and experiences indigenous to different cultures." The WHO Global Traditional Medicine Strategy 2025–2034 is a key framework that aims to help member states build the knowledge base for traditional and complementary medicine (T&CM), strengthen quality assurance, and promote the safe and effective use of T&CM products, practices, and practitioners [10, 11]. The strategy also addresses intellectual property rights and the equitable sharing of benefits from traditional medical knowledge. The WHO recognizes that TM is used by billions of people and seeks to apply rigorous scientific methods to support member states in ensuring the effectiveness, safety, and equity of these systems [12].

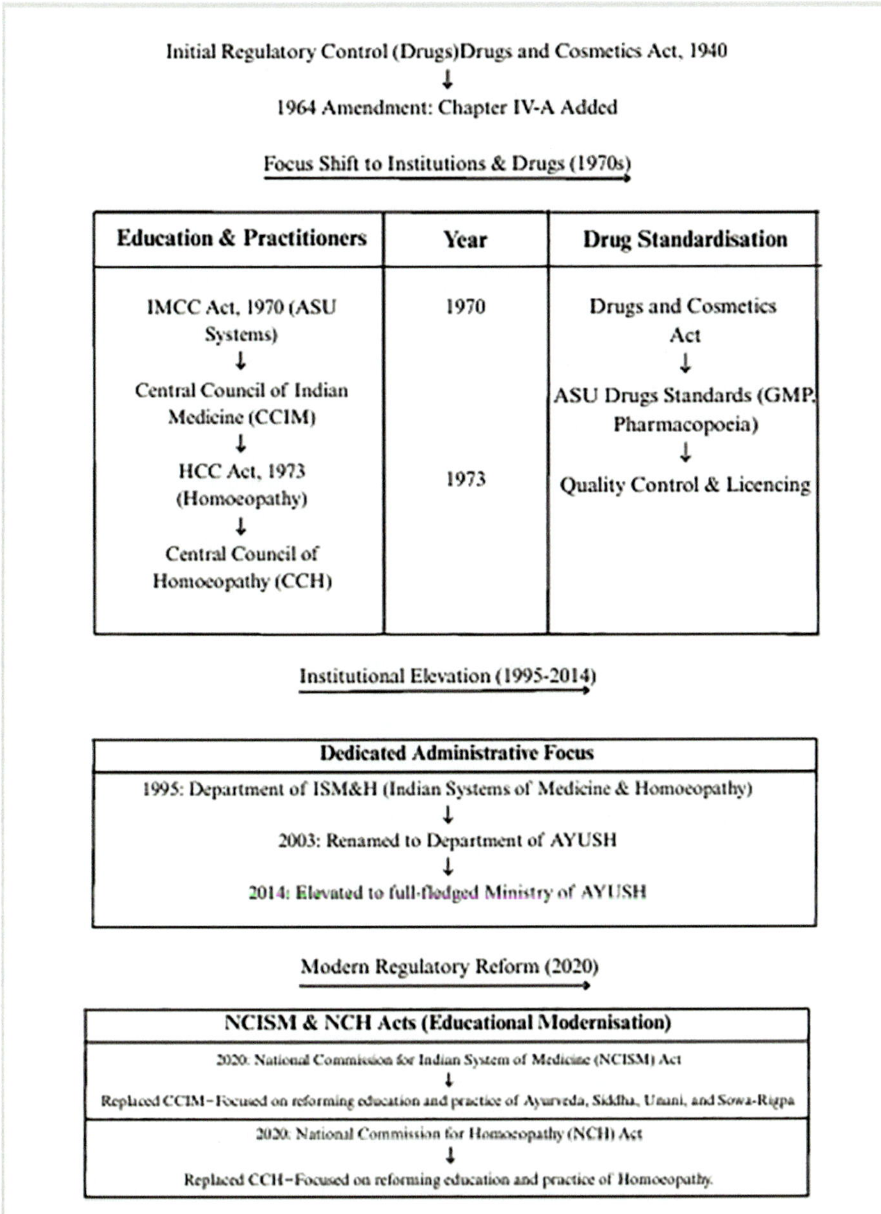

FIGURE 11.1 Evolution of regulatory framework.

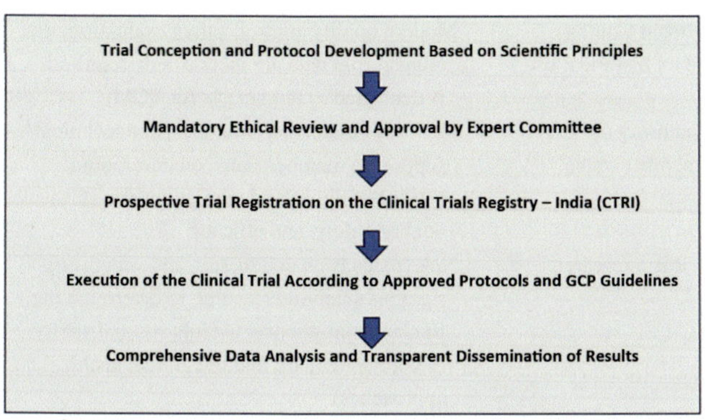

FIGURE 11.2 Flowchart on guidelines for clinical trials of ISM.

11.3.2 Regulatory Models in the EU, the United States, and China

Regulatory approaches to TM vary significantly. The European Union's Traditional Herbal Medicinal Products Directive (THMPD) offers a simplified registration process for products with a long history of traditional use. This framework, which has been successful in facilitating the marketing of numerous products, does not require clinical trials to prove efficacy, but it does mandate rigorous Good Manufacturing Practice (GMP) and safety standards. CAM is also popular among the general public in these countries, although the efficacy is not sufficiently proven according to the standards of evidence-based medicine, it is often available without prescription as over-the-counter products [13, 14].

TABLE 11.1
Key Regulatory Bodies in India

Regulatory Body	Primary Role	Key Responsibilities
Ministry of AYUSH	Nodal Government Ministry	Formulates policies, promotes research and education, and implements schemes related to Ayurvedic, Yoga, Naturopathy, Unani, Siddha, and Homeopathy systems.
CDSCO	Central Drug Regulator	Approves new drugs and regulates clinical trials of all systems of medicine, including ISM. It is responsible for enforcing the Drugs and Cosmetics Act.
DCGI	Head of the CDSCO	The final authority for approving new drugs and clinical trials in India. Oversees the work of the CDSCO.
CCRAS	Central Research Council	Conducts, coordinates, and assists scientific research in Ayurvedic medicine. It develops research protocols for clinical trials.
CCRUM	Central Research Council for Unani Medicine	Engages in scientific research on fundamental and applied aspects of the Unani system, including drug standardization, clinical trials, and surveys of medicinal plants.
CCRH	Central Research Council for Homoeopathy	Undertakes, coordinates, develops, disseminates, and promotes scientific research in the Homeopathy system, focusing on clinical trials, drug proving, and standardization.
CCRYN	Central Research Council for Yoga and Naturopathy	Serves as the apex body for research and development in Yoga and Naturopathy, actively engaged in the promotion, propagation, education, and clinical research of these drugless therapies.

In contrast, the United States primarily regulates herbal products as dietary supplements under the Dietary Supplement Health and Education Act (DSHEA) of 1994. This places the burden of proof for safety on the FDA, which must prove a product is unsafe before it can be removed from the market. For a botanical product to be approved as a "botanical drug," it must undergo the full, rigorous FDA drug approval process, which includes extensive preclinical and multi-phase clinical trials to prove its safety and efficacy.

China operates with a dual regulatory system that officially recognizes and integrates traditional Chinese medicine (TCM) into its healthcare framework. The State Administration of Traditional Chinese Medicine (SATCM) governs this sector, and while there are pathways that rely on historical use and clinical experience, there is also a significant push toward modernization through science [15, 16]. An overview of International regulatory bodies is presented in Table 11.2.

TABLE 11.2
International Regulatory Bodies

Key Regulatory Body	Approval Pathway for Traditional Medicines	Key Requirements
European Medicines Agency (EMA) and national agencies	Traditional Use Registration (THR): Simplified pathway for products with a history of traditional use.	15+ years of traditional use within the EU, safety data, and quality standards. No requirement for clinical efficacy trials.
Food and Drug Administration (FDA)	Regulated as dietary supplements or through the rigorous New Drug Application (NDA) process.	For supplements: No premarket approval required, but must be safe. For drugs: Requires extensive clinical trials for safety and efficacy.
National Medical Products Administration (NMPA)	Specific pathways for Traditional Chinese Medicine (TCM) are based on historical use and modern research.	Modern quality control, safety evaluation, and clinical trial data are increasingly required. A dedicated system exists for TCM.
World Health Organization (WHO)	Does not have an approval pathway or regulatory authority over member states' medicines.	Provides guidelines, strategies, and technical support to member states on developing national policies and regulations for TCM quality, safety, and efficacy.
National Center for Complementary and Integrative Health (NCCIH)	Does not regulate products. It is a research center within the NIH.	N/A (focus is on conducting and supporting rigorous scientific research to determine the fundamental science, usefulness, and safety of complementary and integrative health approaches).
Medicines and Healthcare products Regulatory Agency (MHRA)	Traditional Herbal Registration (THR): Simplified pathway, similar to the EU system (retained post-Brexit).	Evidence of 30 years of traditional use, robust quality and safety data. Used for products for minor conditions without medical supervision. No proof of efficacy is required.

11.3.3 Implications for ISM Clinical Trials

The global regulatory landscape presents both challenges and opportunities for clinical trials of ISM products. The primary implication is the need for harmonization and adaptation [12, 17]. Trials for ISM products that aim to enter markets in the EU or the United States must often conform to Western-style, randomized controlled trial (RCT) designs and adhere to ICH-GCP guidelines. This can be problematic for ISM, which often involves personalized, holistic interventions and complex polyherbal formulations.

Furthermore, international frameworks highlight critical issues of product standardization and quality control. Unlike single-molecule drugs, ISM formulations must demonstrate consistent chemical composition from batch to batch, a requirement that can be difficult to meet due to variations in plant sourcing, cultivation, and processing [17]. The ethical dimension of data sharing and intellectual property rights is also a major concern, as researchers must navigate complex rules to protect traditional knowledge and ensure equitable benefit-sharing with source communities.

11.4 CHALLENGES IN CLINICAL TRIALS OF ISM

11.4.1 Study Design and Methodological Challenges

11.4.1.1 Translating Classical ISM Principles Into Modern Protocols

A primary challenge in conducting clinical trials for ISM is the difficulty of applying the reductionist approach of modern science to a holistic system. ISM principles often involve individualized treatment, where a practitioner may alter a complex polyherbal formulation based on a patient's unique constitution and disease state. This contrasts sharply with the standardized, single-intervention model of a RCT, which is the gold standard for proving drug efficacy. Crafting an effective placebo for a multi-ingredient decoction is also a major methodological obstacle, as it needs to mimic the product's taste, smell, and appearance without having therapeutic effects [17].

11.4.1.2 Standardization, Quality Control, and Integration with Allopathic Methods

Furthermore, ensuring standardization, quality control, and integration with allopathic methods is critical for regulatory acceptance. The chemical composition of an ISM product can vary widely based on factors such as plant origin, soil quality, and harvesting time. This makes it challenging to guarantee that the product used in a trial is identical to the one that would be commercialized, a key requirement for modern drug approval. Integration with allopathic methods requires careful study design to account for potential drug-herb interactions and avoid confounding results, as many patients may be using conventional and traditional therapies simultaneously [17, 18].

11.4.2 Ethical, Safety, and Legal Challenges

11.4.2.1 Patient Safety, Informed Consent, and Pharmacovigilance

Ethical and safety concerns are paramount in ISM clinical trials. While TMs are often perceived as harmless, they can have adverse effects and contraindications, necessitating robust patient safety, informed consent, and pharmacovigilance protocols. The holistic nature of ISM also complicates obtaining informed consent, as it involves a broader approach to health than just treating a specific disease. Comprehensive pharmacovigilance systems must be in place to accurately track and report any adverse events associated with the TM, which is a significant challenge for a largely unregulated sector. For example, a study by Sharma and Singh in the *Journal of Ayurveda and Integrative Medicine* highlighted the need for a more structured approach to pharmacovigilance for herbal drugs to ensure patient safety and build a reliable safety profile [18].

11.4.2.2 Ethics Committee Approvals, Regulatory Overlaps, and Data Integrity

Navigating ethics committee approvals, regulatory overlaps, and data integrity adds another layer of complexity. In India, researchers must seek approval from both the Ministry of AYUSH and the CDSCO, leading to potential regulatory delays and jurisdictional confusion. A comparative review of clinical trial regulations by Ahangar et al. found that this type of dual-agency approval process can hinder research progress and create challenges in harmonizing data standards. Ensuring data integrity is essential for a trial's credibility, as reliable and accurate data are required for publication and global regulatory acceptance. This is especially critical for ISM trials, where the lack of standardized protocols can make data collection and analysis prone to variability [12, 19].

11.4.2.3 Intellectual Property Rights and Protection of Traditional Knowledge

Finally, a major legal challenge is the protection of intellectual property rights and traditional knowledge. The commercialization of traditional formulations raises concerns about biopiracy—the use of genetic resources and indigenous knowledge by Western companies without benefit-sharing with those who generated the traditional knowledge. Frameworks such as the Nagoya Protocol on Access and Benefit-Sharing aim to address these issues; these protocols also seek to engage with traditional communities in a transparent and respectful manner, ensuring proper consent for the use of their knowledge and establishing clear agreements for benefit-sharing. Intellectual Property Rights and Traditional Medicine: Policy Dilemmas at the Interface [20]. For example, the passing of Kuttimathan Kani, a Kani tribal elder, highlights a key issue: while the initial benefit-sharing agreement for the "Jeevani" drug was celebrated as a victory, it ultimately failed to provide lasting economic benefits to the tribe, underscoring the vital role of robust intellectual property rights in protecting indigenous knowledge.

11.5 OPPORTUNITIES FOR ADVANCEMENT OF ISM CLINICAL TRIALS

11.5.1 Leveraging Technology and Digital Platforms

Technology offers significant opportunities to overcome traditional challenges in ISM clinical trials by enhancing efficiency, data integrity, and research scope. Ayur informatics, a portmanteau of Ayurveda and bioinformatics, applies computational tools to analyze complex ISM data, including genetic variations and the multicomponent nature of herbal formulations [21]. This can help researchers better understand the mechanisms of action of TMs and identify potential synergistic effects. For example, machine learning algorithms can analyze vast datasets to predict which herbal combinations might be most effective for a particular condition. Furthermore, digital platforms and applications can streamline data collection and management, improve patient recruitment, and facilitate remote monitoring of participants, which is particularly useful for geographically diverse populations. This not only makes trials more efficient but also expands their reach, allowing for greater diversity in the study population. The use of electronic health records (EHRs) and other digital tools also improves pharmacovigilance by enabling the real-time tracking of adverse events and improving data integrity, thus building a more robust safety profile for ISM products [22].

11.5.2 Harmonization of ISM and Global Clinical Trial Practices

Achieving global acceptance for ISM requires the harmonization of its practices with established international guidelines. A key opportunity lies in adapting the ICH-GCP (International Council for Harmonisation of Technical Requirements for Pharmaceuticals for Human Use—Good Clinical Practice) framework to accommodate ISM's unique features. This would involve developing specific guidelines for traditional medicine that maintain scientific rigor while respecting the holistic and individualized nature of ISM. By creating a unified set of standards, researchers could conduct trials that are not only ethically sound and scientifically valid but also recognized by regulatory bodies worldwide. This harmonization would facilitate the global commercialization of ISM products and foster trust among the broader medical community, as it provides a clear, standardized pathway for evidence-based validation. For instance, harmonizing quality control standards for herbal raw materials could ensure that products meet a consistent level of quality, regardless of their origin [23].

11.5.3 International Collaborations and Government Initiatives

Fostering international collaborations and supporting government initiatives are crucial for the advancement of ISM. Collaborative research projects between Indian institutions and their counterparts in other countries can facilitate knowledge sharing, standardize research methodologies, and accelerate the development of evidence-based ISM products. The WHO, through its Traditional Medicine Strategy, actively promotes such collaborations and encourages member states to integrate traditional medicine into their national healthcare systems. In India, the establishment of the Ministry of AYUSH and the creation of research councils like the CCRAS have provided dedicated funding and a clear regulatory pathway for ISM research. These government initiatives, coupled with international partnerships, are essential for building a robust, credible, and globally recognized body of evidence for the effectiveness and safety of ISM. This dual approach of internal support and external collaboration is vital for overcoming the historical challenges of proving ISM's efficacy and safety in the modern scientific landscape.

11.6 CONCLUSION AND FUTURE PERSPECTIVES

11.6.1 Summary of Key Insights

The integration of ISM into mainstream healthcare represents a pivotal and complex transition from empirical knowledge to evidence-based practice. This journey is underscored by the significant evolution of India's domestic regulatory framework, notably the New Drugs and Clinical Trials Rules, 2019, which have been instrumental in creating a more structured and ethical environment for clinical research on traditional therapies. This domestic progress, however, must navigate a fragmented and often restrictive global landscape. International regulatory models, such as the EU's THMPD and the United States' approach to dietary supplements, present unique challenges that highlight the urgent need for a harmonized, globally recognized standard to overcome regulatory hurdles and facilitate the global market access for ISM products, a goal that is paramount for their wider acceptance.

Beyond the regulatory challenges, our analysis has also identified core methodological and ethical complexities inherent in applying Western-style, reductionist clinical trial protocols to the holistic principles of ISM. The difficulty in standardizing complex herbal formulations and the need to protect traditional knowledge from biopiracy remain significant limitations. However, these challenges also pave the way for a promising future. The strategic leveraging of technology, particularly Ayur informatics, offers a clear pathway for analyzing complex biological data and predicting therapeutic effects. This, combined with dedicated government support through initiatives like the Ministry of AYUSH and the fostering of international collaborations, is pivotal in advancing evidence-based ISM research and building a credible, globally recognized body of evidence for its effectiveness and safety.

11.6.2 Limitations of Current Research

Research into herbal medicinal products is significantly hindered by a lack of sustainable funding, which directly contributes to a shortage of high-quality studies and reliable data. This deficiency has led to a widespread perception that herbal research is often sporadic and haphazard when compared to conventional medicine studies, resulting in data that can be considered outdated or insufficient. A major consequence of this is the lack of adequate scientific evidence to properly assess the efficacy and safety of many herbal products, making it difficult for researchers and clinicians to draw firm conclusions.

Methodological limitations also pose a significant challenge. The traditional reductionist approach, which is highly effective for single-compound drugs, is poorly suited for the complex, multi-chemical nature of herbal products. Furthermore, the gold standard for clinical trials—the placebo-controlled, randomized, double-blind trial—may not be the most appropriate model for evaluating herbal medicines, which are often used in a personalized and holistic context. This mismatch in methodology can make it difficult to capture relevant outcomes and may not accurately reflect the true effects of these traditional remedies.

The scope and design of the research themselves present further limitations. For instance, some reviews rely exclusively on a limited number of major electronic databases, which may inadvertently exclude valuable research on Eastern herbal medicine not indexed in these Western-centric sources. Additionally, studies may not be initially designed to collect detailed information on key variables, such as the specific comparators used (whether other herbal medicines, traditional therapies, or Western medicines), limiting the ability to perform comprehensive comparisons. This can also lead to a lack of risk-of-bias assessments, providing only limited insight into the overall methodological rigor of the research being reviewed [24, 25].

11.6.3 Overview of Ongoing and Recently Completed Trials

Based on a broad analysis of clinical trials registered on ClinicalTrials.gov, the landscape of ongoing and recently completed research is dynamic and extensive. As of March 7, 2025, a review of over 400,000 interventional clinical trials revealed that the majority have been completed, with 204,437 studies reaching their conclusion. Beyond completed trials, the data provides a detailed snapshot of the current status of clinical research, showing that 47,711 trials are actively recruiting participants, while 17,187 have not yet begun recruiting. Additionally, 24,826 trials have been terminated, and 2,733 are enrolling by invitation.

The current landscape of clinical trials has also been significantly shaped by major global health events, with a continuous increase in registrations peaking in 2021 due to the COVID-19 pandemic. A significant number of these trials are focused on cancer, reflecting its global health burden, and the most common study design remains the RCT. However, the field is evolving, with newer, innovative designs like adaptive, decentralized, and real-world evidence trials gaining prominence to make research more efficient and patient-centered. This changing environment, alongside the continuous registration of new studies, highlights the robust and adaptive nature of clinical research [26].

11.6.4 Future Research Directions

The future of ISM clinical research lies in innovative and interdisciplinary approaches that bridge the gap between traditional wisdom and modern science. Future research should explore hybrid research models that combine the gold standard of RCTs with pragmatic, real-world designs that better accommodate the holistic nature of ISM treatments. The integration of technology is also paramount; the use of Ayur informatics, artificial intelligence, and machine learning will be crucial for analyzing complex multicomponent formulations, predicting drug-herb interactions, and identifying promising new leads for targeted clinical investigation. A concerted effort is needed from international bodies like the WHO and key national regulatory agencies to develop globally harmonized guidelines for TM research, creating a clear and unified pathway for ISM products to gain global acceptance. This will require a greater emphasis on developing robust quality control standards and analytical methods to ensure the reproducibility and safety of ISM products for the global market.

REFERENCES

[1] Jaiswal, Y. S., & Williams, L. L. (2017). A Glimpse of Ayurveda: The Forgotten History and Principles of Indian Traditional Medicine. *Journal of Traditional and Complementary Medicine*, *7*(1), 50–53. https://doi.org/10.1016/j.jtcme.2016.02.002

[2] Hajar, R. (2013). The Medicine of Old India. *Heart Views: The Official Journal of the Gulf Heart Association*, *14*(2), 92. https://doi.org/10.4103/1995-705X.115497

[3] Phull, R., & Phull, G. (2019). A Review on Ayurveda, the Indian System of Medicine: Rcih Heritage and Its Scientific Relevance. *International Journal of Research in Ayurveda and Pharmacy*, *10*, 12–15. https://doi.org/10.7897/2277-4343.100226

[4] Ravishankar, B., & Shukla, V. J. (2007). Indian Systems of Medicine: A Brief Profile. *African Journal of Traditional, Complementary, and Alternative Medicines: AJTCAM*, *4*(3), 319–337. https://doi.org/10.4314/ajtcam.v4i3.31226

[5] Singh, H. P., Sharma, S., Chauhan, S., & Kaur, I. (2014). Clinical Trials of Traditional Herbal Medicines in India: Current Status and Challenges. *International Journal of Pharmacognosy*, *1*, 415–421. https://doi.org/10.13040/IJPSR.0975-8232.IJP.1(7).415-421

[6] Poongothai, S., Unnikrishnan, R., Balasubramanian, J., Nair, M. D., & Mohan, V. (2014). Why Are Clinical Trials Necessary in India? *Perspectives in Clinical Research*, *5*(2), 55–59. https://doi.org/10.4103/2229-3485.128018

[7] Singh, N., Madkaikar, N. J., Gokhale, P. M., & Parmar, D. V. (2020). New Drugs and Clinical Trials Rules 2019: Changes in Responsibilities of the Ethics Committee. *Perspectives in Clinical Research*, *11*(1), 37–43. https://doi.org/10.4103/picr.PICR_208_19

[8] Bhat, S., & Guddadar, V. (2020). A Mini Review on Challenges in Regulatory, Development and Clinical Development for Herbal Product. *Gyanshauryam International Scientific Refereed Research Journal*, *3*(6), 50–54. https://doi.org/10.32628/GISRRJ203611

[9] Ng, J. Y., Cramer, H., Lee, M. S., & Moher, D. (2024). Traditional, Complementary, and Integrative Medicine and Artificial Intelligence: Novel Opportunities in Healthcare. *Integrative Medicine Research*, *13*(1), 101024. https://doi.org/10.1016/j.imr.2024.101024

[10] Wasiullah, M., Wasiullah, M., Yadav, P., Yadav, P., Sushil Yadav, S. Y., & Vikash Vishawakarma, V. (2025). An Overview of Regulatory Authorities in India: Rules, Functions, and Impact. *International Journal of Pharmaceutical Research and Applications*, *10*(2), 2665–2670. https://doi.org/10.35629/4494-100226652670

[11] Hoenders, R., Ghelman, R., Portella, C., Simmons, S., Locke, A., Cramer, H., Gallego-Perez, D., & Jong, M. (2024). A Review of the WHO Strategy on Traditional, Complementary, and Integrative Medicine from the Perspective of Academic Consortia for Integrative Medicine and Health. *Frontiers in Medicine*, *11*, 1395698. https://doi.org/10.3389/fmed.2024.1395698

[12] Sharma, M., Grover, M., Sharma, N., Shukla, V. K., & Suryavanshi, S. J. (2025). Comparative Review of Clinical Trial Regulations in Different Countries: Current Scenario and Future Prospect. *Reviews on Recent Clinical Trials*. https://doi.org/10.2174/011574887136532 0250529125930

[13] Efferth, T., Banerjee, M., Abu-Darwish, M. S., Abdelfatah, S., Böckers, M., Bhakta-Guha, D., Bolzani, V., Daak, S., Demirezer, Ö. L., Dawood, M., Efferth, M., El-Seedi, H. R., Fischer, N., Greten, H. J., Hamdoun, S., Hong, C., Horneber, M., Kadioglu, O., Khalid, H. E., Khalid, S. A., Kuete, V., Mahmoud, N., Marin, J., Mbaveng, A., Midiwo, J., Nakagawa, H., Naß, J., Ngassapa, O., Ochwang'i, D., Omosa, L. K., Ooko, E. A., Özenver, N., Poornima, P., Romero, M. R., Saeed, M. E. M., Salgueiro, L., Seo, E-J., Yan, G., Yasin, Z., Saeed, E. M., & Paul, N. W. (2019). Biopiracy Versus One-World Medicine-From Colonial Relicts to Global Collaborative Concepts. *Phytomedicine: International Journal of Phytotherapy and Phytopharmacology*, *53*, 319–331. https://doi.org/10.1016/j.phymed.2018.06.007

[14] Qu, L., Zou, W., Wang, Y., & Wang, M. (2018). European Regulation Model for Herbal Medicine: The Assessment of the EU Monograph and the Safety and Efficacy Evaluation in Marketing Authorization or Registration in Member States. *Phytomedicine: International Journal of Phytotherapy and Phytopharmacology*, *42*, 219–225. https://doi.org/10.1016/j.phymed.2018.03.048

[15] Hua, H., Tang, J.-Y., Zhao, J.-N., Wang, T., Zhang, J.-H., Yu, J.-Y., Yang, C.-M., Ai, Y.-L., & Luo, Q.-X. (2025). From Traditional Medicine to Modern Medicine: The Importance of TCM Regulatory Science (TCMRS) as an Emerging Discipline. *Chinese Medicine*, *20*(1), 92. https://doi.org/10.1186/s13020-025-01152-8

[16] Gudi, N., Kamath, P., Chakraborty, T., Jacob, A. G., Parsekar, S. S., Sarbadhikari, S. N., & John, O. (2022). Regulatory Frameworks for Clinical Trial Data Sharing: Scoping Review. *Journal of Medical Internet Research*, *24*(5), e33591. https://doi.org/10.2196/33591

[17] Aguilera, B., DeGrazia, D., & Rid, A. (2020). Regulating International Clinical Research: An Ethical Framework for Policy-Makers. *BMJ Global Health*, *5*(5). https://doi.org/10.1136/bmjgh-2020-002287

[18] Raha, S. (2011). A Critique of Statistical Hypothesis Testing in Clinical Research. *Journal of Ayurveda and Integrative Medicine*, *2*(3), 105–114. https://doi.org/10.4103/0975-9476.85548

[19] Panda, T., Lala, P. K., Manoharan, K., Jinson, J., & George, M. (2025). Evaluation of the Current Status of Ethics Committees in India Using the National Accreditation Board for Hospitals and Health-Care Providers, Central Drugs Standard Control Organization (CDSCO), and Department of Health Research databases. *Perspectives in Clinical Research*, *16*(2), 75–80. https://doi.org/10.4103/picr.picr_40_24

[20] Timmermans, K. (2003). Intellectual Property Rights and Traditional Medicine: Policy Dilemmas at the Interface. *Social Science & Medicine (1982)*, *57*(4), 745–756. https://doi.org/10.1016/s0277-9536(02)00425-2

[21] Goel, S., Karthik, K. P., Mahapatra, A., Shrikrishna, R., & Gupta, P. (2024). Redefining Ayurveda through the Lens of Bioinformatics: A Conference Report. *International Journal of Ayurveda Research*, *5*, 324–329. https://doi.org/10.4103/ijar.ijar_274_24

[22] Evans, R. S. (2016). Electronic Health Records: Then, Now, and in the Future. *Yearbook of Medical Informatics*, *Suppl 1*(Suppl 1), S48–S61. https://doi.org/10.15265/IYS-2016-s006

[23] Hoffmann, J.-M., Blümle, A., Grossmann, R., Yau, H., Lang, B., & Bradbury, C. (2023). Toward a Global Harmonization of Service Infrastructure in Academic Clinical Trial Units: An International Survey. *Frontiers in Medicine*, *10*, 1252352. https://doi.org/10.3389/fmed.2023.1252352

[24] Pelkonen, O., Xu, Q., & Fan, T.-P. (2014). Why Is Research on Herbal Medicinal Products Important and How Can We Improve Its Quality? *Journal of Traditional and Complementary Medicine*, *4*(1), 1–7. https://doi.org/10.4103/2225-4110.124323

[25] Koonrungsesomboon, N., Sakuludomkan, C., Na Takuathung, M., Klinjan, P., Sawong, S., & Perera, P. K. (2024). Study Design of Herbal Medicine Clinical Trials: A Descriptive Analysis of Published Studies Investigating the Effects of Herbal Medicinal Products on Human Participants. *BMC Complementary Medicine and Therapies*, *24*(1), 391. https://doi.org/10.1186/s12906-024-04697-7

[26] Joshi, G., Bhandari, T. K., Joshi, P., Bhandari, S., Araveeti, S. R., Jain, A., Khadka, S., Trecarten, S., Abdelaziz, A., Garg, H., & Bhandari, M. (2025). The Current Landscape of Clinical Trials. *Journal of Clinical Medicine*, *14*(7). https://doi.org/10.3390/jcm14072519

12 Latest Advancements in Clinical Trials of Herbal Medicines

Chandrakanta[1], *Archana Negi Sah*[1], *and Sweta Bawari*[2]
[1] Amity Institute of Pharmacy, Amity University Campus, Sector-125, Noida, Uttar Pradesh, India
[2] Amity Institute of Pharmacy, Amity University Campus, Sector, Noida, Gautam Buddha Nagar, Uttar Pradesh, India

12.1 INTRODUCTION

National and international drug regulatory frameworks define herbal medicines as herbal products in the medicines category, which can comprise "herbs," "herbal materials," "herbal preparations," and "finished herbal products" or "herbal medicinal products." Certain herbs and herbal materials can be consumed as foods or used as culinary additives in some countries [1].

Herbal medicine is the foundation of complementary and alternative medicine, which is becoming more and more popular worldwide and is progressively making its way into traditional healthcare systems. Gender, social status, and race are no barriers to the use of herbal medicine in developing and developed countries worldwide [2].

Herbal medicine has a significant role in Indian culture and traditional medicine. In addition to being recognized for their economic significance, plant-based medications represent the foundation of the contemporary healthcare system. Research on the production of new drugs heavily relies on traditional medical knowledge and botanicals, as herbs are an immensely rich source of phytotherapeutics, and many phytoconstituents and their analogues, such as taxanes, camptothecin analogues, and artemisinin derivatives, are ubiquitously being used for the treatment of various ailments and diseases [3, 4]. In addition to being employed directly as therapeutic agents, herbal products or constituents are also utilized as lead molecules in the development of novel drugs; e.g., human cancer cells can be treated with *Cinnamomum camphora* as an antimutagenic agent [5, 6]. Other examples of natural components used as therapeutic candidates include digoxin, a diuretic from *Digitalis purpurea*, and digitoxin, which is used to treat heart failure and irregular heart rhythm. Curcumin from *Curcuma longa* is used to treat inflammation [7], and piperine from *Piper* species is used as a bioavailability enhancer [8]. Based on current research and understanding, using raw herbs or herbal products over an extended period of time may have significant health benefits [9].

The mounting adversities associated with the synthetic medicines and population growth are the crucial contributing factors to the rise in popularity of the complementary and alternative medicine, which also includes medicines from plant sources. More broadly, the demand for plant-based pharmaceuticals, nutraceuticals, herbal cosmetics, and herbal medicines is rising globally, and this is a sign of the growing acceptance and trust that people have in herbal remedies [10].

12.2 RECENT ADVANCEMENT IN THE REGULATION OF HERBAL MEDICINE

Herbal medicine laws have changed significantly in the 21st century, with ramifications for both regional and worldwide practices. This includes the creation of efficacy profiles, quality and safety regulations, and the prevention of offensive advertisements related to the regional cultivation of herbal remedies utilized by indigenous people [11].

12.2.1 In China

The China Food and Drug Administration regulates herbal medicines through the Drugs Registration Provision. New draft guidelines are presently being developed. In order to be registered as Traditional Chinese Medicines, herbal remedies for crude medicines and preparations must adhere to the Supplementary Rules. The Traditional Chinese Medicines regulations must be followed by all herbal goods, including natural products, Chinese medicines, and plant-based medications, in order to be registered [12]. Several nations, including China, India, Pakistan, Brazil, and Bahrain, have long histories of traditional medicine practices. These countries have made tremendous progress in introducing diverse herbal medicines into their healthcare systems. This is corroborated by the high prevalence of HM products in both national pharmacopoeias and national essential drug

DOI: 10.1201/9781003522065-12

lists. China and India, the top producers and exporters of medicinal plant products, have exemplary HM regulation among emerging countries [13].

12.2.2 In Bahrain

Herbal medicines can be sold as prescription, non-prescription, or over-the-counter (OTC) via pharmacies, specific outlets such as HM stores, and licensed practitioners. Herbal medications now follow the same restrictions as regular drugs. Bahrain classifies herbal medicines as prescription drugs, OTC medications, herbal treatments, nutritional supplements, and health products [14, 15]. Herbal medications now follow the same restrictions as regular drugs. Bahrain classifies herbal medicines as prescription drugs, OTC medications, herbal treatments, nutritional supplements, and health goods. Herbal medicine manufacturers must follow the same requirements as regular pharmaceuticals. Manufacturers must submit samples of their products to a government-approved laboratory for testing and obtain GMP certification from their local authorities to comply with regulations. In Bahrain, herbal medications must meet the same safety requirements as regular pharmaceuticals. Bahrain requires registration for all complementary and alternative medical items, with the exception of botanical medicines [15].

12.2.3 In Russia

The regulations governing pharmacy-grade drugs under the "Circulation of Medicines in the Russian Federation," which was adopted in September 2010 and amended in 2014 (N-429-F3, 11/22/2014), also govern herbal products of pharmacopoeia grade in Russia. These regulations outline the steps and policies for granting licenses for herbal medicines and registering them with the state; they also cover manufacturing processes and logistics, importing and exporting herbals, selling and advertising guidelines, quality control and standardization, research and development, and trials on both humans and animals [16].

12.2.4 In Australia

Under the Australian National Medicines Policy, herbal medicine is governed in Australia. Herbal medications are divided into three categories under the policy: listed, registered, and exempt. The Australian Therapeutic Goods Administration evaluates the preparations and ingredients based on the first two categories; manufacturers must adhere to Good Manufacturing Practice guidelines when manufacturing. The third group addresses limitations on homeopathic remedies, unprocessed and preliminary materials for conditional use, medications with prescribed therapeutic uses, and medications for individual use. In addition to identifying the seven active ingredients, the labeling of herbal medicines should also specify the intended use and dosage guidelines based on the specified standardized constituents [17].

12.2.5 In India

The Drug and Cosmetic Act, 1940, and the Drug and Cosmetic Rules, which were amended in multiple states, established the country's policy for complementary and alternative medicine in India. The Indian government revised the Drug and Cosmetic Act in 1959 and acknowledged the traditional Indian System of Medicine (ISM). A number of expert committees for various ISM have been formed over the years; the first one was founded in 1962. Act 13 of 1964 added a separate chapter to the act in 1969 that dealt with Ayurvedic, Siddha, and Unani medicines. This chapter was somewhat comparable to the ones for conventional pharmaceuticals. The act was later changed once more in the years 1983, 1987, 1994, and 2002, with a few substitutions. The Drug and Cosmetic Rules, 1945, provided guidelines for the evaluation and analysis of pharmaceuticals under ISM in 2006 and 2008. Since its establishment in 1970, the Central Council of Indian Medicine (CCIM) has been involved in the development and implementation of numerous laws, including those pertaining to the curricula and syllabi of Indian medicine (Ayurveda, Siddha, and Unani). The Sowa Rigpa medical system was integrated into the CCIM in 2012. With the intention of developing the ISM, the Department of Indian Medicine and Homeopathy (ISM & H) was established. The Department of Ayurveda, Yoga and Naturopathy, Unani, Siddha and Homeopathy (AYUSH) replaced this department in 2003. A separate ministry dedicated to AYUSH was established in 2014 [18, 19].

Recent developments in the regulation of herbal medicine show how supervisory procedures have evolved significantly with the goal of improving product quality, efficacy, and customer safety. Comparing these modifications to general pharmacological standards highlights how important it is to incorporate traditional treatments into contemporary healthcare frameworks.

12.3 ADVANCEMENT IN THE INTEGRATION OF HERBAL MEDICINE FOR CLINICAL TRIAL

There are three ways in which herbal medicine can be integrated. It can, first and foremost, be integrated as a fundamental component of a country's healthcare system, being acknowledged as an acceptable and crucial form of medical care within the same framework. Second, it can be used by specific medical professionals in conjunction with contemporary medicine. Third, conventional and contemporary medical practices can coexist as separate branches of the field, eventually combining aspects of both to create an entirely novel discipline [20]. Both contemporary and traditional evidence-based medicine will be integrated into a nation's formal healthcare system; in fact, this is already feasible in many nations, especially Asian nations like China, Japan, Korea, and India [21, 12, 19].

Herbal medicine's incorporation into clinical trials is a developing field of study that combines conventional wisdom with cutting-edge scientific techniques. This shift is a result of a growing understanding of the possible advantages of herbal treatments, as well as the need to conduct thorough clinical studies to assess their safety and efficacy [22].

Many cultures, including traditional Chinese medicine, Ayurveda, and indigenous healing methods, have been using herbal medicine for ages. Historically, because of variations in composition, dosage, and the intricacy of herbal formulations, the scientific community has frequently regarded herbal medicines with suspicion. However, recent advances in pharmacognosy and phytochemistry have brought greater insight into the active chemicals found in these plants, allowing for more structured study methodologies [23].

12.4 RECENT ADVANCEMENTS IN THE STANDARDIZATION OF HERBAL MEDICINE

Herbal medicine originates from plant materials or components that have undergone essential processes, including harvesting, drying, and storage, and are converted into phytopharmaceuticals. Furthermore, the word includes additional plant-derived crude products that have lost their organic structure, such as essential oils, fatty oils, resins, and gums, which play a vital role in the domain of herbal medicine. Herbal raw materials must undergo standardization before they are used in the production of medicinal products to address the variability found in natural sources. Standardization involves a set of procedures aimed at ensuring that the active ingredients, as well as the resulting extracts, are consistent and phyto-equivalent. A basic work design and components involved in standardization methods were given in Figure 12.1. This process guarantees that the effectiveness and safety of the materials and their final products are reproducible, maintaining the quality of the herbal medicine from batch to batch. Standardization ensures that each dose of chemicals has a specific amount, quality level, and therapeutic effect [24, 25]. Various analytical techniques, including high-performance liquid chromatography (HPLC), gas chromatography (GC), and mass spectrometry (MS), are utilized to ensure the quality of both raw materials and end herbal products. High-performance thin-layer chromatography (HPTLC) is distinguished by its simplicity and reliability, making it the most efficient tool for quality control. HPTLC fingerprint analysis is frequently employed for the identification, authenticity, and quality control of herbal medications, rendering it an essential approach in contemporary phytopharmaceutical evaluation [26, 27].

12.4.1 Fingerprinting Approaches for Standardization of Herbal Medicine

Fingerprinting is a vital method for the standardization and quality control of herbal medicine. It ensures consistency, safety, and efficacy by identifying and quantifying various chemical constituents in herbal products. Several chromatographic and spectrophotometric techniques, as well as the assessment of physicochemical qualities, can be used to evolve a pattern for recognizing the presence of distinct ingredients. The existence of the component as a whole can be determined. Fingerprinting techniques can be used, wherever feasible, for the quantitative determination of specific chemicals or bioactive groups of compounds such as

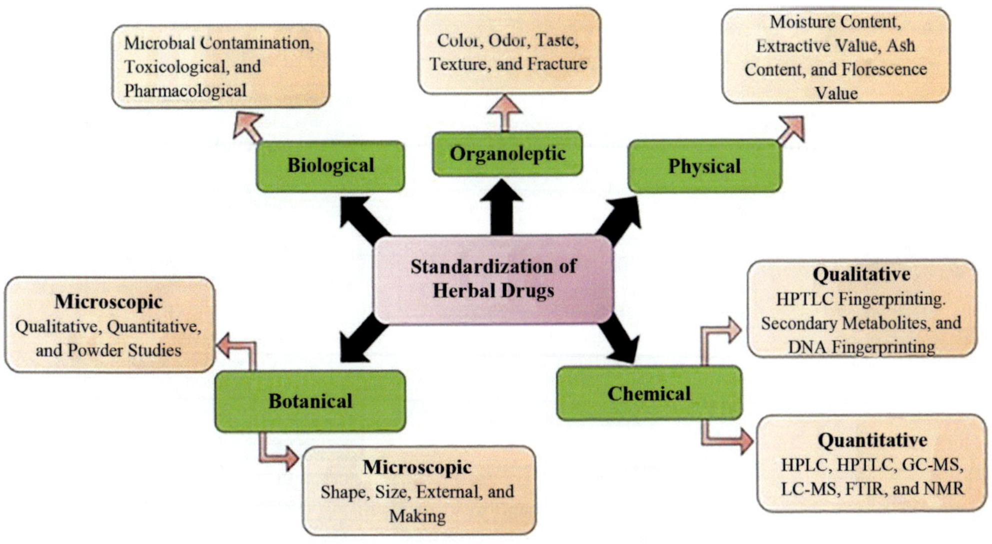

FIGURE 12.1 Basic parameters involved in standardization of herbal medicines.

alkaloids, flavonoids, and polyphenolic and other secondary metabolites [28, 29]. Fingerprinting procedures are employed to ascertain seasonal and geographical origins, taxonomic classification, and the evaluation of extraction processes, quality control, and verification of herbal pharmaceuticals. The fingerprinting method involves methods such as spectroscopic, chromatographic, DNA barcoding, electrophoretic, and direct MS methods. A pictorial representation of fingerprinting methods is shown in Figure 12.2.

12.4.2 Spectroscopic Fingerprinting

Spectroscopic fingerprinting is a technique for identifying and authenticating items, including herbal medications, by creating distinct spectral patterns or "fingerprints" based on their molecular composition. These spectral fingerprints can be checked against reference data to ensure quality, consistency, and validity. Fingerprinting uses a variety of spectroscopic approaches, each of which provides unique insights into the substance's chemical makeup and structure [29]. FTIR and NMR spectroscopies are extensively employed for quality control and herbal authenticity using fingerprint profiling. FTIR is a widely used spectroscopic technique that can detect microgram quantities of materials. Functional groupings of herbal medicine components can absorb IR photons at specific frequencies (or wavenumbers), producing fingerprint spectra suitable for herbal authentication. Chemometrics technologies, particularly exploratory data analysis and pattern identification, can help with data interpretation [30].

12.4.3 Chromatographic Fingerprinting

Chromatographic fingerprinting is a common method for quality control, verification, and standardization of herbal medications. This method maintains consistency, purity, and efficacy by recognizing and quantifying several components in complex combinations. Chromatographic methods are particularly successful for isolating, identifying, and assessing the different chemicals found in herbal products [29]. Given the complexity of plant materials and extracts, chromatographic procedures (e.g., TLC, HPLC, and GC) combined with spectroscopic detectors (UV/diode array detector (DAD), MS, and NMR) are commonly used to standardize herbal medications. Because distinct chemicals may have identical UV spectra, MS or HR-MS/MS is the detector of choice. While LC-NMR can also be used, the cost of analysis with this equipment may be high [31].

12.4.4 Electrophoretic Fingerprinting

Electrophoretic fingerprinting is a technique for separating and identifying molecules such as proteins, nucleic acids, and tiny molecules based on their size, charge, or other physical characteristics. This approach is very useful in quality monitoring and verification of herbal medications, as it allows for comprehensive profiling of physiologically active components [29]. Electrophoretic fingerprinting produces distinct patterns or "fingerprints" for various samples, which can subsequently be compared to reference profiles to assure authenticity, purity, and consistency. Fingerprinting with electrophoresis has significant advantages for the investigation of complicated herbal medicines due to its separation capability [29]. Electrophoretic techniques have been used for decades to help with molecular and chemical fingerprinting. These techniques can effectively differentiate charged compounds, such as nucleic acids and metabolites, resulting in fingerprints [32].

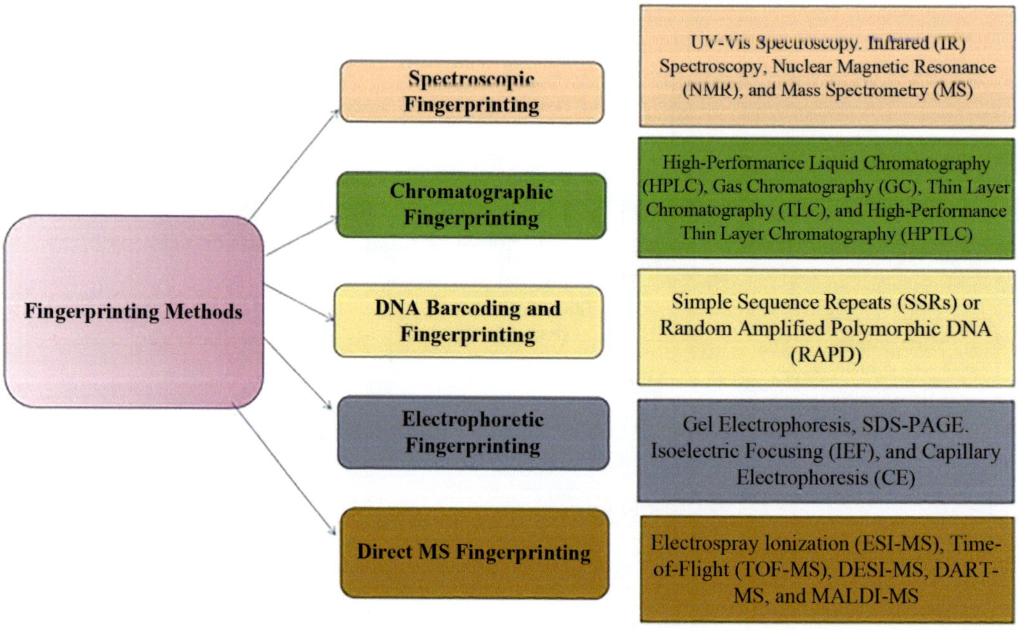

FIGURE 12.2 Fingerprinting methods used for the standardization of herbal drugs.

12.4.5 Direct MS Fingerprinting

Direct MS fingerprinting is a sophisticated analytical approach that allows for the quick identification and characterization of complex combinations, such as herbal medications, without the need for lengthy sample preparation or chromatographic separation. MS creates a comprehensive chemical "fingerprint" of a sample's molecular composition by examining the mass-to-charge ratio (m/z) of its ions [33, 29]. This technology is highly sensitive and accurate, and it can analyze a large range of substances at the same time. Different types of MS fingerprinting techniques are Matrix-Assisted Laser Desorption/Ionization (MALDI-MS), Electrospray Ionization Mass Spectrometry (ESI-MS), Desorption Electrospray Ionization (DESI-MS), and Direct Analysis in Real Time Mass Spectrometry (DART-MS) [34].

12.4.6 DNA Barcoding and Fingerprinting

DNA barcoding and DNA fingerprinting are molecular techniques used to identify, authenticate, and monitor the quality of herbal medications. These approaches give genetic "fingerprints" of plants, guaranteeing the correct species is used while preventing adulteration or misidentification [29]. Both strategies are crucial for standardizing herbal medications. The methods include DNA extraction, PCR amplification, gel electrophoresis, capillary electrophoresis, DNA sequencing, RAPD, AFLP, SSR, ISSR, and RFLP techniques [35].

12.4.7 Chemometric Approach to Data Analysis in Fingerprint

Chemometric approaches are mathematical and statistical tools for extracting important information from complicated datasets, such as those produced by fingerprinting techniques (e.g., spectroscopic, chromatographic, and electrophoretic data) in herbal medication standardization. These methods aid in pattern recognition, sample classification, and compound quantification, resulting in improved quality control, authenticity, and discrimination of herbal medications [29, 36]. Recent studies on chemometric methods in fingerprinting are summarized in Table 12.1.

Fingerprinting procedures are used to standardize herbal products in a complete and reproducible manner, ensuring quality, safety, and efficacy. These techniques—whether spectroscopic, chromatographic, electrophoretic, DNA-based, or chemometric—are critical for accurately identifying active ingredients, detecting adulteration, and maintaining overall quality control of herbal medicines. Combining these fingerprinting procedures ensures the validity and consistency of herbal medications in both raw materials and finished products. HPLC fingerprinting has been employed for the evaluation and standardization of various plants such as *Zingiber officinalis*, *Curcuma longa*, and *Phyllanthus emblica*. some of the recent studies regarding fingerprinting techniques are discussed in Table 12.2.

12.5 OPPORTUNITIES AND CHALLENGES FOR CLINICAL TRIALS OF HERBAL MEDICINES

Clinical trials of herbal medicines present opportunities for innovation and integration into mainstream healthcare; there are also obstacles to be overcome, including the requirement for strong scientific evidence, product standardization, and regulatory barriers [61, 62]. Clinical trials of herbal medicines present the following opportunities.

12.5.1 Opportunities

12.5.1.1 Opportunities in Globalization

Because of people's curiosity about herbal medicines and their remarkable acceptance for their helpful capabilities, with little to no side effects, against a variety of difficult

TABLE 12.1
Summary of Chemometric Methods in Fingerprinting

Method	Purpose	Application
Principal Component Analysis (PCA)	Reduce dimensionality, find patterns	Classify samples, identify major sources of variability
Cluster Analysis (CA)	Group samples based on similarity	Classify samples by species, origin, and quality
Partial Least Squares (PLS)	Relate fingerprint data to target properties	Quantify compounds, predict bioactivity
Discriminant Analysis (DA)	Classify samples into predefined groups	Authenticate herbal products, differentiate species
Support Vector Machine (SVM)	Classify complex, nonlinear data	Authenticate herbal drugs, identify adulteration
Multivariate Curve Resolution (MCR)	Resolve overlapping signals, quantify components	Analyze mixtures, identify individual compounds
Artificial Neural Networks (ANN)	Pattern recognition, classification	Classify and predict herbal sample quality
Principal Component Regression (PCR)	Dimensionality reduction with regression	Predict concentrations of compounds or quality markers
Soft Independent Modeling of Class Analogy (SIMCA)	Build PCA models for classification	Authenticate herbal samples, quality control
Hierarchical Modeling	Analyze structured data with nested relationships	Identify variations between batches or sources

TABLE 12.2
Different Fingerprinting Techniques Involved in The Standardization of Herbal Drugs and Formulations

S. No.	Fingerprinting Techniques/ Methods	Medicinal Plants/Herbal Preparations	Validatory Methods for Evaluation	Observations	References
1.	Ultra-high-performance liquid chromatography with diode array detector (UHPLC-DAD)	Jiawei Huoxiang Zhengqi Pill	Quantification using tandem mass spectrometry (UPLC-MS/MS) and quality evaluation	109 compounds were identified and analyzed	Lu et al., 2024. [37]
2.	HPLC fingerprint and electrochemical fingerprint (ECFP)	Tianma Toutong Tablet (*Gastrodia elata* Bl., *Angelica dahurica* Benth., *Ligusticum chuanxiong* Hort., *Schizonepeta tenuifolia* Briq., *Angelica sinensis* Diels., and *Boswellia carterii* Birdw)	Cosine angle method, SQFM, and CSF model	Estimation of components and improvement of formulation quality	Wang et al., 2024. [38]
3.	High-performance liquid chromatography (HPLC) UHPLC-QTRAP-MS/MS	*Lysimachiae* Herba (Jinqiancao)	Precision, stability, and repeatability of RSDs	Pharmacological effects on cholestatic liver injury and screening of active compounds	Zhou et al., 2024. [39]
4.	High-performance liquid chromatography method	*Centella* Herba	Similarity analysis, hierarchical cluster analysis, and principal PCA with multivariate econometric analysis	Determination of five triterpenoid active components	Yu et al., 2024. [40]
5.	Ultra-performance liquid chromatography coupled with Fourier transform ion cyclotron resonance mass spectrometry (UPLC-FT-ICR-MS)	Changan powder	Relative standard deviation (RSD) of precision and stability	Quick and effective identification of chemical components	Wang et al., 2024. [41]
6.	Ultra-high-performance liquid chromatography (UPLC) chemometric analysis and UPLC-Q-TOF/MS analysis	*Phyllanthus emblica*	Peak resolution, shapes, and response values, and similarity assessment using multipoint correction and the median method	Identification of eight different compounds	Xu and Chen, 2024. [42]
7.	High performance liquid chromatography (HPLC) and Fourier transform Infrared spectroscopy (FTIR)	"Yiqing" tablet (*Scutellaria baicalensis* Georgi, *Rheum officinale* Baill, and *Coptis chinensis)*	SQFM method; accurate and fast multicomponent quantification; multi-markers assay by mono-linear method (MAML); partial least squares regression (PLSR) model	Samples had good quality consistency within the acceptable quality range	Liu et al., 2024. [43]
8.	Near-infrared spectrometry (NIR)	Ginger (*Zingiber officinale* Rosc)	Optimal calibration model with root mean square error of calibration	6-gingerol content and provides geographical traceability-based identification	Naskar et al., 2024 [44]
9.	HPLC-DAD (high-performance liquid chromatography coupled with diode array detection)	*Cinnamomum tamala*	The LOD and LOQ values of the compounds and the RSD percentage	Five bioactive compounds, *viz.*, coumarin, cinnamyl alcohol, cinnamic acid, cinnamaldehyde, and cinnamyl acetate, were identified and quantified in samples	Champati et al., 2024 [45]

(Continued)

TABLE 12.2 ***(Continued)***
Different Fingerprinting Techniques Involved in The Standardization of Herbal Drugs and Formulations

S. No.	Fingerprinting Techniques/ Methods	Medicinal Plants/Herbal Preparations	Validatory Methods for Evaluation	Observations	References
10.	Species-specific start codon-targeted SCoT, polymorphism and CAAT-box-derived polymorphism (CBDP)-based (SCAR) markers	*Senna alexandrina* Mill.	Translational tool to bridge the distance between the species	Distinct amplicons were observed in Indian senna only, while no such amplicon was observed in the other four adulterant species	Chouksey et al., 2024 [46]
11.	HPTLC and LC-MS fingerprint	Kakoli *Roscoea purpurea* Sm., *Roscoea capitata* Sm., *Roscoea alpina* Royle.	Fragmentation pattern and similarity score	*R. purpurea* Sm. could be the most likely source of Kakoli	Acharya et al., 2024 [47]
12.	Fourier transform near-infrared (FT-NIR)	*Curcuma longa*, *Ocimum sanctum*, and *Withania somnifera*	Metabolic fingerprinting strategy. Extremely good discrimination, accuracy, sensitivity, and specificity	FT-NIR-based fast metabolic fingerprinting can authenticate several essential therapeutic plants	Khan et al., 2023 [48]
13.	LC-ESI-QTOF-MS/MS analysis.	Anemone including *A. obtusiloba*, *A. falconeri*, and *A. tetrasepala*	Distribution among the analyzed species was studied using different statistical tools	Different classes of compounds identified belong to diterpenoids, alkaloids, phenols, and flavonoids, etc.	Syed et al., 2021 [49]
14.	Fourier transform infrared (FTIR) fingerprinting chemometrics:	*Apium graveolens*	Analyze and classify according to their geographical origin using principal component analysis (PCA), CA, and discriminant analysis (DA)	Analytical approaches for A. graveolens geographical provenance will improve herbal material quality control and ensure product safety and efficacy	Kartini et al., 2020 [50]
15.	Electrochemical fingerprint	*Eupatorium fortunei*, *E. japonicum*, *E. chinense*, *E. lindleyanum*, *Ageratina adenophora*, *Cynanchum bungei*, *Fallopia multiflora*, *F. aubertii*, *Dioscorea cirrhosa*, and *Stephania cephalantha*	Taxonomic relationships and degree of differences in electroactive compounds can reflect genetic differences	The results show that fresh sample electrochemical fingerprints can accurately reflect species phylogenetic positions	Fu et al., 2021 [51]
16.	High-performance TLC (HPTLC), HPLC, and GC-MS	Safoof-e-Pathar phori (*Didymocarpus pedicellata*, *Dolichos biflorus*, *Rheum emodi*, *Raphanus sativus*, potassium carbonate, and potassium nitrate)	The relative retention times of the characteristic peaks in the HPLC fingerprinting and GC-MS were used for comparative checks	Fingerprint profiling of different compounds present in single drugs	Ahmad et al., 2020 [52]
17.	HPSEC-MALL-RID, FTIR, and HPLC	Zishen Yutai Pills (ZYP)	Correlation coefficient, Similarity Evaluation System	Reliable and effective for ZYP and polysaccharide-enriched herbal medication quality control	Li et al., 2020 [53]

(Continued)

TABLE 12.2 *(Continued)*
Different Fingerprinting Techniques Involved in The Standardization of Herbal Drugs and Formulations

S. No.	Fingerprinting Techniques/ Methods	Medicinal Plants/Herbal Preparations	Validatory Methods for Evaluation	Observations	References
18.	Mechano-chemical extraction (MCE)/Real-time mass spectrometry (DART-MS)	Herbal polysaccharides in *Dendrobium officinale*	Decomposed into small fragment ions less than *m/z* 350 instantaneously using a DART ion source	Mechanochemical extraction and DART-MS for rapid polysaccharide fingerprint analysis in medicinal plants	Wang et al., 2021 [54]
19.	(UPLC), (FTIR), and (QAMS) methods	Yankening Tablets (YKNT)	The contents of berberine and baicalin measured by the calibration curve method (CCM) and the QAMS method were compared, and *t*-test results were analyzed	Hierarchical cluster analysis categorized all samples into four groups. This research greatly improved TCM and food quality evaluation	Wang et al., 2021 [55]
20.	Electrochemical fingerprint, HPLC	Gastrodia elata	PCA and HCA suggest that electrochemical fingerprint analysis can serve as a complementary tool to conventional HPLC analysis for a more comprehensive quality assessment of G. elata	Improved consistency in ginger-processed samples	Liao et al., 2024 [56]
21.	Comprehensive two-dimensional gas chromatography (GC × GC).	Curcuma *wenyujin*	Affinity propagation (AP) clustering for peak detection Contour edge tracking	Quality assessment of Curcuma products	Yang et al., 2024 [57]
22.	High-performance liquid chromatography (HPLC) fingerprints combined with chemometric analysis	*Coptidis rhizome*	Cluster analysis, PCA, Pearson correlation analysis, gray relation analysis, and PLSR	Antioxidant activity evaluation, fingerprint-efficacy relationship analysis, and simultaneous quantification of multiple components	Tian et al., 2024 [58]
23.	High-performance liquid chromatography (HPLC) fingerprinting	*Angelica sinensis*	Model prediction performance metric, relative closeness of the TOPSIS method, and analysis of similarity	Reference for the quality control of Angelica sinensis	Liang., 2024 [59]
24.	LC-MS, GC-MS, FT-NIR, and combined chemometrics	*Ocimum* species	PCA, partial least squares discriminant analysis (PLS-DA), DD-SIMCA, random forest, and k-nearest neighbors	Rapid screening and confirmatory approaches as a highly efficient means to identify the species and variety of *Ocimum*	Rai et al., 2023 [60]

health-related conditions, the traditional healthcare system is becoming more and more popular globally. Currently, 60% of the global population uses conventional and herbal medicines as their main means of treating fever caused by malaria. When it comes to treating various illnesses, 80% of Africans, 30–50% of Chinese people, 48% of Australians, 70% of Canadians, 80% of Germans, 42% of Americans, 39% of Belgians, and 76% of French people prefer herbal or alternative medications [63]. Here are some examples of herbal medicines being used globally: the herb ginseng is extensively utilized in traditional Chinese medicine for its adaptogenic properties, which boost vitality and lower stress [64]. A growing number of industrialized nations have also embraced herbal remedies, and the United

Kingdom, Australia, Europe, and North America are all seeing an enormous rise in the use of these remedies [65].

12.5.1.2 Opportunities in Research and Education

Nowadays, preclinical or clinical studies, toxicological assessment, standardization, and the production of herbal products are the main areas of focus for research on traditional remedies. Numerous governmental and private research institutes, organizations, and colleges are actively researching, developing, and promoting conventional herbal treatments [61, 62].

Here are some examples of ongoing research on herbal medicine. Research is exploring the anti-inflammatory and antioxidant properties of curcumin for chronic diseases [66]. Another ongoing research is evaluating the efficacy of echinacea in preventing and treating respiratory tract infections [67].

In order to rediscover herbal plants as a source of novel pharmaceuticals, scientists and multinational pharmaceutical companies have combined traditional wisdom with the most recent scientific knowledge and technology [62]. The evolution of drug discovery from herbal plants has been altered by new methods and technologies, which are now essential resources for the study of traditional remedies. Novel phytoconstituents identified in traditional formulations or plants have been separated and developed with the aid of state-of-the-art research instruments and methodologies. As a result, these cutting-edge techniques were crucial for evaluating and upholding the standards of conventional preparations, as well as for the discovery of new drugs, the identification of mechanisms, pharmacokinetic profiles, and toxicity profiles. They were also employed in semisynthetic and synthetic processes for the synthesis of natural constituents [68].

12.5.2 Challenges

Standardization of herbal medicine is a major challenge for clinical trials of herbal medicine. Biological activity and consistency in composition are crucial for the safe and efficient administration of herbal medications [69]. Nonetheless, there are many reasons why herbal medications usually fall short of this requirement: challenges with plant identification; genetic variability; variations in growth environments; differences in harvesting and extract processing techniques; and a lack of knowledge regarding active pharmacologic principles [70]. For the purpose of standardizing herbal products, chromatographic techniques and marker chemicals can be employed to ensure batch-to-batch uniformity; however, this is not enough to ensure consistent pharmacologic action or stability. Researchers would face significant challenges if herbal drugs were not standardized, as they would be unable to utilize commercially accessible herbal items for their investigations [71].

Quality control is another challenge in clinical trials of herbal medicine. Quality control is necessary to make sure an herbal treatment works as intended and produces the desired effects. Since quality is a crucial factor in determining safety, quality assurance is vital even in the absence of efficacy data. There is a major issue with plant adulteration. Medicinal plants, hazardous metals, microbes, microbial poisons, insecticides, and fumigants are a few examples of common adulterants. Ephedrine, chlorpheniramine, methyltestosterone, and phenacetin were the most commonly discovered substances; lead, mercury, or arsenic was present in 10% to 15% of the samples [69].

Drug interactions and various safety concerns can also play a significant role in clinical trials of herbal medicines. Herbal medicines are generally thought to be safer than synthetic drugs [72]. While this may be true, case studies indicate that serious side effects and relevant interactions with other drugs can occur. For example, the herb *Ephedra*, which was marketed as a dietary aid, has been linked to at least a dozen deaths, heart attacks, and strokes. Other well-known safety issues include the hepatotoxicity of kava and the renal effects of aristolochic acid. In addition, drug interactions with herbal drugs are a serious concern. For instance, hypericum extracts have been shown to decrease the concentration of a variety of other drugs through enzyme induction [73].

Maintaining the same color, smell, and taste of the herbal medication in the active and control groups is exceedingly challenging, impractical, and occasionally unachievable. Comparable challenges arise when using placebos as well because, unlike herbal research drugs, they cannot be manufactured to have the same distinct flavor or potent perfume. Distinct tastes are impossible to replicate while creating a placebo. A multi-herbal system–integrated approach fails to distinguish between the patient and the disease [74].

12.6 CONCLUSION

Herbal medicine has been a significant factor in preserving people's health since ancient times. Recent studies have shown that the traditional medical systems of India, China, Japan, and the United States have a strong scientific foundation and are successful. The WHO has produced quality control manuals and recommendations for GACP and GCP for medicinal plants and materials. Recent advances in herbal medicine clinical trials reflect an increasing understanding of its potential advantages, as well as a commitment to rigorous scientific investigation. Important advancements consist of the following: Improved methods for standardizing herbal products have increased the consistency and dependability of trial results by applying sophisticated analytical techniques like MS and HPLC. The methodological rigor of studies on herbal medicine has been strengthened by the use of adaptive trial designs and multisite cooperation, enabling more dynamic evaluation of safety and efficacy. A more evidence-based approach and easier integration into traditional medical practices are made possible by the increased focus on comprehending the biochemical mechanisms of herbal medications. Credibility is being increased by regulatory authorities' attempts to create more precise criteria for the use of herbal medicines in clinical trials. This is creating a more organized research environment. More pertinent and effective trial designs have resulted from an emphasis on patient-reported outcomes and customized medicine, guaranteeing the patients' needs and experiences are at the center of research endeavors.

Integrative therapeutic approaches are becoming more popular, especially in the management of chronic illnesses, as a result of growing research into the synergistic benefits of herbal remedies and mainstream medicine. With the ultimate goal of improving patient care and results, these developments are essential for expanding our knowledge of herbal medicine from a scientific perspective and for its clinical application. In order to fully realize the promise of herbal medicines in contemporary healthcare, it is imperative that researchers, physicians, regulatory authorities, and the herbal medicine community continue to collaborate.

REFERENCES

[1] U.S. Food and Drug Administration (2016). Botanical drug development: Guidance for industry. Silver Spring, MD: U.S. Food and Drug Administration.

[2] Xutian, S., Zhang, J., & Louise, W. (2009). New exploration and understanding of traditional Chinese medicine. *American Journal of Chinese Medicine*, 37(3), 411–426.

[3] Cragg, G. M., Newman, D. J. (2013). Natural products: A continuing source of novel drug leads. *Biochimica et Biophysica Acta*, 1830, 3670–3695.

[4] Kinghorn, A. D., Pan, L., Fletcher, J. N., & Chai, H. (2011). The relevance of higher plants in lead compound discovery programs. *Journal of Natural Products*, 74, 1539–1555.

[5] Chen, W., Vermaak, I., & Viljoen, A. (2013). Camphor-a fumigant during the black death and a coveted fragrant wood in ancient Egypt and Babylon: A review. Molecules, 18, 5434–5454.

[6] Hamidpour, R., Hamidpour, S., Hamidpour, M., & Shahlari, M. (2013). Camphor (Cinnamomum camphora), a traditional remedy with the history of treating several diseases. International Journal of Case Reports and Images, 4(2), 86–89.

[7] Srivastava, K. C., Bordia, A., & Verma, S. K. (1995). Curcumin, a major component of food spice turmeric (Curcuma longa) inhibits aggregation and alters eicosanoid metabolism in human blood platelets. Prostaglandins, Leukotrienes & Essential Fatty Acids, 52(4), 223–227.

[8] Chopra, B., Dhingra, A. K., & Prasad, D. N. (2020b). Modification in the natural bioactive molecule: Piperine; a continuing source for the drug development. Current Bioactive Compounds, 16(6), 714–725.

[9] Singh, H. (2007). Prospects and challenges for harnessing opportunities in medicinal plants sector in India. Law, Environment and Development Journal, 2(2), 196–211.

[10] Bandaranayake, W. M. (2006). Quality control, screening, toxicity, and regulation of herbal drugs. in Modern phytomedicine: Turning medicinal plants into drugs, eds. I. Ahmad, F. Aqil, & M. Owais. Weinheim: Wiley-VCH GmbH & Co. KGaA, 25–57.

[11] Nielsen, B. (2020). Regulatory frameworks for herbal medicines: Current trends and future directions. International Journal of Herbal Medicine, 8(2), 22–30.

[12] Teng, L., Zu, Q., Li, G., Yu, T., Job, K. M., Yang, X., Di, L., Sherwin, C. M., & Enioutina, E. Y. (2016). Herbal medicines: Challenges in the modern world. Part 3. China and Japan. Expert Review of Clinical Pharmacology, 9, 1225–1233.

[13] Picking, D. (2024). The global regulatory framework for medicinal plants. In Pharmacognosy, 1, 769–782. https://doi.org/10.1016/B978-0-443-18657-8.00023-2

[14] World Health Organization (2019). WHO global report on traditional and complementary medicine 2019. Geneva: World Health Organization.

[15] Alostad, A. H., Steinke, D. T., & Schafheutle, E. I. (2019). A qualitative exploration of Bahrain and Kuwait herbal medicine registration systems: Policy implementation and readiness to change. JPPP, 12, 32. https://doi.org/10.1186/s40545-019-0189-7

[16] Sammons, H. M., Gubarev, M. I., Krepkova, L. V., Bortnikova, V. V., Corrick, F., Job, K. M., Sherwin, C. M., & Enioutina, E. Y. (2016). Herbal medicines: Challenges in the modern world. Part 2. European Union and Russia. Expert Review of Clinical Pharmacology, 9, 1117–1127.

[17] Barnes, J., McLachlan, A. J., Sherwin, C. M., & Enioutina, E. Y. (2016). Herbal medicines: Challenges in the modern world. Part 1. Australia and New Zealand. Expert Review of Clinical Pharmacology, 9, 905–915.

[18] Ministry of AYUSH, CCRAS (2018). Guidelines for standardization, quality control, and safety evaluation of Ayurvedic formulations. New Delhi: Ministry of AYUSH, vol. 1, 1–11.

[19] Sahoo, N., & Manchikanti, P. (2013). Herbal drug regulation and commercialization: An Indian industry perspective. The Journal of Alternative and Complementary Medicine, 19, 957–963.

[20] World Health Organization and Western Pacific Region (2000). Traditional and modern medicine: Harmonizing the two approaches. Manila: World Health Organization. https://iris.who.int/bitstream/handle/10665/207162/19991126_CHN_eng.pdf?sequence=1

[21] Dong, J. C. (2013). The relationship between traditional Chinese medicine and modern medicine. Evidence-Based Complementary and Alternative Medicine, 2013, 153148.

[22] Liu, Y., Xu, Y., Chen, Y., & Li, X. (2022). Synergistic effects of herbal medicines with conventional treatments. Frontiers in Pharmacology, 13, 856332.

[23] Zhao, J., Zhang, X., Liu, Y., Yang, L., & Wang, Y. (2020). Quality control of herbal medicines. Frontiers in Pharmacology, 11, 660.

[24] Lamichhane, S., Sahariah, B. J., Das, B., Khatiwara, D., Sola, P., Adhikari, R. P., . . . & Alam, T. (2023). Herbal drug standardization: A systemic review. Journal of Drug Delivery and Therapeutics, 13(4), 149–153.

[25] Kaggwa, B., Anywar, G., Munanura, E. I., Wangalwa, R., Kyeyune, H., Okella, H., & Engeu, O. P. (2023). Application of the herbal chemical marker ranking system (Herb MaRS) to the standardization of herbal raw materials: A case study. BMC Complementary Medicine and Therapies, 23(1), 348.

[26] Balekundri, A., & Mannur, V. (2020). Quality control of the traditional herbs and herbal products: A review. Future Journal of Pharmaceutical Sciences, 6, 1–9.

[27] Fitzgerald, M., Heinrich, M., & Booker, A. (2020). Medicinal plant analysis: A historical and regional discussion

of emergent complex techniques. Frontiers in Pharmacology, 10, 1480.
[28] Lamichhane, S., Sahariah, B. J., Das, B., Khatiwara, D., Sola, P., Adhikari, R. P., & Alam, T. (2023). Herbal drug standardization: A systemic review. Journal of Drug Delivery and Therapeutics, 13(4), 149–153.
[29] Noviana, E., Indrayanto, G., & Rohman, A. (2022). Advances in fingerprint analysis for standardization and quality control of herbal medicines. Frontiers in Pharmacology, 13, 853023.
[30] Sahoo, M. R., & Umashankara, M. S. (2023). FTIR based metabolomics profiling and fingerprinting of some medicinal plants: An attempt to develop an approach for quality control and standardization of herbal materials. Pharmacognosy Research, 15(1).
[31] Cuadros-Rodríguez, L., Ortega-Gavilán, F., Martín-Torres, S., Arroyo-Cerezo, A., & Jiménez-Carvelo, A. M. (2021). Chromatographic fingerprinting and food identity/quality: Potentials and challenges. Journal of Agricultural and Food Chemistry, 69(48), 14428–14434.
[32] Gackowski, M., Przybylska, A., Kruszewski, S., Koba, M., Mądra-Gackowska, K., & Bogacz, A. (2021). Recent applications of capillary electrophoresis in the determination of active compounds in medicinal plants and pharmaceutical formulations. Molecules, 26(14), 4141.
[33] Wang, Y., & Liu, S. (2024). Recent application of direct analysis in real time mass spectrometry in plant materials analysis with emphasis on traditional Chinese herbal medicine. Mass Spectrometry Reviews, 43(5), 1150–1171.
[34] Li, Y., Shen, Y., Yao, C. L., & Guo, D. A. (2020). Quality assessment of herbal medicines based on chemical fingerprints combined with chemometrics approach: A review. Journal of Pharmaceutical and Biomedical Analysis, 185, 113215.
[35] Gesto-Borroto, R., Medina-Jiménez, K., Lorence, A., & Villarreal, M. L. (2021). Application of DNA barcoding for quality control of herbal drugs and their phytopharmaceuticals. Revista Brasileira de Farmacognosia, 31, 127–141.
[36] Li, H., Cao, J., Wu, X., Deng, Y., Ning, N., Geng, C., & Wang, Y. (2020). Multiple fingerprint profiling for quality evaluation of polysaccharides and related biological activity analysis of Chinese patent drugs: Zishen Yutai Pills as a case study. Journal of Ethnopharmacology, 260, 113045.
[37] Lu, Y., Lei, T., Chen, X., Ning, N., Huang, Q., Wu, X., & Cao, J. (2024). A comprehensive strategy based on ultra-high performance liquid chromatography with diode array detector fingerprinting and multi-component ultra-performance liquid chromatography with tandem mass spectrometry technology for quality control of Jiawei Huoxiang Zhengqi Pill. Journal of Separation Science, 47(13), 2400308.
[38] Wang, S., Yang, T., Guo, P., Lan, L., & Sun, G. (2024). A new method to comprehensively evaluate the quality of Tianma Toutong tablets by multiple fingerprints combined with quantitative analysis and prescription analysis. Journal of Pharmaceutical and Biomedical Analysis, 242, 116008.
[39] Zhou, Y., Chen, C., Yuan, J., Xue, J., Chen, H., Liu, X., Cheng, J. (2024). A study for quality evaluation of Lysimachiae herba from different origins based on fingerprint-activity relationship modeling and multi-component content determination. Journal of Ethnopharmacology, 325, 117840.
[40] Yu, D., Zhongjing, G., Zhimin, Z., Depo, Y., & xinjun, X. (2024). Establishment of chemical fingerprint and determination of main active ingredients in Centellae Herba by HPLC combined with multivariate chemometrics analysis. Separation Science Plus, 2400021.
[41] Wang, Z., Zhang, Y., Fu, Z., Jin, T., Zhao, C., & Zhao, M. (2024). A comprehensive strategy for quality evaluation of Changan powder by fingerprinting combined with rapid qualitative and quantitative multi-ingredients profiling. Phytochemical Analysis, 35(4), 840–859.
[42] Xu, Y. H., & Chen, J. (2024). Discovery of quality markers of Phyllanthus emblica by integrating chromatographic fingerprint, serum pharmacochemistry and network pharmacology. Journal of Pharmaceutical and Biomedical Analysis, 249, 116346.
[43] Liu, X., Yang, T., Chen, L., Lan, L., Sun, G., & Guo, P. (2024). A strategy takes "Yiqing" tablets as an example to carry out simpler multi-component quantification and use fingerprint technology for comprehensive quality consistency evaluation. Journal of Pharmaceutical and Biomedical Analysis, 238, 115809.
[44] Naskar, S., Sing, D., Banerjee, S., Shcherbakova, A., Bandyopadhyay, A., Kar, A., & Bandyopadhyay, R. (2024). Rapid quality assessment and traceability of ginger powder from Northeast India and Indian market based on near infrared spectroscopic fingerprinting. Phytochemical Analysis, 36(2), 369–376.
[45] Champati, B. B., Das, P. K., Sahoo, C., Ray, A., Jena, S., Sahoo, A., & Panda, P. C. (2024). Chemical fingerprinting and multicomponent quantitative analysis for quality control of Cinnamomum tamala collected from Western Himalaya by HPLC-DAD. Heliyon, 10(9).
[46] Chouksey, S., Ashfaq, M. A., Kaira, P., Farhat, S., Pandey, M., Kumar, C. A., & Nagaraja Reddy, R. R. (2024). Development of highly discriminatory SCoT-and CBDP-based SCAR fingerprint for authentication of Indian senna (Senna alexandrina Mill.) formerly Cassia angustifolia Vahl. Frontiers in Plant Science, 15, 1424665.
[47] Acharya, B., Mishra, R., Thakur, P., Singh, S., Narsingh, S., & Varshney, A. (2024). Pharmacognostic, phytochemical, and chromatographic fingerprinting of three probable species accepted as Kakoli: A member of Astavarga. Indian Journal of Traditional Knowledge (IJTK), 23(8), 747–759.
[48] Khan, S., Rai, A. K., Singh, A., Singh, S., Dubey, B. K., Lal, R. K., & Ch, R. (2023). Rapid metabolic fingerprinting with the aid of chemometric models to identify authenticity of natural medicines: Turmeric, Ocimum, and Withania somnifera study. Journal of Pharmaceutical Analysis, 13(9), 1041–1057.
[49] Syed, M., Khan, M. N., Khadim, A., Shadab, H., Perveen, A., El-Seedi, H. R., & Musharraf, S. G. (2021). Chemical fingerprinting of three Anemone species and an adulteration study to detect cross mixing of medicinal plants by HPLC-HR-ESI-MS/MS method. Journal of King Saud University-Science, 33(5), 101461.
[50] Kartini, K., Putri, L. A. D., & Hadiyat, M. A. (2020). FTIR-based fingerprinting and discriminant analysis of Apium graveolens from different locations. Journal of Applied Pharmaceutical Science, 10(12), 62–67.
[51] Fu, L., Zheng, Y., Wang, A., Zhang, P., Ding, S., Wu, W., & Zhao, S. (2021). Identification of medicinal herbs in Asteraceae and Polygonaceae using an electrochemical fingerprint

recorded using screen-printed electrode. Journal of Herbal Medicine, 30, 100512.

[52] Ahmad, W., Parveen, R., Mujeeb, M., & Zaidi, S. A. (2020). Comparative fingerprint profiling of Unani polyherbomineral (Safoof-e-Pathar phori) formulation by HPTLC, HPLC, and GC-MS. Journal of AOAC International, 103(3), 659–668.

[53] Li, Z., Zhang, X., Liao, J., Fan, X., & Cheng, Y. (2021). An ultra-robust fingerprinting method for quality assessment of traditional Chinese medicine using multiple reaction monitoring mass spectrometry. Journal of Pharmaceutical Analysis, 11(1), 88–95.

[54] Wang, X., Jiang, Q., Li, H., & Chen, D. D. (2021). Rapid fingerprint analysis for herbal polysaccharides using direct analysis in real-time ionization mass spectrometry. Rapid Communications in Mass Spectrometry, 35(16), e9139.

[55] Wang, X., Liu, X., Wang, J., Wang, G., Zhang, Y., Lan, L., & Sun, G. (2021). Study on multiple fingerprint profiles control and quantitative analysis of multi-components by single marker method combined with chemometrics based on Yankening tablets. Spectrochimica Acta Part A: Molecular and Biomolecular Spectroscopy, 253, 119554.

[56] Liao, Y., Wang, P., Zhang, R., Chen, G., & Fu, J. (2024). Investigation of quality control of Gastrodia elata using electrochemical fingerprint technology. International Journal of Electrochemical Science, 19(8), 100702.

[57] Yang, X., Sima, Y., Luo, X., Li, Y., & He, M. (2024). Analysis of GC× GC fingerprints from medicinal materials using a novel contour detection algorithm: A case of Curcuma wenyujin. Journal of Pharmaceutical Analysis, 14(4), 100936.

[58] Tian, M., Nie, L., Yin, Y., Zhou, H., Meng, Z., Cao, G., & Zang, H. (2024). Study on quality analysis of different species of Coptidis rhizome based on fingerprint-effect relationship. Phytochemical Analysis, 35(1), 77–86.

[59] Liang, Y. (2024) Liquid chromatographic fingerprinting technique for quality control of Angelica sinensis and its preparation. Applied Mathematics and Nonlinear Sciences, 9(1), 1–13.

[60] Rai, A. K., Khan, S., Kumar, A., Dubey, B. K., Lal, R. K., Tiwari, A., & Ch, R. (2023). Comprehensive metabolomic fingerprinting combined with chemometrics identifies species- and variety-specific variation of medicinal herbs: An Ocimum study. Metabolites, 13(1), 122.

[61] Wang, Y., Zhang, L., & Liu, Y. (2021). Clinical trials of herbal medicines: Opportunities and challenges. Journal of Ethnopharmacology, 268, 113642.

[62] Ernst, E. (2018). The role of herbal medicine in the future of healthcare: Challenges and opportunities. European Journal of Integrative Medicine, 18, 66–70.

[63] Elachouri, M. (2018). Ethnobotany/ethnopharmacology, and bioprospecting: Issues on knowledge and uses of medicinal plants by Moroccan people, in Natural products and drug discovery. Elsevier, 105–118.

[64] Reay, J. L., Kennedy, D. O., & Scholey, A. B. (2021). The efficacy of Panax ginseng on cancer-related fatigue: A randomized controlled trial. Supportive Care in Cancer, 29(6), 3045–3052.

[65] Ekor, M. J. (2013). The growing use of herbal medicines: Issues relating to adverse reactions and challenges in monitoring safety. Frontiers in Pharmacology, 4, 177.

[66] Anand, P., Thomas, S. G., Kunnumakkara, A. B., Sundaram, C., Harikumar, K. B., & Behl, T. (2022). Curcumin: A review of its effects on human health. Foods, 11(11), 1520.

[67] Woo, P. M., Wong, C. C., & Lau, K. C. (2023). Echinacea for the prevention and treatment of respiratory tract infections: A systematic review. Phytotherapy Research, 37(2), 671–684.

[68] Koparde, A. A., Doijad, R. C., & Magdum, C. S. (2019). Natural products in drug discovery, in Pharmacognosy-medicinal plants, eds. S. Perveen & A. Al-Taweel. IntechOpen, 269–270.

[69] Wang, H., Chen, Y., Wang, L., Liu, Q., Yang, S., & Wang, C. (2023). Advancing herbal medicine: Enhancing product quality and safety through robust quality control practices. Frontiers Pharmacology, 14, 1265178. https://doi.org/10.3389/fphar.2023.1265178

[70] Marcus, D. M., & Grollman, A. P. (2000). Botanical medicines—the need for new regulations. New England Journal of Medicine, 347, 2073–2076.

[71] Lee, J.-S., & Lee, H.-J. (2019). Regulatory challenges in herbal medicine clinical trials. Journal of Ethnopharmacology, 245, 112164.

[72] Fong, W., & Liao, Y. (2020). Ethnopharmacological variability in clinical trials of herbal medicine. Phytomedicine, 68, 153157.

[73] Stein, M. C. (2002). Are herbal products dietary supplements or drugs? An important question for public safety. Clinical Pharmacology & Therapeutics, 71, 411–413.

[74] Chen, Y., Liu, J., Wang, R., & Zhang, L. (2024). Integrating herbal medicine with conventional treatments: Barriers and solutions. Integrative Medicine Research, 13, 100877.

13 Clinical Trials of the Indian System of Medicine Based on Case Studies

Khushi Gupta, Avadh Biharee, Aatif Subhan, and Suresh Thareja
Department of Pharmaceutical Sciences and Natural Products,
Central University of Punjab, Bathinda, Punjab, India

13.1 INTRODUCTION

The Indian System of Medicine (ISM) encompasses a rich tapestry of traditional healing practices, with Ayurveda, Siddha, and Unani medicine at the forefront [1]. Each of these systems is deeply rooted in historical and cultural contexts, reflecting a unique synthesis of philosophy, observation, and practice that has evolved over centuries [2]. Ayurveda, known for its holistic approach to health and wellness, emphasizes the balance of bodily energies and the use of natural substances for healing [3]. Ayurveda is characterized by its comprehensive framework that addresses not only physical health but also mental and spiritual well-being. Siddha medicine, with its origins in South India, integrates a profound understanding of the human body and mind, often utilizing herbal and mineral-based formulations [4]. Siddha is distinguished by its intricate methodologies and a focus on the internal balance of elements within the body, mirroring some of the principles found in Ayurveda while also showcasing unique therapeutic modalities. Unani medicine, influenced by ancient Greek teachings and further developed in the Islamic world, focuses on the humoral theory of health, advocating for individualized treatment based on the body's constitution [5]. Unani categorizes individuals according to their dominant humors and employs a range of therapeutic interventions tailored to restore balance. By leveraging a combination of herbal remedies, dietary adjustments, and various regimental therapies, Unani practitioners aim to achieve holistic wellness. As the demand for evidence-based medicine intensifies, the incorporation of traditional systems into clinical research becomes essential [6]. Clinical trials not only provide a systematic approach to assess therapeutic outcomes but also address the need for standardization in treatment methodologies [7]. This chapter establishes the importance of such evaluations, emphasizing their importance for legitimizing traditional practices within contemporary healthcare frameworks. Assessment of clinical trials pertaining to traditional systems of medicine, particularly Ayurveda, Siddha, and Unani, through various case studies will shed light on the methodologies employed, the types of interventions tested, and the populations involved. Insights into the efficacy and safety of the treatment under investigation will be delineated based on the trial outcomes, followed by implications for future research.

13.2 DEVELOPMENT AND EVOLUTION OF ISM

13.2.1 Historical Background

The development and evolution of the ISM are deeply rooted in an extensive historical and cultural continuum that spans millennia [8]. This tradition of medical practice, often called ethnomedicine, originates as far back as 5000 BCE, reflecting an intricate interplay between spiritual, philosophical, and practical dimensions of healthcare [9]. The earliest recorded evidence of medicinal practices in India is embedded in ancient Vedic literature. The "Rigveda," which dates from approximately 1700 to 1100 BCE, along with the "Yajurveda" (1400–1000 BCE) and the "Atharvaveda" (1200 BCE), outlines early medicinal knowledge and therapeutic principles [10] (Figure 13.1).

These Vedic texts are not merely religious scriptures but also include detailed descriptions of medicinal plants, treatments, and rituals used for healing, indicating a sophisticated understanding of medicinal botany and therapeutic practices from a very early period [11]. The transition from Vedic to classical Indian medicine is marked by the emergence of pivotal texts such as the "Sushruta Samhita" (circa 660 BCE) and the "Charaka Samhita" (circa 990 BCE) [12]. The "Sushruta Samhita" is renowned for its comprehensive treatises on surgical techniques and instruments, while the "Charaka Samhita" focuses on internal medicine, emphasizing a holistic approach to health and disease [13]. These texts codify the principles of Ayurveda, which are grounded in the concept of balance among the body's three doshas, viz. Vata, Pitta, and Kapha. Further evolution of Indian medicine continued through the integration of various therapeutic practices and texts [14]. For instance, the "Dhanwantari Nighantu," compiled around 1800 CE, illustrates the continued evolution and systematization of medicinal knowledge through multiple periods of Indian history. An extensive inventory of medicinal plants and their uses, reflecting indigenous and assimilated knowledge, has been delineated within these textbooks [15]. The dynamic

DOI: 10.1201/9781003522065-13

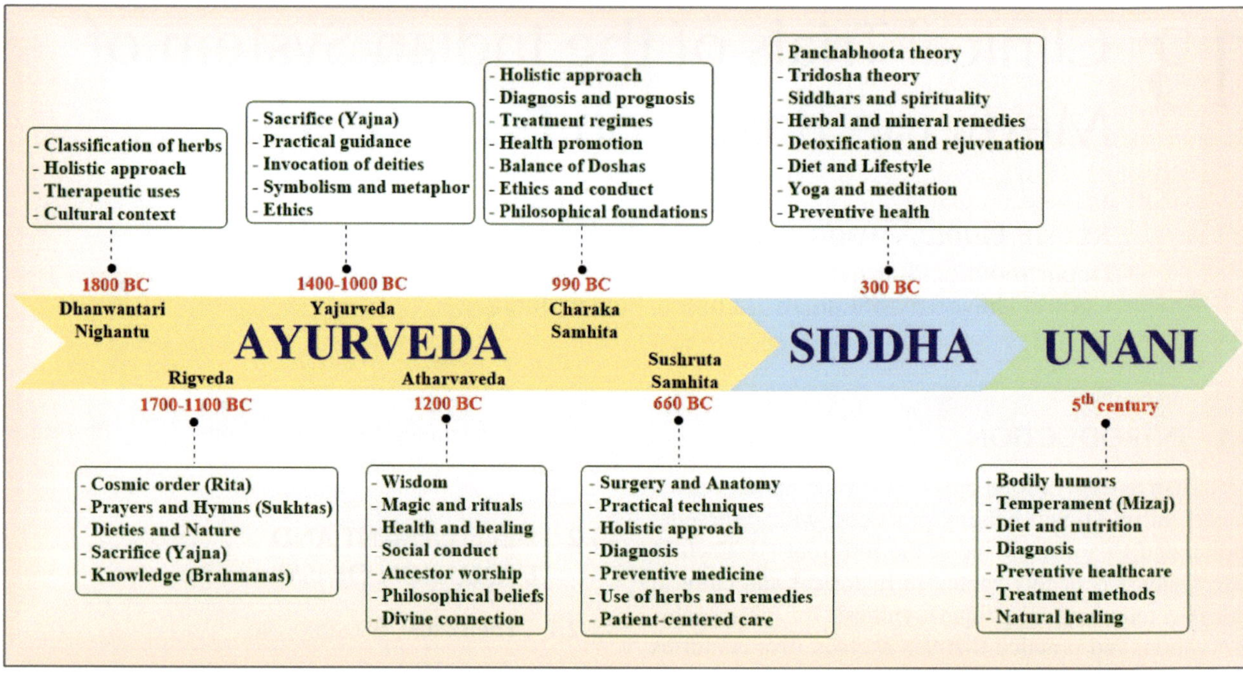

FIGURE 13.1 Historical backdrop of the Indian system of medicine.

nature of Indian medicine is also evident in its assimilation of external medicinal systems. The traditional Indian medicinal system encompasses a diverse array of practices in Ayurveda, Unani, and Siddha medicine, each contributing to a comprehensive medical system [16]. This plurality reflects the historical amalgamation of various medical traditions within the Indian subcontinent.

13.2.2 Foundation Pillars of ISM

13.2.2.1 Ayurvedic System of Medicine

Ayurveda is recognized as one of the most ancient systems of medicine, with its origins tracing back to 2500–500 BCE in India [17]. The term "Ayurveda," derived from the Sanskrit words "Ayus" (life) and "Veda" (science or knowledge), encapsulates its essence as the "science of life" [18]. The foundational texts, primarily the Charaka Samhita, Sushruta Samhita, and Ashtanga Hridaya, emerged around the first millennium BCE, codifying practices and philosophical insights that emphasize the interplay between mind, body, and environment. The classical texts that underpin Ayurvedic medicine include the "Charaka Samhita" and the "Sushruta Samhita," attributed to the scholars Charaka and Sushruta, respectively, around 300 BCE [13]. These texts provide comprehensive guidelines on therapeutic practices and surgical techniques. Additionally, works such as the "Ashtanga Hridaya" and the "Madhava Nidhana" further enrich the Ayurvedic literature [19].

During the Samhita period (6th century BCE–7th century CE), Ayurveda underwent systematic development, leading to its current status as a well-regarded system of medicine in India and globally. Central to Ayurvedic practice are concepts such as the Pancha Mahabhutas (five significant elements: air, water, fire, earth, and ether), the three doshas (*Vata*, *Pitta*, and *Kapha*), *Pancha Karma* (detoxification methods), and the seven *Dhatus* (tissues) [20]. Ayurveda posits that health arises from a balance among these elements and doshas [21]. At the same time, life is regarded as a synthesis of the physical body (*sarira*), senses (*indriya*), mind (*sattva*), and soul (*atma*) [22]. The interplay between macrocosmic and microcosmic principles is fundamental to understanding health and disease in Ayurveda [23]. Ayurveda emphasizes *Dhatus* (seven bodily tissues) and *Malakriya* (waste products) for detoxification, aligning with modern metabolic practices [24]. The concept of Agni (digestive fire) is crucial for health, producing Ojas (vitality and immunity). Additionally, the *Gunatraya* (Sattva, Rajas, Tamas) connects mental health to overall well-being, highlighting the impact of environmental factors and lifestyle choices on health, in line with epigenetic principles [19].

13.2.2.2 Siddha System of Medicine

Siddha medicine has its roots in the Dravidian culture and dates back to between 10,000 and 4000 BCE [16]. The Siddha system is intertwined with the Saiva philosophy of Hinduism, and the term "Siddha" signifies "holy harmony" [25]. This tradition is attributed to 18 Siddhars, who are believed to possess extraordinary intellectual abilities. Siddha medicine incorporates various physiological and psychological factors, such as age, race, diet, and environment, into its diagnostic and therapeutic approaches [26]. Foundational texts such as the Siddha Sangam, Piyamoli, and Agasthiyar's Thirumanthiram codify the principles and methodologies of this medical tradition, effectively integrating spiritual insights with empirical health practices. A distinctive feature of Siddha medicine is its use of processed minerals and metals in treatments, including substances like lead, sulfur, silver, and arsenic [27]. These components are meticulously prepared to avoid significant side effects. Practitioners employ complex preparation techniques such as calcination (*pudichu*), distillation (*kottai*), and fermentation (*kudi maruthuvam*) to enhance the therapeutic efficacy and bioavailability of these remedies. Diagnosis within the Siddha system is a holistic process that involves *nadi* (pulse) diagnosis, tongue examination, and a thorough assessment of clinical signs and symptoms. Treatment modalities are multifaceted and include internal medicine (orally administered herbal formulations), external applications (ointments and poultices), and therapeutic interventions such as panchakarma (detoxification therapies). Furthermore, Siddha medicine recognizes the intrinsic relationship between physical health and mental well-being, incorporating practices such as yoga, meditation, and pranayama (breathing exercises) to foster mental clarity and emotional equilibrium. The Siddha medical system has seen a resurgence in recent decades, supported by the establishment of institutions such as the National Institute of Siddha in Chennai and the Central Council for Research in Siddha (CCRS), which fosters research and educational advancements in the field [28].

13.2.2.3 Unani System of Medicine

Unani medicine traces its origins to ancient Mesopotamia and Egypt but is fundamentally rooted in Greek medical philosophy, particularly the work of Hippocrates (460–377 BCE) [29]. The Unani system was further developed by Arab scholars and integrated with indigenous practices during its introduction to India [30]. Notable contributions to Unani literature include Al-Qanoon by Sheikh Bu-Ali Sina (Avicenna) and Al-Havi by Razi (Rhazes), which are foundational texts in Unani medicine [31]. During the medieval period, particularly under the patronage of various Indian dynasties, such as the Delhi Sultans, Khiljis, Tughlaqs, and Mughal emperors, Unani medicine gained prominence [32]. The theoretical framework of Unani medicine is based on the four humors (blood, yellow bile, mucus, and black bile) and the four temperaments (hot, cold, dry, and moist) [33]. Diagnostic methodologies within Unani medicine involve a comprehensive assessment that includes pulse examination, urine analysis, and an evaluation of the patient's temperament (*mizaj*) to identify underlying imbalances [34]. Therapeutic interventions encompass phytotherapy, which employs a diverse array of medicinal plants, dietary adjustments, and regimental therapies such as *hijama* (wet cupping) and venesection. Unani practitioners utilize a combination of observational techniques and analytical assessments for diagnosis, including *nadi pariksha* (pulse examination), *mutala'a* (urine analysis), and visual assessments of the skin and overall physical condition [35]. The interplay of these evaluations enables a comprehensive understanding of the patient's health status, facilitating the identification of humor imbalances.

13.3 CLINICAL TRIAL METHODOLOGIES

13.3.1 Research Design and Approaches

The research methodologies employed in clinical trials for Indian systems of medicine, such as Ayurveda, Siddha, and Unani, demonstrate a sophisticated integration of traditional practices with contemporary scientific standards [36]. These methodologies are essential for systematically evaluating the efficacy, safety, and mechanistic aspects of conventional interventions [37]. The following discussion outlines the primary research trial methodologies used in this context. Commonly employed research designs and approaches in ISM are illustrated in Figure 13.2.

13.3.1.1 Design of Clinical Trials

a. **Protocol Development**: Clinical trials in ISM begin with the meticulous development of study protocols. Protocols address research questions derived from traditional practices and texts, incorporating contemporary scientific methodologies. These protocols outline objectives, study design, intervention details, outcome measures, and statistical analysis plans. Traditional practices are interpreted in the context of modern scientific research, ensuring that the study design is both rigorous and relevant [38].
b. **Randomized Controlled Trials (RCTs)**: RCTs are considered the gold standard for evaluating therapeutic interventions. RCTs are employed to assess the efficacy and safety of traditional treatments by randomly assigning participants to either the intervention group or a control group, which may receive a placebo or standard care. Randomization minimizes selection bias and ensures that observed effects can be attributed to the intervention. For example, an RCT may be designed to evaluate the impact of a specific Ayurvedic formulation on a chronic condition compared to a placebo [39].
c. **Comparative Effectiveness Research (CER)**: Involves comparing the effectiveness of different interventions, including traditional and conventional treatments. In ISM, CER can compare the outcomes of Ayurvedic, Siddha, or Unani treatments with those of allopathic interventions or other traditional therapies. This approach helps elucidate the relative efficacy and safety of various treatment modalities [40].

13.3.1.2 Observational Studies

a. **Cohort Studies**: Follow a group of individuals receiving a particular treatment over time to observe outcomes and associations with health conditions. In the context of traditional medicine, cohort studies might track patients receiving herbal therapies to evaluate long-term health benefits and adverse effects [41].

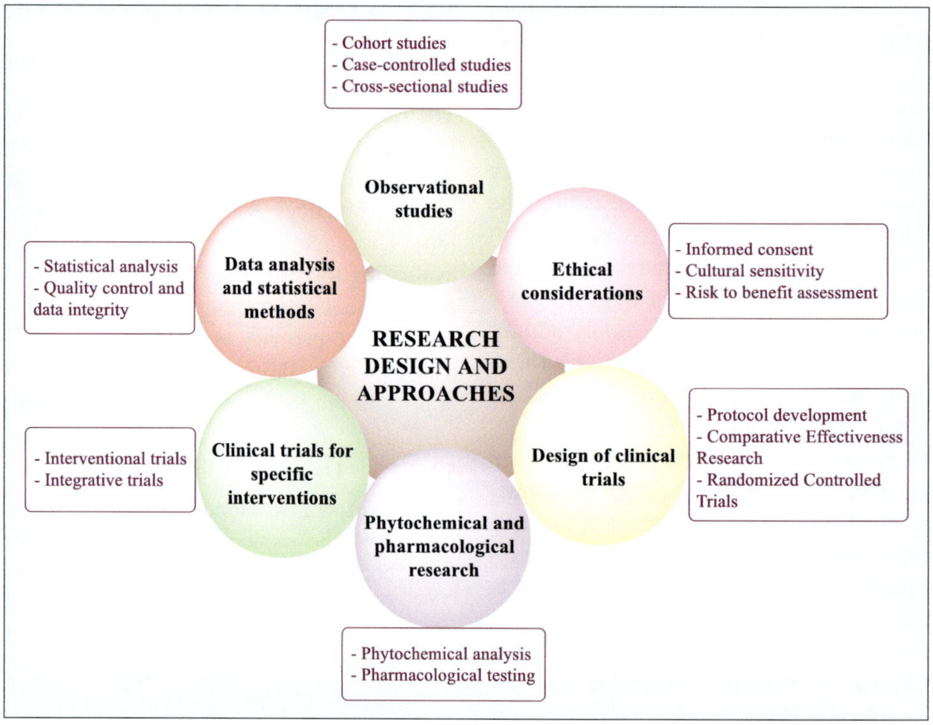

FIGURE 13.2 Commonly employed research designs and approaches in ISM.

b. **Case-Control Studies**: Compare individuals with a particular condition (cases) with those without the condition (controls) to identify potential risk factors or treatment effects. For instance, a case-control study could assess the efficacy of Siddha treatments in patients with a specific disease by comparing them with controls who did not receive the treatment [42].
c. **Cross-Sectional Studies:** Analyze data collected at a single point in time to understand the prevalence of conditions or the effects of treatments within a population. These studies help assess the immediate impact of traditional therapies and their association with health outcomes [43].

13.3.1.3 Phytochemical and Pharmacological Research

a. **Phytochemical Analysis:** Involves identifying and quantifying bioactive compounds in traditional medicinal plants. Techniques such as chromatography and mass spectrometry are employed to elucidate the chemical constituents of herbal formulations used in ISM [44].
b. **Pharmacological Testing:** Evaluates the biological activity of traditional remedies. Preclinical studies, including *in vitro* and *in vivo* assays, assess the therapeutic potential and mechanisms of action of medicinal plants and formulations. For example, *in vivo* studies might test an Ayurvedic herb's efficacy in animal disease models [45].

13.3.1.4 Clinical Trials for Specific Interventions

a. **Interventional Trials:** Evaluate the effects of specific traditional treatments or practices. These trials are designed to test the therapeutic claims of conventional medicine, such as the efficacy of panchakarma in detoxification or the impact of Unani formulations on metabolic disorders [46].
b. **Integrative Trials:** Assess the combination of traditional and conventional treatments. These trials evaluate how traditional therapies can be integrated with modern medical practices to enhance patient outcomes. For instance, an integrative trial might study the combined effects of Ayurvedic and allopathic treatments for managing diabetes [47].

13.3.1.5 Data Analysis and Statistical Methods

a. **Statistical Analysis:** Employed to analyze clinical trial data. Techniques such as regression analysis, survival analysis, and meta-analysis are used to interpret the results and draw conclusions about the efficacy and safety of traditional interventions. Using appropriate statistical methods ensures the robustness and reliability of trial outcomes [48].
b. **Quality Control and Data Integrity:** Maintaining high standards of quality control and data integrity is crucial in clinical trials. This includes rigorous documentation, adherence to Good Clinical Practice (GCP) guidelines, and regular monitoring to ensure compliance with the study protocol and ethical standards [49].

13.3.1.6 Ethical Considerations

a. **Informed consent:** Informed consent must be comprehensive and culturally appropriate, ensuring participants understand the trial's purpose, procedures, risks, and benefits. Researchers should communicate in local languages and consider participants' literacy levels, allowing for ongoing consent and the option to withdraw at any time [50].
b. **Cultural sensitivity:** Trials must respect local beliefs and practices, engaging with communities and traditional practitioners to ensure interventions align with cultural norms. Involving community leaders enhances trust and supports participant recruitment and retention [51].
c. **Risk to benefit assessment:** A thorough risk-benefit assessment is essential to safeguard participants. Researchers should minimize potential harms and transparently communicate risks and benefits. Ensuring equitable access to trial benefits for marginalized populations is also crucial for ethical integrity [52].

13.4 METHODOLOGICAL CONSIDERATIONS

Methodological considerations in clinical trial methodolo gies for the Indian systems of medicine, such as Ayurveda and Unani, require a tailored approach to effectively evaluate their unique therapeutic modalities. Key aspects include the following:

a. **Study Design:** Employing rigorous designs, such as RCTs and cohort studies, is crucial. RCTs help minimize biases inherent in traditional practices while ensuring robust data on efficacy and safety [53].
b. **Randomization:** Implementing appropriate randomization techniques is essential to prevent selection bias and ensure balanced distribution of confounding variables across intervention and control groups [54].
c. **Blinding:** Employing single or double blinding can help mitigate bias in both treatment administration and outcome assessment, which is particularly important given the subjective nature of many Indian medicine assessments [55].
d. **Sample Size and Power Calculations:** Adequate sample size determination is vital to ensure sufficient power to detect clinically significant differences, addressing both statistical and clinical relevance [56].
e. **Inclusion and Exclusion Criteria:** Clearly defined participant selection criteria should consider the holistic nature of Indian systems of medicine, including patient characteristics, disease stages, and comorbidities, to enhance the relevance and applicability of findings [57].
f. **Outcome Measures:** Selecting culturally and contextually appropriate outcome measures is essential. This includes both objective assessments and subjective measures that reflect patient-reported outcomes, aligning with holistic health principles [58].
g. **Standardization of Interventions:** Given the variability in herbal formulations and treatment protocols, standardizing interventions is critical to ensure reproducibility and reliability of results [59].
h. **Ethical Considerations:** Adhering to ethical guidelines, including informed consent and participant safety, is paramount. This includes respecting cultural beliefs and practices associated with traditional medicine [60].
i. **Statistical Analysis:** Employing appropriate statistical techniques that account for the complexity of data, including possible clustering effects in community-based trials, is necessary for robust analysis [61].
j. **Multicentric Trials:** Conducting multicentric studies can enhance the generalizability of findings across diverse populations and healthcare settings, reflecting the varied practices within Indian systems of medicine [62].

By addressing these methodological considerations, researchers can enhance the credibility and applicability of clinical trials in ISM, facilitating their integration into mainstream healthcare and contributing to evidence-based practices.

13.5 CASE STUDIES IN CLINICAL TRIALS

13.5.1 Clinical Trials Based on Ayurveda

Clinical trials in the Ayurvedic system of medicine represent a significant endeavor to bridge traditional therapeutic practices with contemporary scientific methodologies. These studies are designed to rigorously assess the efficacy and safety of Ayurvedic interventions, employing RCT designs and other robust statistical frameworks [63]. By incorporating pharmacological evaluations, biomarker analyses, and CER, researchers aim to elucidate the biochemical mechanisms underpinning Ayurvedic treatments.

Singh et al. (2024) conducted a prospective, open-label, multicentric clinical trial evaluating the potential of Ayurvedic compound formulations (Kantakaryavaleha and Eladi gutika) in tablet and semi-solid dosage forms for the management of bronchial asthma. The evaluation was conducted based on the Asthma Control Questionnaire (ACQ). A significant enhancement ($P < 0.001$) in the ACQ score was observed, from 1.61 (0.45) to 0.61 (0.60), and in FEV1, from 0.86 (0.05) to 0.93 (0.09), after 84 days [64].

Timilsina et al. (2023) led a phase III randomized controlled clinical trial and a preclinical study investigating the efficacy of the Ayurvedic formulation Arjuna Ksheerapaka Churna (AKC) for dyslipidemia. The study was carried out at Rajiv Gandhi Government Postgraduate Ayurvedic

College and Hospital, Paprola, Himachal Pradesh, India, wherein AKC was administered to 100 patients. The study reported that AKC was effective in regulating dyslipidemia compared with rosuvastatin, leading to decreased levels of serum aminotransferases. However, minor discrepancies in certain parameter levels were observed between the pre-clinical and clinical outcomes, such as variations in metabolic rates, differences in dietary habits between animals on high-fat diets and humans adhering to normal lifestyles, physiological and anatomical dissimilarities, etc. Moreover, the influence of pharmacogenetics and pharmacogenomics has been proposed as significant contributors to the disparities observed in the trial. The trial was registered on June 26, 2020, under clinical trial no. CTRI/2020/06/026173, while the findings were published in the *Indian Journal of Traditional Knowledge* [65].

M. Hedaoo et al. (2023) conducted a quasi-experimental study to assess the anti-arthritic effects of an Ayurvedic formulation, Abha Guggulu (A.G.), in osteoarthritic patients. According to the WOMAC index, the results demonstrated a substantial reduction in pain ($p = 0.002$) after administering 1.5 gm of tablets twice daily for one month. No adverse effects were seen, suggesting the safety of A.G. for the management of osteoarthritis and associated complications. The study has been completed [66].

Chitre et al. (2022) conducted a multicentric, randomized, double-blind, placebo-controlled phase III clinical study of an Ayurvedic polyherbal tablet, BV-4051, investigating its anti-inflammatory action. BV-4051 was effective in reducing disease duration and symptom severity, ultimately preventing the incidence of COVID-19 complications, and boosting immunity. Recent studies reveal that the phase III clinical trial of this Ayurvedic formulation (BV-4051), involving 208 randomized subjects from four sites in India and aimed at evaluating the efficacy of BV-4051 compared to a placebo when administered alongside standard care for treating moderate SARS-CoV-2 infections, has been completed [67].

Gupte et al. (2020) orchestrated a clinical study to evaluate the potential of the herbal formulation HFO-02 (CTRI/2016/07/007067) for managing obesity in overweight individuals. No ADRs were observed in the study. HFO-02 alleviated anthropometric parameters, such as body circumferences and skinfold thickness, in overweight individuals. The trial found that the herbal formulation did not lead to statistically notable weight loss or changes in BMI among test subjects, although there were some improvements in circumferential measurements. The study was published in 2019, supporting its completion, and no adverse events were reported during the study [68].

Steels et al. (2018) carried out a double-blind, randomized, placebo-controlled trial to investigate the potential of an Ayurvedic botanical combination (*Tinospora cardifolia, Asparagus racemosus, Withania somnifera,* and *Commiphora mukul*) in reducing vasomotor and menopausal symptoms in healthy women. The findings revealed that the active treatment group showed significant improvements compared to placebo after 12 weeks in the total MENQOL score, the vasomotor domain (hot flashes, night sweats), and the psychosocial, physical, and sexual domains. However, there were no significant changes between groups in serum hormone levels or health indices. The Integrated Health Group Pvt Ltd. study in New Zealand aims to evaluate the effectiveness of this Ayurvedic botanical formulation in alleviating menopausal symptoms such as hot flashes, mood swings, and sleep disturbances [69].

Modak et al. (2017) conducted a phase III, double-blind, randomized controlled study to compare the effects of the Ayurvedic formulation Nartana with the NSAID diclofenac in mild-to-moderate osteoarthritis. The participants were randomly assigned to receive either Nartana (450 mg capsule twice daily for 15 days and then once daily) with a diclofenac-matching placebo or diclofenac (50 mg tablet twice daily for 15 days and then 100 mg once daily) with a Nartana-matching placebo for 12 weeks. The findings revealed that Nartana reduced joint pain, stiffness, and physical prowess and relieved pain with efficacy similar to diclofenac, suggesting the potential of Nartana in alleviating symptoms associated with osteoarthritis [70].

Awasthi et al. (2015) conducted a 24-week randomized, active-controlled clinical study detailing the antidiabetic potential of an Ayurvedic polyherbal formulation (PHF) constituted by *Berberis aristata*, *Cyperus rotundus*, *Cedrus deodara*, *Emblica officinalis*, *Terminalia chebula,* and *Terminalia bellirica* (500 mg capsule). After 24 weeks of treatment, the group under the PHF experienced a mean decrease in fasting and postprandial blood glucose levels of 25.52% and 24.22%, respectively. In contrast, reductions of 31.46% and 24% in the Metformin group was also observed. Additionally, a decrease in HbA1c levels was comparable, with an estimated treatment difference of 0.01. However, the reduction in total cholesterol was highly significant in the PHF group, with an estimated mean difference of 61.3. The clinical trial was registered under CTRI/2014/03/004490 with an aim to assess the effectiveness of a standardized PHF for individuals with type 2 diabetes mellitus. [71].

Chopra et al. (2013) orchestrated a randomized, double-blind, controlled equivalence drug trial for knee osteoarthritis (CTRI/2008/091/000063). The six-month controlled study concluded that the Ayurvedic formulation SGCG significantly reduced knee pain and improved knee function, with effects similar to those of glucosamine and celecoxib. The confirmatory study had a statistical power of 80%, and the α-level was set at 0.05 (two-tailed). The Ayurvedic formulation exhibited positive outcomes in pain and equivalent function to that of standard drugs. However, a longer study period would be required to ascertain its long-term efficacy and safety, especially with reference to hepatic adverse effects [72].

Chopra et al. (2012) investigated the anti-arthritic potential of an Ayurvedic polyherb formulation (comprising *Tinospora cordifolia, Zingiber officinale, Withania somnifera,* and *Tribulus terrestris*) and a monoherb formulation (constituting *Semecarpus anacardium*). The 24-week study involved an investigator-blind, parallel efficacy, three-arm, multicenter drug trial, with 120 patients receiving polyherb capsules and monoherb powder (approximately 500 mg each), using the ACR 20 index for improvement. The findings revealed that the monoherb formulation was inferior (36%) to both polyherb formulation (44%) and hydroxychloroquine sulfate (HCQs) (51%). The Ayurvedic polyherb drug could be an effective and safe alternative to HCQS in controlling active rheumatoid arthritis, with better results employing a longer evaluation period [73].

Sharma et al. (2010) investigated the antidiabetic potential of GlucoCare Ayurvedic capsules through a double-blind, placebo-controlled clinical evaluation. GlucoCare demonstrated positive effects in alleviating the symptoms and bringing about overall improvement. Nineteen of 25 patients suffering from NIDDM showed an adequate response to the treatment. GlucoCare exhibited notable enhancement in symptoms and glycemic control in 50 patients, with a favorable safety profile. Further research with a substantial sample size is warranted to authenticate these findings [74].

The findings from such trials hold the potential to inform evidence-based approaches to personalized healthcare, ultimately enriching the global discourse on integrative medicine (Table 13.1). As research continues to evolve, it is imperative to maintain a dialogue between traditional wisdom and scientific inquiry to harness the full potential of Ayurveda in contemporary health systems.

13.5.2 Clinical Trials Based on Siddha Medicine

Siddha medicine, one of the oldest traditional healing systems originating from South India, is characterized by its holistic approach to health and its emphasis on the balance between body, mind, and spirit [63]. Clinical trials within this framework aim to rigorously evaluate the efficacy and safety of Siddha interventions, which often include herbal formulations, mineral-based medicines, and therapeutic practices.

Pavithra et al. (2022) orchestrated a preclinical safety evaluation of an herbal-mineral formulation, Guru Parpam, administered orally to manage various diseases. The study revealed that the drug is safe, with a wide safety margin established through both acute and subacute toxicity assessments. However, further preclinical validation is required before the drug can be applied clinically to treat chronic conditions in humans [75].

Srivastava et al. (2021) carried out a double-blind, placebo-controlled clinical trial evaluating the potential

TABLE 13.1
Case Studies on Clinical Trials of Ayurvedic Formulations

Ayurvedic Formulation	Dosage Form	Study Type	Activity Reported	References
Abha Guggulu (A.G.)	Tablet (1.5 g/twice daily)	Quasi-experimental study (CTRI/2019/09/021354)	Arthritis	Hedaoo et al., 2023
Nartana	Capsule (450mg)	Phase 3, double-blind, double-dummy, randomized controlled study	Osteoarthritis	Modak et al., 2017
Polyherb formulation and monoherb containing Bhallataka Parpati (BPRT)	Polyherb capsules Monoherb powder (500 mg each)	Randomized investigator-blind controlled study	Rheumatoid arthritis	Chopra et al., 2012
Zingiber officinale, *Tinospora cordifolia*, *Phyllanthus emblica*, and *B. serrata*	Shunthi-guduchi formulation capsule (400 mg)	Randomized, double-blind, controlled equivalence drug trial (CTRI/2008/091/000063)	Symptomatic knee osteoarthritis	Chopra et al., 2013
Polyherbal formulation (PHF) (*B. aristata*, *C. rotundus*, *C. deodara*, *E. officinalis*, *T. chebula*, *and T. bellirica*)	Capsule (500 mg)	Randomized, active-controlled clinical study (CTRI/2014/03/004490)	Diabetes	Awasthi et al., 2015
Ayurvedic medicine composed of 75 mg *T. cordifolia*, 100 mg *A. racemosus*, 100 mg *W. somnifera*, and 225 mg *C. mukul*	Capsule (500 mg)	Double-blind, randomized, placebo-controlled trial	Vasomotor and other menopause-associated symptoms	Steels et al., 2018
BV-4051 polyherbal formulation	Tablet	Multicenter, randomized, double-blind, placebo-controlled phase III clinical study	Inflammation	Chitre et al., 2023
GlucoCare	Capsule	Double-blind, placebo-controlled clinical evaluation	Diabetes	Sharma & Patki, 2010
Kantakaryavaleha & Eladi gutika	Tablets, semi-solid dosage	Prospective, open-label, multicentric clinical trial	Bronchial asthma	Singh et al., 2024
HFO 02	Tablet HFO-02 (500 mg)	Clinical trial registration number: CTRI/2016/07/007067	Obesity	Gupte et al., 2020
Arjuna Ksheerapaka	Churna 6 g/day	Randomized controlled clinical trial and preclinical efficacy CTRI/2020/06/026173	Dyslipidemia	Timilsina et al., 2024

to manage mild-to-moderate symptomatic COVID-19 patients. Nilavembu Kudineer (NVK) and Kaba Sura Kudineer (KSK) treatments resulted in a significant reduction in hospital stay duration, SARS-CoV-2 viral load, and the time taken for patients to achieve asymptomatic status compared to the placebo group (decaffeinated tea). Notably, the KSK group exhibited superior primary outcomes to the NVK group [76].

Kumar et al. (2021) investigated the effectiveness of the Siddha PHF, Vipro™, for managing uncomplicated respiratory infection. At the end of the study, a complete blood count (CBC) and nasal/throat swab for culture were carried out. Vipro™ effectively alleviated clinical symptoms comparable to those achieved with standard treatment [77].

Iswarya et al. (2020) conducted an open-label clinical trial detailing the potential of the Siddha formulation *Pereechangai Nei,* which is a medicated ghee, for the management of type 2 diabetes mellitus. The trial drug demonstrated significant clinical improvement in 29 cases, representing 72.5% of the participants. Twenty percent of the cases exhibited moderate results, while 7.5% showed a poor prognosis based on the HbA1c level study [78].

Nagarajan et al. (2019) conducted a comparative clinical study of the Siddha PHF *Thetran ilagam.* The trial drug was given at 5 gm twice a day with cow's milk, paired with yoga therapy, which led to the confirmation of the efficacy of the drug, demonstrating substantial enhancements in sperm count and sexual health. The test formulation is associated with the clinical trial CTRI/2020/05/025215 and is currently in active recruitment. This trial attempts to assess the effectiveness and safety profile of the intervention in test subjects [79].

Madesh et al. (2018) conducted a clinical study on the Siddha PHF Azhinjiyathikashayam to manage type 2 diabetes mellitus. Azhinjiyathikashayam (60 mL) successfully lowered blood sugar levels in patients without eliciting adverse reactions during prolonged administration. As a safe, herbal-based therapeutic option, it has the potential to prevent complications associated with chronic diabetes [80].

Chellakkan et al. (2018) conducted the clinical evaluation of *Amirthathi Churnam and Nandukal Chunnam,* which were administered as *churnam* (powder) and *chunnam* (calcined nanoparticle powder). The assessment revealed that the trial drug was used as a lithotriptic agent to decrease the calculi's size, resulting in promising outcomes in eliminating urinary stones. Besides, symptoms of burning micturition, dysuria, hematuria, abdominal pain from loin to groin, back pain, nausea, and vomiting reduced significantly in all the patients, with nonrecurrence of the disease upon drug withdrawal. While the initial results are promising, more research is needed to confirm the efficacy and safety of Amirthathi Churnam and Nandukal Chunnam for the treatment of nephrolithiasis. Larger, RCT are necessary to establish their effectiveness and to identify any potential side effects [81].

Kingsley et al. (2018) conducted an open-labeled pilot study investigating the anti-arthritic potential of the Siddha formulation *Seenthil chooranam and Serankottai nei.* The findings revealed that among the 20 patients, 85% reported a significant reduction in joint pain and swelling and marked enhancement in joint mobility. The Siddha formulation is concerned with clinical trial CTRI/2013/12/004203 and is currently under investigation to study the formulation's efficacy in managing rheumatoid arthritis [82].

The insights gained from these studies are instrumental in establishing evidence-based guidelines that support the efficacy and safety of Siddha therapies (Table 13.2). Furthermore, as Siddha medicine gains empirical support, it

TABLE 13.2
Case Studies on Clinical Trials Involving Siddha Formulations

Siddha Formulation	Dosage Form	Study Type	Activity Reported	References
Pereechangai nei	Medicated ghee formulation, 5 mL dose	Open-label clinical trial	Diabetes	Iswarya & Lakshmikantham
Guru parpam	Herbo-mineral formulation, orally administered	Preclinical safety evaluation	Multiple diseases	Pavithra & Anbu, 2022
Polyherbal formulation—*thetran ilagam*	5 gm twice a day with cow's milk	Comparative clinical trial (Ref/2018/03/018682)	Male infertility	Karthik Nagarajan et al., 2019
	Herbo-mineral Siddha polyherbal formulation, 60 mL	Clinical evaluation	Diabetes	Madesh et al., 2018
Nilavembu Kudineer and *Kaba Sura Kudineer*	Polyherbal decoctions	Double-blind, placebo-controlled clinical trial	Mild-to-moderate symptomatic COVID-19	Srivastava et al., 2021
Vipro	Siddha polyherbal formulation	Single-center, prospective, randomized, open-labeled clinical study	Uncomplicated respiratory infection	Abidharini et al., 2023
Seenthil chooranam and *Serankottai nei*	Oral formulation	Open-labeled pilot study–CTRI/2013/12/004203	Arthritis	Kingsley et al., 2018
Amirthathi Churnam and *Nandukal Chunnam*	*Churnam* (powder) and *chunnam* (calcined nanoparticle powder)	N.A.	Nephrolithiasis	Chellakkan et al., 2018

presents opportunities for integration with modern medical practices, fostering a more comprehensive approach to health that addresses both preventive and therapeutic needs.

13.5.3 Clinical Trials Based on Unani Medicine

Unani medicine, an ancient therapeutic system rooted in the teachings of Hippocrates and Galen, further developed by Islamic scholars, emphasizes the concept of balancing the body's four humors *viz.*, blood, phlegm, yellow bile, and black bile. This holistic approach integrates physical, mental, and environmental health, using a range of natural remedies, including herbal formulations, dietary interventions, and therapeutic practices [83]. Clinical trials of Unani medicine play a crucial role in evaluating its efficacy and safety, helping to integrate traditional practices with modern scientific standards (Table 13.3).

Khanum et al. (2024) carried out a randomized, single-blind, placebo-controlled clinical trial to investigate the effectiveness of the polyherbal Unani formulation Arq-e-Gaozaban for managing diastolic dysfunction in the left ventricle. The study concluded that the test formulation exhibited superior effectiveness in alleviating the dyspnea. Compared to the control group, the test formulation significantly improved diastolic function and easy fatigability [84].

Qaiyyum et al. (2023) investigated the potential of a topical Unani formulation *via* a randomized controlled clinical trial to treat melasma (acquired facial hypermelanism). The study revealed that the Unani formulation and the control drug were equivalent in effectiveness and safety for managing melasma [85].

Rahi et al. (2022) conducted a randomized controlled clinical trial on a Unani oral formulation containing the herbs Aftimoon and ma'uljubn to manage psoriasis. The study revealed that no adverse effects were noted throughout the study, and all blood safety parameters remained within normal limits. A decrease in TNF-α was also observed, with the test group showing a more significant reduction (P = 0.08) than the control group (P = 0.10). The trial suggested that these Unani interventions may be feasible alternative remedies for chronic plaque psoriasis. Further investigation in substantial multicenter studies is required to confirm efficacy and safety in the broader population [86].

Saima et al. (2022) conducted a randomized controlled trial evaluating the efficacy and safety profile of Unani herbo-mineral preparations Qurs-e-Asfar (Q.A.) and Rogan-e-Narjeel for the management of dermatophytosis. The results revealed notable improvement in itching (approximately 86%) compared to 78% in the control group, as well as improvement in erythema and scaling. The findings concluded that Qurs-e-Asfar (Q.A.) and Rogan-e-Narjeel efficiently treated dermatophytosis. The findings unveiled that these Unani interventions are effective and safe for dermatophytosis treatment, leading to their potential as possible choices in clinical practice [87].

Qurratul et al. (2022) orchestrated a prospective randomized controlled trial to evaluate the safety and effectiveness of polyherbal Unani formulation for managing dyslipidemia. The study concluded that Jawarish Falafili

TABLE 13.3
Case Studies on Clinical Trials of Unani Formulations

Unani Formulation	Dosage Form	Study Type	Activity Reported	References
Aftimoon and ma'uljubn	Oral formulation	Randomized controlled clinical trial	Psoriasis	Rahi et al., 2022
Unani formulation	Fine powder	Randomized, single-blind, standard-controlled clinical study	Diabetes	Khan et al., 2020
Itrifal-e-Sagheer	Sugar-based semi-solid formulation	Randomized, single-blind, controlled clinical trial on safety and efficacy (CTRI/2017/09/009783)	Dyslipidemia	Alam et al., 2020
UNIM-401 and UNIM-403	Capsule	Non-inferiority, randomized, active-controlled clinical trial	Psoriasis	Khanna et al., 2018
Qatoor Ramad	Eye drop	Comparative, double-blind, randomized, placebo-controlled clinical trial	Conjunctivitis	Siddiqui et al., 2002
Zimade Muhasa	Powder	Single-blind, randomized, controlled clinical trial	Acne	Tabasum et al., 2014
Unani formulation	Capsule	Randomized, double-dummy clinical trial	Lower back pain	Aafreen et al., 2019
Qurs-e-Asfar (QA) and Rogan-e-Narjeel	Unani herbo-mineral preparation	Randomized, active-controlled, and open-label clinical study	Dermatophytosis	Naseem et al., 2023
Unani formulation	Topical	Randomized controlled clinical trial	Acquired facial hypermelanism	Qaiyyum et al., 2023
Jawarish Falafili	Polyherbal Unani formulation	Prospective, randomized, controlled trial	Dyslipidemia	Ain et al., 2022
Arq-e-Gaozaban	Capsule along with 35 mL of an extract of *B. officinalis*	Randomized, single-blind, placebo-controlled clinical trial	Left ventricular diastolic dysfunction in hypertensive patients	Khanum et al., 2024

and atorvastatin were equivalent in managing dyslipidemia. The outcomes assisted the efficacy of the PHF as a sustainable alternative treatment for dyslipidemia, requiring further research and validation in more extensive studies [88].

Khan et al. (2020) investigated the antidiabetic potential of the Unani formulation in type 2 diabetes mellitus through a randomized, single-blind, standard-controlled clinical study. The trial formulation included Hulba (*Trigonella foenum-graecum*) 10 gm, Tabasheer (*Bambusa indica*) 10 gm, Shilajit (*Asphaltum*) 5 gm, Magaze Khasta Jamun (*Syzygium cumini*) 15 gm, and Afsanteen Rumi (*Artemisia absinthium*) 10 gm. The findings revealed a significant reduction in subjective symptoms such as polyuria, polydipsia, polyphagia, and tiredness. Both the test and control groups also showed a significant decrease in fasting blood sugar (FBS), postprandial blood sugar (PPBS), and HbA1c levels [89].

Alam et al. (2020) conducted a randomized, single-blind, controlled clinical trial evaluating the safety profile and effectiveness of Itrifal-e-Sagheer for managing dyslipidemia. The trial drug notably alleviated the symptoms of palpitations, dyspnea, and weight gain ($p < 0.05$). Compared with those receiving the control drug treatment, there was a statistically significant decrease in the lipid profile of patients in the test group ($p < 0.05$). The trial findings suggest that Itrifal-e-Sagheer is potentially effective and safe for the management of dyslipidemia, with notable enhancement in lipid profiles among test subjects [90].

Siddiqui et al. (2019) studied the efficacy of a Unani formulation in lower back pain (Waja-ul-Khasirah) through a randomized, double-dummy clinical trial. The inter- and intra-group findings revealed notable improvements in lower back pain, stiffness, and QoL ($p < 0.001$), concluding that the trial formulation proved efficacious and well tolerated [91].

Khanna et al. (2018) orchestrated a non-inferiority randomized controlled clinical trial for the management of psoriasis. The findings revealed that 287 patients were randomized; 84 out of 147 in the Unani group and 67 out of 140 in the PUVA sol group were included, with a treatment duration of 12 weeks. ITT analysis showed that patients on Unani medication showed outcomes equivalent, if not superior to, those who received PUVA sol. The oral UNIM-401 and topical UNIM-403 were effective and well tolerated by patients with moderate-to-severe CPP. The drugs were registered under IEC/NP 223/2010 and CTRI 2010/091/001253, and the DCGI approved the drug's marketing in India after completion of the required clinical trials and regulatory approval process [92].

Tabasum et al. (2014) designed and carried out a single-blind, randomized, controlled clinical trial to investigate the effectiveness of Zimade Muhasa in managing *Acne vulgaris*. The trial lasted six weeks and unveiled that a formulation equivalent to 5% benzoyl peroxide reduced acne lesions and was well tolerated by the test subjects. As of 2024, the Unani anti-acne formulation has been evaluated in clinical trials, exhibiting notable efficacy in managing acne vulgaris [93].

Siddiqui et al. (2002) investigated the potential of the ophthalmic formulation Qatoor Ramad by conducting a comparative double-blind, randomized, placebo-controlled clinical trial to manage conjunctivitis. The results revealed that Qatoor Ramad was effective, with no side effects, and well tolerated in almost all test subjects. The findings suggest that Qatoor Ramad eye drops exhibit effective potential and a favorable safety profile in managing conjunctivitis, specifically the mucopurulent type, compared to placebo [94].

The body of clinical trials on Unani medicine underscores its potential efficacy and safety across various health conditions, including cardiovascular, dermatological, and metabolic disorders. These studies not only affirm the therapeutic benefits of Unani formulations but also highlight the necessity for further research to solidify their role as viable alternatives in contemporary medical practice.

13.6 CHALLENGES AND BOTTLENECKS IN CLINICAL TRIALS OF ISM

Conducting clinical trials within the framework of Indian systems of medicine presents several significant challenges and bottlenecks, as stated later:

a. **Regulatory Framework:** The regulatory landscape for clinical trials in India is evolving but remains complex. Existing guidelines may not adequately address the unique characteristics of traditional systems of medicine. This can lead to ambiguities in trial approval processes, prolonged timelines, and difficulties in compliance [95].
b. **Standardization of Interventions:** ISM often employ individualized treatments based on holistic assessments. This variability poses challenges in standardizing interventions for clinical trials, complicating the establishment of control groups and the replicability of results. The lack of universally accepted protocols can hinder the scientific rigor of studies [96].
c. **Quality Control and Safety:** Ensuring the quality and safety of herbal formulations and therapies used in trials is critical. Variability in the sourcing, preparation, and dosage of traditional medicines can affect efficacy and safety profiles. There is often insufficient research on the pharmacokinetics and toxicology of these interventions, necessitating robust quality assurance measures [97].
d. **Research Funding and Resources:** Securing funding for clinical trials in traditional medicine can be challenging, as many funding bodies prioritize research in conventional biomedical fields. Limited financial resources can restrict the scope of studies and impact participant recruitment, data collection, and overall study quality [98].
e. **Participant Recruitment and Retention:** Recruiting participants who are familiar with and willing to engage in trials of traditional medicine

can be difficult. Cultural beliefs and skepticism toward clinical research may deter potential participants. Additionally, retention rates may suffer due to the prolonged nature of some trials, which can be demanding for participants [99].

f. **Integration with Conventional Medicine:** There is often a perceived divide between conventional and traditional medical practices. This can lead to challenges in integrating findings from clinical trials of Indian systems of medicine into mainstream healthcare. Addressing this gap requires a paradigm shift in how both systems are viewed and utilized [100].

g. **Data Management and Analysis:** The holistic approach of Indian systems of medicine complicates data collection and analysis. Trials must account for several variables, including lifestyle, diet, and environmental factors, which can introduce confounding elements. Establishing robust data management systems that can handle these complexities is crucial for valid outcomes [61].

h. **Ethical Considerations:** Ensuring ethical compliance in trials involving traditional medicine requires sensitivity to cultural values and practices. The informed consent process must be adapted to address diverse literacy levels and cultural contexts, which can be challenging in heterogeneous populations [60].

Although there is significant potential for clinical trials in Indian systems of medicine, addressing these challenges and bottlenecks is essential for ensuring scientific rigor, participant safety, and the successful integration of findings into broader health systems.

13.7 IMPLICATIONS FOR FUTURE RESEARCH

The integration of ISM, such as Ayurveda, Siddha, and Unani, into clinical research necessitates a rigorous, multifaceted approach to address existing challenges. Standardization of herbal formulations and therapeutic protocols is imperative to mitigate variability, thereby ensuring reproducibility and reliability of clinical outcomes. Future investigations should prioritize elucidating the pharmacodynamics and pharmacokinetics of natural compounds, alongside their interactions with conventional therapeutics, to foster evidence-based integration into mainstream healthcare. CER is critical for evaluating ISM against conventional modalities, enabling identification of optimal patient populations for integrative therapies. Moreover, interdisciplinary collaboration encompassing pharmacology, biochemistry, and traditional medicine is essential for advancing mechanistic understanding and the clinical application of ISM.

Establishing robust regulatory frameworks is crucial for ensuring the safety and efficacy of ISM interventions, with particular emphasis on ethical considerations and compliance tailored to traditional practices. Incorporating cultural dimensions into research design enhances the relevance of clinical trials by addressing patient perspectives and adherence factors. Longitudinal studies investigating the sustained efficacy of ISM in managing chronic conditions, coupled with the collection of real-world evidence, can provide insights into the practical application of these therapies. Advancing clinical trials of ISM requires a systematic, multidimensional strategy that encompasses standardization, mechanistic research, regulatory compliance, cultural competence, and long-term efficacy assessments to elucidate their potential in enhancing global health outcomes.

13.8 CONCLUSION

The clinical trials conducted on the ISM demonstrate promising potential for integrating traditional practices with modern medical approaches. Case studies have revealed that various ISM therapies, such as Ayurveda, Siddha, and Unani, can effectively address a range of health conditions, often with fewer side effects compared to conventional treatments. Key findings indicate significant improvements in patients' health, particularly for chronic diseases like diabetes, arthritis, and stress-related disorders, underscoring the holistic nature of ISM. Moreover, these therapies generally present a favorable safety profile, aligning with ISM's principles of natural and balanced treatments. High levels of patient satisfaction have been reported, attributed to personalized treatment plans that consider the physical, mental, and emotional aspects of health. ISM can complement conventional medicine, fostering a more integrated healthcare model that enhances treatment outcomes. However, there is an essential need for standardized protocols and larger-scale studies to validate these findings and ensure consistency in treatment practices among practitioners. Overall, the clinical trials and case studies on ISM highlight its significant potential in modern healthcare, and continued research could pave the way for a more inclusive approach, benefiting both patients and practitioners.

ACKNOWLEDGMENTS

ST is sincerely grateful to ICMR for providing financial assistance through an extramural grant (EMDR/SG/13/2023-2887). The authors acknowledge the Department of Science and Technology (DST), Government of India, for providing the Departmental DST-FIST grant (SR/FST/LSI-656/2016) to the Department of Pharmaceutical Sciences and Natural Products, Central University of Punjab, India.

AUTHORS' CONTRIBUTIONS

S.T. conceptualized the present book chapter. K.G., A.S., and A.B. collected the data and wrote the initial draft of the book chapter. Review and editing of the final book chapter were carried out by K.G., A.B., and A.S. under the supervision of S.T.

REFERENCES

[1] Taneja, S. (2009). Therapeutic products from botanical sources: The Indian scenario. *Planta Medica*, 75(4), S–25.

[2] Elendu, C. (2024). The evolution of ancient healing practices: From shamanism to Hippocratic medicine: A review. *Medicine*, 103(28), e39005.

[3] Verma, S.K., Pandey, M., Sharma, A., et al. (2024). Exploring Ayurveda: Principles and their application in modern medicine. *Bulletin of the National Research Centre*, 48(1), 77.

[4] Subbarayappa, B.V. (1997). Siddha medicine: An overview. *The Lancet*, 350(9094), 1841–1844.

[5] Parveen, A., Ahmad, M., Parveen, B., et al. (2022). The traditional system of Unani medicine, its origin, evolution and Indianisation: A critical appraisal. *Indian Journal of Traditional Knowledge*, 21(3), 511–521.

[6] Sen, S., & Chakraborty, R. (2017). Revival, modernization and integration of Indian traditional herbal medicine in clinical practice: Importance, challenges and future. *Journal of Traditional and Complementary Medicine*, 7(2), 234–244.

[7] Williamson, P.R., Altman, D.G., Blazeby, J.M., et al. (2012). Developing core outcome sets for clinical trials: Issues to consider. *Trials*, 13, 1–8.

[8] Mukherjee, P.K., & Wahile, A. (2006). Integrated approaches towards drug development from Ayurveda and other Indian system of medicines. *Journal of Ethnopharmacology*, 103(1), 25–35.

[9] Gogoi, I., Dowara, M., & Chetia, P. (2024). Traditional medicinal plants and their ethnomedicinal values, in *Traditional resources and tools for modern drug discovery: Ethnomedicine and pharmacology*. Singapore: Springer Nature Singapore. 377–399.

[10] Mukherjee, P.K., Harwansh, R.K., Bahadur, S., et al. (2017). Development of Ayurveda-tradition to trend. *Journal of Ethnopharmacology*, 197, 10–24.

[11] Jaiswal, Y.S., & Williams, L.L. (2017). A glimpse of Ayurveda: The forgotten history and principles of Indian traditional medicine. *Journal of Traditional and Complementary Medicine*, 7(1), 50–53.

[12] Borah, S., Barman, K., Mazumdar, A., et al. (2024). Ethnotherapeutic practices in India. *African Journal of Biomedical Sciences*, 6(10), 4913–4923.

[13] Shankar, D. (2021). Roadmap for Ayurveda education in modern India, in *Handbook of education systems in South Asia*. Singapore: Springer Singapore,. 101–119.

[14] Khare, C. (2019). *Evidence-based Ayurveda: Defining a new scientific path*. Routledge. 1–301

[15] Ranade, A., & Acharya, R. (2015). Contribution of Dhanwantari Nighantu towards drug safety: A critical review. *Global Journal of Research on Medicinal Plants & Indigenous Medicine*, 4(2), 20.

[16] Karunamoorthi, K., Jegajeevanram, K., Xavier, J., et al. (2012). Tamil traditional medicinal system-siddha: An indigenous health practice in the international perspectives. *CellMed*, 2(2), 12.1–12.11.

[17] Lepcha, I., Patra, B., & Saha, D. (2023). *Medicinal plants in Ayurveda and Unani systems including Siddha in the Indian scenario*. Shaswat Publication. 85–118.

[18] Matthews, S. (2020). Ayurveda, in *Introduction to complementary medicine*. Routledge. 15–32.

[19] Kizhakkeveettil, A., Parla, J., Patwardhan, K., et al. (2024). History, present and prospect of Ayurveda, in *History, present and prospect of world traditional medicine*. World Scientific. 1–72.

[20] Debnath, P.K., Banerjee, S., Debnath, P., et al. (2015). Ayurveda–opportunities for developing safe and effective treatment choices for the future, in *Evidence-based validation of herbal medicine*. Elsevier. 427–454.

[21] Lad, V. (2018). Ayurvedic medicine: An integrative approach, in *Nutrition and integrative medicine*. CRC Press. 329–346.

[22] Kapur, M., & Kapur, M. (2016). Basic principles of ayurveda. *Psychological Perspectives on Childcare in Indian Indigenous Health Systems*, 15–29.

[23] Kumar, J. (2014). Ayurveda and early Indian medicine, in *Science and religion*. Routledge. 184–196.

[24] Jha, M., Sah, M.K., Singh, P.K., et al. (2021). Role of Ayurveda in public health: A critical review. *International Journal of Research in AYUSH and Pharmaceutical Sciences*, 595–599.

[25] White, D.G. (2022). *The alchemical body: Siddha traditions in medieval India*. University of Chicago Press. 1–555.

[26] Muthiah, K., Ganesan, K., Ponnaiah, M., et al. (2019). Concepts of body constitution in traditional Siddha texts: A literature review. *Journal of Ayurveda and Integrative Medicine*, 10(2), 131–134.

[27] Subramaniyan Parimalam, S., Badilescu, S., Bhat, R., et al. (2022). The Siddhars: The great artisans of gold medicines in medieval South India. *Advances in Traditional Medicine*, 22(3), 509–517.

[28] Niva, N., Sasirekha, R., & Anbu, N. (2020). Contribution of Tamil poets and philosophers for the development of Siddha system of medicine. *Journal of Research in Biomedical Sciences*, 3(1), 13–19.

[29] Ahmad, A., Naqvi, S.B.K., & Sultana, N. (2024). Exploring the roots: The historical journey and evolution of the Unani system of medicine. *Journal of Emerging Technologies and Innovative Research*, 11(6), 432–432.

[30] Islam, A. (2018). Origin and development of Unani Medicine: An analytical study. *Intellectual Discourse*, 26(1), 23–49.

[31] Shahin, S., Babji, S., Haq, M.S., et al. (2023). Clinical study and efficacy of Unani treatment in acute cystitis (Warm-E-Masana Haad). *International Journal of Unani and Integrative Medicine*, 7(3), 44–58.

[32] Islam, A. (2018). Unani methods of cure in the Indian subcontinent: An analytical study. *International Journal of Engineering & Technology*, 7(2.29), 480–487.

[33] Ahmad, B., & Akhtar, J. (2007). Unani system of medicine. *Pharmacognosy Reviews*, 1(2), 210.

[34] Shaikh, N., Bamne, F., Ali, A., et al. (2024). Herbal medicine: Exploring its scope across belief systems of the Indian medicine, in *Herbal medicine phytochemistry: Applications and trends*. Springer. 1279–1304.

[35] Parveen, A., Parveen, R., Akhatar, A., et al. (2020). Concepts and quality considerations in Unani system of medicine. *Journal of AOAC International*, 103(3), 609–633.

[36] Sen, S., & Chakraborty, R. (2015). Toward the integration and advancement of herbal medicine: A focus on traditional Indian medicine. *Botanics: Targets and Therapy*, 5, 33–44.
[37] Friedman, L.M., Furberg, C., DeMets, D.L., et al. (2010). *Fundamentals of clinical trials*. Vol. 4. Springer. 1–18.
[38] Meyer, E.L., Mesenbrink, P., Dunger-Baldauf, C., et al. (2020). The evolution of master protocol clinical trial designs: A systematic literature review. *Clinical Therapeutics*, 42(7), 1330–1360.
[39] Zabor, E.C., Kaizer, A.M., & Hobbs, B.P. (2020). Randomized controlled trials. *Chest*, 158(1), S79–S87.
[40] Concato, J., Lawler, E.V., Lew, R.A., et al. (2010). Observational methods in comparative effectiveness research. *The American Journal of Medicine*, 123(12), e16–e23.
[41] Wang, X., & Kattan, M.W. (2020). Cohort studies: Design, analysis, and reporting. *Chest*, 158(1), S72–S78.
[42] Dey, T., Mukherjee, A., & Chakraborty, S. (2020). A practical overview of case-control studies in clinical practice. *Chest*, 158(1), S57–S64.
[43] Wang, X., & Cheng, Z. (2020). Cross-sectional studies: Strengths, weaknesses, and recommendations. *Chest*, 158(1), S65–S71.
[44] Abraham, A., Samuel, S., & Mathew, L. (2020). Phytochemical analysis of Pathyashadangam kwath and its standardization by HPLC and HPTLC. *Journal of Ayurveda and Integrative Medicine*, 11(2), 153–158.
[45] Kaushik, P., Ahlawat, P., Singh, K., et al. (2023). Chemical constituents, pharmacological activities, and uses of common ayurvedic medicinal plants: A future source of new drugs. *Advances in Traditional Medicine*, 23(3), 673–714.
[46] Moher, D., & Chan, A.W. (2014). SPIRIT (standard protocol items: Recommendations for interventional trials), in *Guidelines for reporting health research: A user's manual*. John Wiley & Sons. 56–67.
[47] Mirunaleni, P., Narmadha Jothinathan, A., Paargavi, S., et al. (2018). Effect of integrative approach using siddha medicines, Isha yoga and dietary modifications in treatment of madhumegam (Diabetes mellitus) in holistic approach observational study. *International Journal of Current Research in Chemistry Pharmaceutical Sciences*, 5, 9–12.
[48] Zink, R.C., Marchenko, O., Sanchez-Kam, M., et al. (2018). Sources of safety data and statistical strategies for design and analysis: Clinical trials. *Therapeutic Innovation & Regulatory Science*, 52(2), 141–158.
[49] Prasanna, B., Kothapalli, P., & Vasanthan, M. (2024). The role of quality assurance in clinical trials: Safeguarding data integrity and compliance. *Cureus*, 16(8), 2–7.
[50] Falagas, M.E., Korbila, I.P., Giannopoulou, K.P., et al. (2009). Informed consent: How much and what do patients understand? *The American Journal of Surgery*, 198(3), 420–435.
[51] Hamer, D.H., Herlihy, J.M., Musokotwane, K., et al. (2015). Engagement of the community, traditional leaders, and public health system in the design and implementation of a large community-based, cluster-randomized trial of umbilical cord care in Zambia. *The American Journal of Tropical Medicine and Hygiene*, 92(3), 666.
[52] Schnarr, A., & Mertz, M. (2022). States of uncertainty, risk–benefit assessment and early clinical research: A conceptual investigation. *Science and Engineering Ethics*, 28(6), 68.
[53] Raman, S.R., Landi, S.N., Lund, J.L., et al. (2021). Approaches to mitigate bias in the design and analysis of pRCTs, in *Pragmatic randomized clinical trials*. Elsevier. 131–149.
[54] Sanderson, E., Glymour, M.M., Holmes, M.V., et al. (2022). Mendelian randomization. *Nature Reviews Methods Primers*, 2(1), 6.
[55] Kiani, A.K., Naureen, Z., Pheby, D., et al. (2022). Methodology for clinical research. *Journal of Preventive Medicine and Hygiene*, 63(2 Suppl 3), E267.
[56] Julious, S.A. (2023). *Sample sizes for clinical trials*. New York: Chapman and Hall/CRC. 420.
[57] Meline, T. (2006). Selecting studies for systemic review: Inclusion and exclusion criteria. *Contemporary Issues in Communication Science and Disorders*, 33, 21–27.
[58] Williams, K., Sansoni, J., Darcy, M., et al. (2016). Patient-reported outcome measures, in *Literature review*. Sydney: Australian Commission on Safety and Quality in Health Care. 1–95.
[59] Mirza, K., Sharma, A., Naved, T., et al. (2020). New technologies and interventions for the standardization of Unani drugs. *International Journal of Unani and Integrative Medicine*, 4(1), 22–25.
[60] Vanclay, F., Baines, J.T., & Taylor, C.N. (2013). Principles for ethical research involving humans: Ethical professional practice in impact assessment Part I. *Impact Assessment and Project Appraisal*, 31(4), 243–253.
[61] Atienza, A.A., & King, A.C. (2002). Community-based health intervention trials: An overview of methodological issues. *Epidemiologic Reviews*, 24(1), 72–79.
[62] Kahan, B.C., & Morris, T.P. (2013). Analysis of multicentre trials with continuous outcomes: When and how should we account for centre effects? *Statistics in Medicine*, 32(7), 1136–1149.
[63] Patwardhan, B., Mutalik, G., & Tillu, G. (2015). *Integrative approaches for health: Biomedical research, Ayurveda and Yoga*. Academic Press. 79–106.
[64] Singh, H., Babu, G., Sharma, B.S., et al. (2024). Safety and efficacy of Ayurveda compound formulations (Kantakaryavaleha and Eladi gutika) in the management of mild stable bronchial asthma (Tamaka shwasa): A prospective, open-label, multicentric clinical trial. *International Journal of Ayurveda Research*, 5(1), 31–39.
[65] Timilsina, K., Mishra, A., Patial, V., et al. (2024). A randomized controlled clinical trial and preclinical efficacy of an Ayurvedic formulation Arjuna Ksheerapaka Churna for dyslipidemia: Arjuna Ksheerapaka Churna for dyslipidemia. *Indian Journal of Traditional Knowledge*, 23(6), 519–529.
[66] Hedaoo, M., Patil-Bhole, T., Sharma, R., et al. (2023). Exploratory quasi-experimental study of anti-arthritic activity of Ayurvedic polyherbal formulation, Abha Guggulu in osteoarthritis patients. *Drug Metabolism and Personalized Therapy*, 38(3), 281–288.
[67] Chitre, D., Nadkarni, S., Jagtap, N., et al. (2023). Phase III randomized clinical trial of BV-4051, an Ayurvedic polyherbal formulation in moderate SARS-CoV-2 infections and its impact on inflammatory biomarkers. *Phytotherapy Research*, 37(4), 1232–1241.
[68] Gupte, P., Harke, S., Deo, V., et al. (2020). A clinical study to evaluate the efficacy of herbal formulation for obesity

(HFO-02) in overweight individuals. *Journal of Ayurveda and Integrative Medicine*, 11(2), 159–162.

[69] Steels, E., Steele, M., Harold, M., et al. (2018). A double-blind, randomized, placebo-controlled trial evaluating safety and efficacy of an ayurvedic botanical formulation in reducing menopausal symptoms in otherwise healthy women. *Journal of Herbal Medicine*, 11, 30–35.

[70] Modak, M., Barde, M., Modak, M.D., et al. (2017). Efficacy of an Ayurvedic formulation for mild-to-moderate osteoarthritis: A phase 3, randomized controlled study. *Alternative Therapies in Health & Medicine*, 23(1), 1–26.

[71] Awasthi, H., Nath, R., Usman, K., et al. (2015). Effects of a standardized Ayurvedic formulation on diabetes control in newly diagnosed Type-2 diabetics; a randomized active controlled clinical study. *Complementary Therapies in Medicine*, 23(4), 555–561.

[72] Chopra, A., Saluja, M., Tillu, G., et al. (2013). Ayurvedic medicine offers a good alternative to glucosamine and celecoxib in the treatment of symptomatic knee osteoarthritis: A randomized, double-blind, controlled equivalence drug trial. *Rheumatology*, 52(8), 1408–1417.

[73] Chopra, A., Saluja, M., Tillu, G., et al. (2012). Comparable efficacy of standardized Ayurveda formulation and hydroxychloroquine sulfate (HCQS) in the treatment of rheumatoid arthritis (RA): A randomized investigator-blind controlled study. Clinical Rheumatology, 31, 259–269.

[74] Sharma, R., & Patki, P.S. (2010). Double-blind, placebo-controlled clinical evaluation of an Ayurvedic formulation (GlucoCare capsules) in non-insulin dependent diabetes mellitus. *Journal of Ayurveda and Integrative Medicine*, 1(1), 45.

[75] Pavithra, K., & Anbu, N. (2022). Preclinical safety evaluation of siddha formulation Guru Parpam by acute and sub-acute repeated oral toxicity studies in rodents. *International Journal of Translational Research Indian Medicine*, 4(3), 10–20.

[76] Srivastava, A., Rengaraju, M., Srivastava, S., et al. (2021). Efficacy of two siddha polyherbal decoctions, Nilavembu Kudineer and Kaba Sura Kudineer, along with standard allopathy treatment in the management of mild to moderate symptomatic COVID 19 patients-a double-blind, placebo-controlled, clinical trial. *Trials*, 22(1), 570.

[77] Abidharini, J.D., Souparnika, B.R., Elizabeth, J., et al. (2023). Herbal formulations in fighting against the SARS-CoV-2 infection, in *Ethnopharmacology and drug discovery for COVID-19: Anti-SARS-CoV-2 agents from herbal medicines and natural products*. Springer. 85–113.

[78] Iswarya, P., & Lakshmikantham, T. (2020). Clinical evaluation of siddha formulation Pereechangai Nei in the management of Madhumegam (Type 2 diabetes mellitus). *Journal of Dental and Medical Sciences*, 19, 55–64.

[79] Karthik Nagarajan, S., Nagalakshmi, S., Kumar, N.M., et al. (2019). A comparative clinical trial of siddha poly herbal formulation thetran ilagam (internal) and yogam therapy in the management of aan maladu (male infertility). *World Journal of Pharmaceutical Research*, 8, 878–893.

[80] Madesh, M., Manickavasakam, K., & Banumathi, V. (2018). Clinical evaluation of a siddha poly herbal formulation Azhinjiyathikashayam in Madhumegam (Type-2 diabetes) patients. *International Journal of Ayurveda and Pharma Research*, 6(3), 17–23.

[81] Chellakkan, E., Nainarpandian, C., Blessed, F., et al. (2018). Amirthathi Churnam and Nandukal Chunnam of siddha formulation and clinical evaluation of nephrolithiasis. *Journal of Complementary Medicine and Alternative Healthcare*, 6(2), 1–7.

[82] Kingsley, J., Arun Kumar, G., Sivakkumar, S., et al. (2018). A pilot study on "combination of siddha formulation" in the treatment of "vali azhal keel vayu" (rheumatoid arthritis). *World Journal of Pharmaceutical Research*, 7, 1299–1307.

[83] Husain, M.K., Khalid, M., Penchala Pratap, G., et al. (2017). Relevance of traditional Unani (Greco-Arab) system of medicine in cancer: An update. *Anticancer Plants: Clinical Trials and Nanotechnology*, 3, 273–302.

[84] Khanum, A., Khan, J.A., Shahid, A., et al. (2024). Effect of a polyherbal Unani formulation on left ventricular diastolic dysfunction in hypertensive patients: A randomized single blind placebo controlled clinical trial. *Drug Metabolism and Personalized Therapy*, 39(3), 137–144.

[85] Qaiyyum, I.A., Nawab, M., & Kazmi, M.H. (2023). A randomized controlled clinical trial to evaluate safety and efficacy of a Unani formulation in the management of Kalaf (Melasma). *Journal of Complementary and Integrative Medicine*, 20(1), 233–240.

[86] Rahi, T., Islam, N.U., Pandith, R.A., et al. (2022). Evaluation of oral and topical herbal formulations used in Unani medicine for treatment of psoriasis and its effect on inflammatory biomarkers: A randomized controlled clinical trial. *European Journal of Integrative Medicine*, 54, 102154.

[87] Naseem, S., Nawab, M., Fatima, S.H., et al. (2023). Assessing the efficacy and safety of Unani pharmacopoeial formulations in dermatophytosis (quba): A randomized controlled trial. *Drug Metabolism and Personalized Therapy*, 38(2), 169–177.

[88] Ain, Q., Nawab, M., Ahmad, T., et al. (2022). Evaluating the safety and efficacy of a polyherbal Unani formulation in dyslipidaemia: A prospective randomized controlled trial. *Journal of Ethnopharmacology*, 289, 115036.

[89] Khan, N.A., Alam, N., Eqbal, K., et al. (2020). Clinical efficacy of Unani formulation in type II diabetes: A randomized, single blind standard control clinical study. *International Journal of Physical Education, Sports and Health*, 7(5), 138–143.

[90] Alam, S., Alam, M., Quamri, M., et al. (2020). A randomized single blind controlled clinical trial on safety and efficacy of a Unani formulation (Itrifal-e-Sagheer) in dyslipidemia. *CellMed*, 10(1), 8.1–8.7.

[91] Aafreen, S., Siddiqui, M., Quamri, M., et al. (2019). Efficacy of Unani formulations in Waja-ul-Khasirah (low back pain): A randomised double dummy clinical trial. *International Journal of Research and Review*, 6(9), 88–96.

[92] Khanna, N., Nazli, T., Siddiqui, K.M., et al. (2018). A non-inferiority randomized controlled clinical trial comparing Unani formulation & psoralen plus ultraviolet: A sol in chronic plaque psoriasis. *Indian Journal of Medical Research*, 147(1), 66–72.

[93] Tabasum, H., Ahmad, T., Anjum, F., et al. (2014). The effect of Unani antiacne formulation (Zimade Muhasa) on acne vulgaris: A single-blind, randomized, controlled clinical trial. *Journal of Pakistan Association of Dermatologists*, 24(4), 319–326.

[94] Siddiqui, T., Zafar, S., & Iqbal, N. (2002). Comparative double-blind randomized placebo-controlled clinical trial of a herbal eye drop formulation (Qatoor Ramad) of Unani medicine in conjunctivitis. *Journal of Ethnopharmacology*, 83(1–2), 13–17.

[95] Barnes, M., Flaherty, J., Caron, M., et al. (2018). The evolving regulatory landscape for clinical trials in India. *Food and Drug Law Journal*, 73(4), 601–623.

[96] Piantadosi, S. (2024). *Clinical trials: A methodologic perspective*. John Wiley & Sons. 10–32.

[97] Kumari, A., Bajwa, N., Singh, P.A., et al. (2024). Quality by design in relation to clinical trials, in *Introduction to Quality by Design (QbD) from theory to practice*. Singapore: Springer Nature Singapore. 353–417.

[98] Subbiah, V. (2023). The next generation of evidence-based medicine. *Nature Medicine*, 29(1), 49–58.

[99] Houghton, C., Dowling, M., Meskell, P., et al. (2020). Factors that impact on recruitment to randomised trials in health care: A qualitative evidence synthesis. *Cochrane Database of Systematic Reviews*, (10).

[100] Jansen, C., Baker, J., Kodaira, E., et al. (2021). Medicine in motion: Opportunities, challenges and data analytics-based solutions for traditional medicine integration into western medical practice. *Journal of Ethnopharmacology*, 267, 113477.

14 Clinical Trials of the Traditional Chinese Medicine System Based on Case Studies

Mithun Singh Rajput[1] *and Gopal Natesan*[2]

[1] Department of Pharmacology, School of Pharmacy and Technology Management, SVKM's Narsee Monjee Institute of Management Studies (NMIMS), Deemed-to-be University, Indore, MP, India

[2] Department of Pharmaceutical Chemistry and Analysis, Institute of Pharmacy, Nirma University, SG Highway, Ahmedabad, Gujarat, India

14.1 INTRODUCTION

Traditional, complementary, and alternative medicine has become more prevalent in global healthcare in recent decades. Medication therapies and nonmedication therapies—for example, manual remedies or acupuncture—that are administered without the use of medicine are examples of traditional, complementary, and alternative modalities [1]. Numerous patients have pursued a holistic method as an approach to the well-being of the whole body in lieu of the symptom-based conventional tactic. This has led to a rise in interest in traditional, complementary, and alternative medicine when individuals search for reliable data regarding procedures and treatments in order to investigate additional healthcare options. The market for complementary, alternative, and traditional medicine is predicted to develop at a rapid rate and reach a valuation of 163.3 billion dollars by 2025 [2].

The foundation of traditional Chinese medicine (TCM), which was developed more than three thousand years ago, is based on the concept of modification of the body's vital energy, or *qi* (pronounced "chee"). Certain anatomical sites along the body's surface allow one to access and control this energy, which flows along meridians. Herbal therapy, *qigong* and *t'ai chi*, diet, moxibustion, acupuncture, reflexology, and acupressure are all included in TCM. Disease is defined by TCM as an imbalance in the stream of *qi*. Notably, TCM never separates the emotions, mind, personality, or body [3]. The primary goal of TCM is disease prevention, while disease diagnosis and treatment are also crucial aspects of this medical practice. Because TCM places a strong emphasis on disease prevention, with appropriate training, its ideas can be readily applied to the nursing profession, preserving health promotion as a useful practice area.

The discovery of TCM is regarded as a significant medical advancement, and natural medicine is thought to be humanity's gift from nature. TCM has been practiced widely during the COVID-19 pandemic [4]. Clinical trials are considered the benchmark in the medical field for determining the efficacy and safety of therapies as well as for generating evidence-based knowledge. Clinical trials of novel TCMs must strictly adhere to the standards of good clinical practice in order to ensure that the standardized process of clinical drug trials is followed, that reliable and scientific data are collected, and that the safety, rights, and interests of trial participants are protected [5]. Chinese scholars have conducted significant research using data from registered clinical trials. For instance, reviewing changes in mainland China's cancer drug clinical trials between 2009 and 2018 [6], summarizing the country's progress in 2020 [7], examining pediatric clinical trials carried out during the last ten years in mainland China [8], and examining how anti-lung cancer clinical drug trials in mainland China have changed over the years [9].

After going over the aforementioned and numerous other studies of this kind, it was believed that concentrating on TCM clinical trials using a case study methodology helped to clarify the difficulties and procedures associated with assessing TCM treatments in a clinical setting. The chapter delves into the challenges of TCM clinical trials that uphold contemporary scientific standards and TCM principles. It also presents successful analyses of case studies that have helped close the knowledge gap between contemporary medicine and traditional methods, offering predictions for the role of TCM in international healthcare.

14.2 FUNDAMENTALS OF TCM AND ITS THERAPEUTIC FRAMEWORK

14.2.1 Core Concepts: Yin-Yang, Qi, and the Five Elements

Conventionally, health is attributed as the state of equilibrium between the body and its internal and external environments. Emotional state, genetic factors, the neuroendocrine system,

DOI: 10.1201/9781003522065-14

congenital influence, and personality are examples of internal environments. Climate, diet, geophysical forces, and outside pressure are examples of the external surroundings [10]. Diseases result from the body's efforts to maintain homeostasis when there is an imbalance between the internal and external surroundings or among internal variables [10]. This relationship is explained in terms of TCM concepts by the Yin-Yang theory, which is based on the philosophical premise that two energies are equal to "*qi*," and the five-element theory, which is based on the idea that everything in the world is the change of five basic elements: wood, fire, earth, metal, and water. Every element has a relationship to a certain organ system, which is represented by five Zang (Yang) and five Fu (Yin). The liver, kidney, spleen, heart, and lung are the five main Yin organs. According to Zhang (2020), the small intestine, gallbladder, large intestine, triple heater, and urinary bladder are the five main Yang organs [11]. With inhibition and augmentation on both sides, these constituents are constantly in motion and changing. The mutual interaction between the five components of TCM aids in the diagnosis and mitigation of medical conditions [10].

The development of these two philosophical stances is to comprehend and explain natural phenomena. According to Xie and Preast (2007), these two hypotheses play a significant role in TCM pathology, physiology, diagnosis, and disease therapy [10]. Their mutual dependence shows that neither of the two elements could exist on its own and that each is necessary for the other to exist. According to the Yin-Yang hypothesis, they work together to shield the body from pathogenic elements like stress, germs, and viruses. Four pathogenic alterations of Yin-Yang imbalance result from this situation, either through excess or deficiency of Yin and Yang. Illness and disease are caused by pathological alterations [11].

14.2.2 Comparing TCM and Western Medicine Approaches

Two different approaches to health and healing are represented by TCM and Western medicine. Those have a unique history, set of concepts, and approaches [12]. TCM aims to treat the underlying causes of sickness and is founded on a holistic understanding of the human body. TCM makes a diagnosis based on the patient's symptoms, appearance, and skin, tongue, and eye colors. It then aims to treat the underlying systemic issue while avoiding any possible negative consequences. In contrast, Western medicine is based on scientific research and primarily addresses disease symptoms. Instead of viewing the body as a single, interconnected system, in Western medicine, the target or target organ is treated as an independent entity from the rest of the body, and symptoms are treated. Western medicine employs lab tests to diagnose patients and concentrates on treating symptoms rather than addressing harmful consequences on the body [12]. Also, these practices differ based upon the mechanism of action. In Western medicine, molecular medications are the main emphasis, as this facilitates the identification of targeted molecules and associated signal pathways in studies. As a result, chemical synthesis is used in the development of most medications. TCM, on the other hand, attempts to treat the biosystem as a whole rather than focusing on specific targets with specific chemicals. The objective is to modify the body's general immunological system or bioenvironment in order to produce long-lasting results. In TCM, ingredients typically function as a single formula and are therefore inseparable [12]. Figure 14.1 indicates various modalities associated with TCM and Western medicine, with core concepts and treatment approaches.

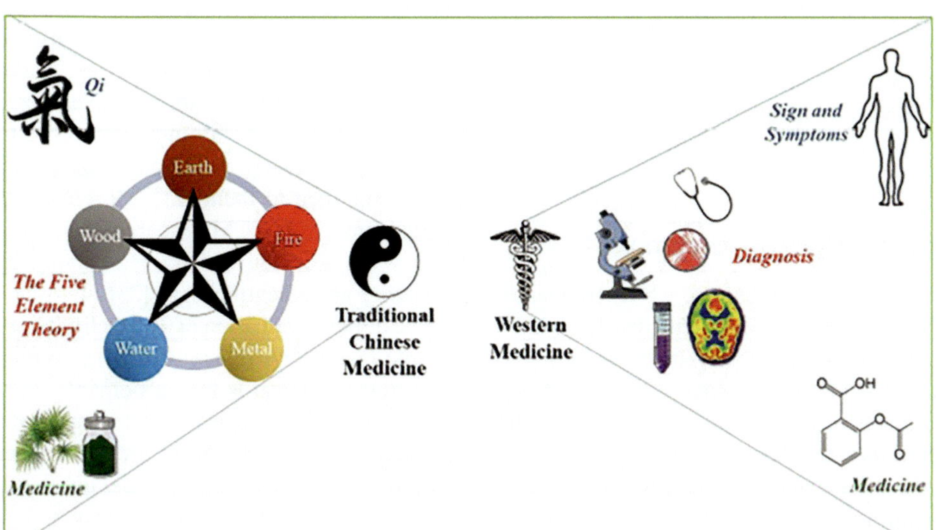

FIGURE 14.1 TCM and Western medicine: core concepts and treatment approaches.

14.3 CHALLENGES IN CONDUCTING CLINICAL TRIALS FOR TCM

14.3.1 Standardization of TCM Practices

TCM has expanded to over 100 nations and developed into a global industry; however, TCM's development is hampered, and its progress toward internationalization and modernization is adversely affected by the absence of pertinent standards [13]. The creation and implementation of standards for the many connections, activities, and outcomes of TCM encompass the fields of medicine, management, science, culture, education, and industry [13]. In order to produce beneficial social and economic consequences, it also promotes international conveyance and modernization of TCM, ensures its safety and quality, and directs and finances various initiatives that aim to advance TCM holistically. By applying the concepts of fusion, simplification, harmonization, and optimization, and by making extensive use of them, all of these objectives are accomplished.

Certain advancements have been made in the ongoing investigation into TCM standardization since the 1980s. However, the existing TCM standardization still lags behind the evolving needs of TCM due to its lengthy, difficult, and complex character [13]. The following are difficulties in standardizing TCM practices:

- TCM standardization is still quite in its primary phases of advancement and has not been well publicized or implemented, which has led to its uncertain identity [14].
- As of right now, published standardization has covered only a small portion of TCM requirements, stressing the basics and how things work without going into detail [15].
- The implementation of many current standardizations in clinical practice is hampered by the fact that they mostly rely upon the literature and lack clinical evidence and evidence-based practice [16].
- There is a lack of skilled specialists in this field, as evidenced by the implementation and oversight scheme of TCM formalization, as well as imperfect surveys and comments regarding its application [17].

14.3.2 Variability in Herbal Formulations

TCM makes use of herbal formulations, which frequently consist of a combination of different herbs that have been meticulously blended in exact ratios. Within the formula, these herbs are divided into four unique roles: the envoy, which guides and harmonizes the acts of the formula; the helper, which reinforces the therapy; the deputy, which offers supportive functions; and the chief, which provides the major therapeutic impact [18]. In this way, TCM uses a polypharmacological strategy for treating multifactorial disorders that affect numerous targets at once. Three stages are involved in multi-substance-acting herbal therapy: the simultaneous action of an herb's several bioactive chemicals, the use of several herbs as a single treatment (known in Chinese as "FangjiPeiwu"), and the concurrent administration of herbal medicine and synthetic drugs as distinct medications. Combination therapy is a very easy approach to achieve polypharmacology; nevertheless, in order to achieve the best therapeutic benefit, great pharmacokinetic compatibility between the co-administered medications is required [19].

Because herbal medicine prescriptions are highly individualized and frequently customized to each patient's specific situation, applying the clinical trial approach to this field has limits [20]. The intricacy of herbal blends, which is impacted by variables including plant species, geographical region, and growing techniques, makes it extremely difficult to standardize dosages for various herbal products. The lack of established procedures for making and delivering herbal medications can lead to dose irregularities and compromise the validity of study findings [20].

14.3.3 Regulatory Hurdles and Ethical Considerations

China's TCM clinical research ethics review procedure is very much the same as it is in Western medicine. It does not, however, specifically take into account the distinctive qualities and cultural facets of TCM. The foundation of TCM includes beliefs such as prescription compatibility, syndrome distinction, and Yin-Yang and the five elements. These components give rise to certain ethical issues that the available literature does not adequately address [21]. There are numerous obstacles to overcome in the ethical evaluation of TCM clinical studies. Due to its tailored treatment approaches, the scientific validity TCM research is difficult to verify. Recent years have seen the emergence of real-world studies due to TCM's individualization and nonrepeatability. TCM may not always benefit from standard techniques such as randomized controlled trials [16]. Because TCM therapies involve the use of herbal medicines and other traditional practices, evaluating the risks associated with them requires a specialized methodology [22]. Furthermore, obtaining informed consent might be difficult, particularly when participants need TCM principles explained to them [21]. Moreover, there are ethical conundrums with TCM research's use of placebos, especially when it comes to striking a balance between cultural sensitivity and scientific rigor [23].

The ethical evaluation of TCM clinical trials should take specific factors into account in order to maximize benefits and minimize hazards for participants. In order to protect participant rights and interests, sample content should be calculated and minimized to prevent an excessive number of participants from receiving therapy that is ineffective or has negative side effects. It is necessary to make sure that the supporting materials or raw materials used for the production of TCM placebos are safe. This will ensure that the product is of the highest caliber and meets all applicable international safety and quality requirements. Adverse events—including significant ones—should be closely monitored, and standard operating procedures should be established that include fair termination requirements and suitable first aid techniques. It is conceivable to use an accompaniment design scheme in which standard therapy is employed as the primary intervention in both the experimental group and the control group, and TCM decoction and placebo are used, respectively, to minimize harm [23].

14.4 DESIGNING CLINICAL TRIALS FOR TCM

14.4.1 Methodological Approaches: Randomized Controlled Trials versus Pragmatic Trials

With the advent of evidence-based medicine, new concepts and techniques—like systematic reviews and randomized controlled trials—are being used for the production of clinical trial data [24]. However, evaluating the therapeutic benefits of TCM therapies is invariably complicated by the glaring disparities in the theoretical foundations and fundamental philosophies between Western medicine and TCM. A crucial tactic for TCM modernization is the creation of clinical research procedures that concentrate on the clinical impacts of TCM. These methodologies are essential to TCM's survival and innovation [25].

In an evidence-based medicine substructure, classical RCTs, or explanatory RCTs, are typically considered the benchmark. Cohort studies are combined with additional research methodologies to create a hierarchy of evidence [26]. RCTs, however, have difficulty accurately capturing the aspects of TCM therapies because of the intricacy of TCM interventions and their distinctions from Western medicine. The methodological community has spent the last few decades primarily focusing on therapeutic efficacy, which may have resulted in less attention being paid to other equally essential problems such as effectiveness or safety in actual clinical settings. Furthermore, the integration of intricate theories and useful aspects of TCM therapies is frequently hampered by this research pattern, which places a heavy emphasis on standard RCTs. The nature of TCM therapies may not be reflected in the final results. Consequently, it is highly desirable to use an improved strategy [26].

In integrated medicine, acupuncture and Chinese herbal remedies are often utilized therapeutic methods. These interventions are highly standardized and well-recognized in China and elsewhere. Even so, there are still differences between the appropriate procedures for evaluating the therapeutic effects of these modalities and those used in Western medicine. The main concern with acupuncture and Chinese patent medicine is how TCM characteristics are included in the evidence while treatment outcomes are consistently evaluated. Despite having excellent internal validity, classical randomized controlled trials (RCTs) are never sufficient because of their limitations in terms of study characteristics (such as setting, demographics, and therapies) and their consequent inability to accurately reflect TCM properties. The use of pragmatic randomized controlled trials, or pRCTs, is essential in assessing the practical effectiveness of acupuncture and Chinese patent medicines. Given that it combines the advantages of explanatory trials (such as randomization) with additional TCM intervention elements, this approach might be the ideal selection for evaluating the efficacy of TCM medicines clinically [27] (Figure 14.2).

Additionally, certain integrated techniques can be designed to incorporate TCM ideas and features into important study characteristics while combining the advantages of practice-based research and traditional RCTs. In this way, the difficulties in evaluating the therapeutic benefits of TCM therapies can be successfully overcome [26].

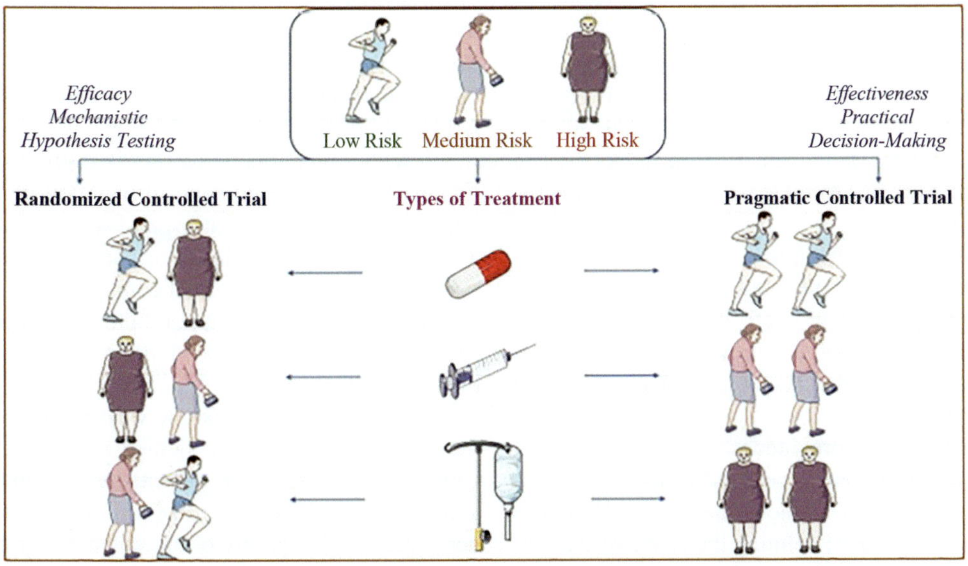

FIGURE 14.2 Comparison among randomized controlled trials and pragmatic controlled trials.

14.4.2 Integrating TCM Diagnosis and Treatment Customization

Standardization of significant diagnosis and treatment variables that may have an impact on TCM clinical trials is one tactic for tackling the complexity of TCM treatments. When evaluating the clinical effects of TCM interventions, a progressive standardization strategy—that is, standardizing fully or largely standardizable factors first, followed by partially standardizable factors—may be used due to the complexity of TCM interventions and the difficulties in standardizing TCM features (such as syndromes). For instance, three distinct disease diagnosis models may be used to evaluate the clinical outcomes of TCM interventions: diagnosis based on TCM-based syndrome distinction, diagnosis based on Western medicine theory, or a mix of both [28]. The first is frequently standardized; however, it might not accurately reflect the intricacy of TCM interventions. Disease diagnoses made exclusively on the basis of TCM theory are frequently less consistent and variable. In this situation, a diagnosis might be made mostly on the basis of Western medicine, and the feature of TCM interventions might be strengthened by adding illness distinction based on TCM theory [26].

14.4.3 Outcome Measurements: Bridging TCM and Western Medicine Criteria

Certain clinical testings are ineligible to be included because of differences in how clinical trial outcomes are reported between research integrating Western medicine and TCM. As a result, even legitimate studies are less likely to have an influence because the excluded studies are not as widely shared. Creating core outcome sets for integrated medicine in particular healthcare domains could potentially solve this issue [29]. Standardizing the outcomes in integrative medicine clinical trials is the aim of the core outcome set creation process [29]. If required, the researchers can suggest a period of time (or "when to measure") for either short- or long-term results. For practice and trials, deciding on measuring procedures (or "how to measure") is equally essential, especially when including subjective outcomes in the main outcome sets. Heterogeneity is introduced when identical outcomes are measured in several ways during the course of clinical studies.

The efficiency of interventions is also predisposed by the caliber of measuring tools. Researchers may be able to optimize their measuring methods with the use of instructions that are existing to help with the selection of measurement instruments for outcomes contained in core outcome sets [30].

Qiu et al. (2021) have provided a suitable example of such strategies for developing a core outcome set for clinical trials merging Western medicine and TCM [29]. The process involves four stages: defining the scope of integrative medicine's core result sets in the first stage, proving the need for them in the second stage, and creating a protocol and registering the core outcome sets in the third step. An additional three steps make up the final stage: (a) comprehensively compiling a list of results (such as safety, effectiveness, and TCM symptoms) by means of qualitative or cross-sectional studies, systematic review, and examination of medication labels and health records; (b) consolidating and arranging results within domains; and (c) conducting two Delphi survey rounds and holding consensus sessions with different stakeholders. The final core outcome set will comprise generic core outcome sets and a core TCM syndrome set. Standardizing outcome reporting and reducing publication bias in the future may be facilitated by the development of core outcome sets for clinical trials involving integrative medicine [29].

14.5 ANALYSIS OF CASE STUDIES OF TCM CLINICAL TRIALS

A growing number of TCM clinical trials have been reported in the last ten years both domestically in China and elsewhere. They have included a wide range of illnesses, such as those that were classified using TCM patterns and those that weren't. The majority of these RCTs' findings demonstrate the effectiveness of TCM in curing illnesses or reducing symptoms, while some report no effect or ineffectiveness. One of the most important aspects of TCM is pattern differentiation of diseases. A growing number of clinical studies involving TCM pattern differentiation of diseases have been published in recent years, suggesting a general trend toward more clinical trials involving TCM with pattern categorization.

14.5.1 TCM-Based Approaches in Oncology

Lv et al. (2024) conducted a systematic evaluation of the clinical trials that have been registered thus far regarding the anticancer impact of TCM monomers [31]. Sixty-nine of the 131 TCM monomers were used in the initiation of a total of 1,982 studies. Five hundred and nineteen interventional anti-tumor trials including 26 monomers were conducted, with vinblastine and camptothecin being the most frequently utilized. These 519 trials examined 45 tumors in all, with lymphoma being the most often studied. Additionally, there is an unequal distribution of clinical trials among funders, collaborators, and sites. The United States and the NIH were the location and sponsor/trial sponsor, respectively, with the greatest counts of completed studies. The majority of the anti-tumor monomers have not yet been used in clinical trials, which remains an issue that needs to be resolved [31].

Future studies may look into ways to improve this condition such as combination therapy and structural changes. China's clinical research on anti-tumor monomers is now a little behind. The extent of improvement, however, is dependent on the quality of investigations at Chinese academies, businesses, and various organizations, as well as the robust backing of national policies.

14.5.2 Intervention trials of TCM for Neurological Disorders

Wei et al. (2022) conducted an analysis of the present state of TCM clinical trial registration for neurological illnesses. Searches of the Chinese Clinical Trial Registry (ChiCTR) and ClinicalTrials.gov platforms turned up a total of 180 interventional clinical trials of TCM treatment for specific illnesses like tension-type headache, ischemic stroke, vascular cognitive impairment, and hemorrhagic stroke [32]. China was the main sponsor of the studies that were included, and hospitals were a prevalent category of primary sponsors. The study design primarily involved randomized allocation, with parallel assignment of the interventional model, double-blind masking, and a sample size of more than four hundred. Most of the trials were conducted at a single location. Interestingly, 50 studies were in phase III of the trials. Chinese herbal preparations were the most frequently used intervention, along with acupuncture coming in second. By checking the registration number on the China National Knowledge Infrastructure Database and PubMed, 38 registered trials were located, consisting of 25 protocol studies and 14 research results (with one study publishing both the protocol and the outcomes) [32].

The irregular and inadequate reporting, the delayed updates and publication, the lack of international collaborations, and the inadequate information on the unique characteristics of traditional medicine are among the issues recognized in the interventional clinical registration trials of TCM treatment on neurological diseases. To improve and standardize the registration of trials using conventional medicine, more work needs to be done on the aforementioned fronts.

14.5.3 Chinese Herbal Medicine for Respiratory Disorders

Ji et al. (2022) conducted a search of CNKI, Wanfang, and SinoMed for RCTs of TCM for pneumonia. There were 1,245 RCTs in all, with 84 Chinese patent medications—45 oral and 39 injectable—involved. The RCTs included 13 different types of interventions/controls, with Chinese patent medicine + Western medication versus Western medicine being the most commonly employed. Physicochemical detection and symptoms/signs were the most commonly used outcome indicators [33].

Similarly, analyses of RCTs using TCM to treat influenza were also conducted in order to give a foundational understanding for clinical decision-making and associated research [34]. Included were 92 RCTs for influenza that were published between 2005 and 2021, with an average sample size of roughly 145, and 17 RCTs with a sample size of more than two hundred. There were 27 Chinese patent medications involved, comprising 11 oral and 6 injectable forms. The most often utilized combination was TCM plus Western medication plus conventional treatment, as opposed to Western medicine plus conventional treatment. A total of 14 intervention protocols were outlined. Moreover, the majority of RCTs reported treatment and intervention courses, but not all of them did. Forty-five outcome indicators were extracted and used 434 times. These indicators included long-term prognosis, symptoms and signs, quality of life, physicochemical detection, TCM symptoms and syndromes, safety incidents, and economic evaluation. The signs and symptoms that occurred most frequently were noted [34].

In the studies, reporting of registration and research ethics was ambiguous. Numerous methodological flaws in the planning and execution of the included RCTs have the potential to affect the validity and applicability of the trials' findings. Crucial standards, such as the choices of basic outcome indicators and disease classification techniques, were also ambiguous. RCTs ought to prioritize the protocol's accuracy and scientificity, bolstering process quality control and hastening the standardization of research on critical links.

14.5.4 Randomized Controlled Trials in Heart Failure in TCM

For the treatment of heart failure, Wu et al. (2022) examined RCTs of TCM. CNKI, Web of Science, VIP, PubMed, Wanfang, and SinoMed, based on the collection in the TCM Clinical Evidence Database System (EVDS), were reviewed. Analysis and evaluation were conducted on the methodological quality, disease type, assessment indices, publication date, treatment plan, sample size, and intervention/control setting [35]. There were 1,631 RCTs in all, with 1,622 in Chinese and nine in English. Just 56 RCTs had a sample size larger than two hundred participants. There were 78 different kinds of Chinese patent medications involved, comprising 39 different kinds of oral medications and 19 different kinds of injectable medications. Thirty-four intervention/control procedures were found, with Chinese patent medicine plus conventional treatment outperforming conventional treatment in the majority of cases. The majority of RCTs described the treatment plan, which typically lasted 14–56 days. The majority of the evaluation indices consisted of physical and chemical examinations as well as symptoms and indications; the most widely utilized measurement index was left ventricular ejection fraction. The most commonly utilized enumeration index was clinical efficacy (response rate). According to methodology, the majority of study participants had a high risk of blindness. Just 13 RCTs with recorded data were available [35].

To enhance the quality of evidence and prevent research waste, increased standardization of clinical study design, execution, and quality control is required.

14.5.5 TCM Intervention in Combating COVID-19

Even with numerous clinical trials underway, there is currently no authorized medication or therapy specifically designed for COVID-19. Numerous clinical trial studies filed on http://www.chictr.org.cn and www.clinicaltrials.gov have been examined by TCM in an effort to find a successful COVID-19 therapy and prevention strategy [36]. A total of 46 TCM interventions were found, which mostly defined the COVID-19 infection prevention tactics and potential therapies. It was shown that natural substances (such as vitamin C and heparin) are useful natural products and therapies based on TCM for boosting the immune system and fighting COVID-19. All analytical investigations, however, contain significant methodological faults and significant bias concerns. Therefore, the therapeutic application of these naturally occurring, accumulating heparin and vitamin C is still not supported by clinical data. Currently, a few more rigorously designed randomized clinical trials with a larger number of patients are being conducted on Chinese medicine. However, it's challenging to find trustworthy and high-quality clinical evidence on COVID-19 treatments due to factors like small sample sizes, subpar drug quality, and lengthy completion times [36].

Another investigation searched all TCM clinical studies for COVID-19 that were found in registry platforms across the globe [37]. A review and summary of pertinent data, registration temporal trend data, interventions, design, and goals were conducted. A total of one hundred and 61 TCM trials were found using three registries. Of them, 114 trials evaluated treatment efficacy, 94 were RCTs, while the remaining trials examined epidemiology, prevention, and rehabilitation of TCM disorders. The TCM therapies Jinye Baidu granules, moxibustion, and Huoxiang Zhengqi pill had the potential to be further evaluated in terms of prevention. Toujie Quwen granules, Lianhua Qingwen capsules, Huashi Baidu decoction, Xuebijing injection, Qingfei Paidu decoction, and Xiyanping injection were to be examined for their therapeutic benefits and ability to relieve symptoms in order to cure COVID-19. Tai Chi and Liuzijue were to be evaluated for their potential to improve patients' lung function during rehabilitation [37].

More research is necessary to determine the evidence base for promising TCM interventions, especially when it comes to populations outside of China. On the other hand, TCM and herbal remedies can be employed as an efficient COVID-19 preventive medication. After completing fruitful clinical trials, countries all over the world can support these promising cures.

14.6 META-ANALYSIS AND SYSTEMATIC REVIEWS OF TCM CLINICAL TRIALS

As a more trustworthy and legitimate method of summarizing research information, systematic reviews are increasingly likely to take the place of conventional narrative reviews and expert commentary. Systematic reviews may or may not include a statistical synthesis known as a meta-analysis, depending on whether the studies are comparable enough that combining their results is meaningful. Chinese academics started to promote the creation of systematic reviews and meta-analyses of TCM in the age of evidence-based medicine [38]. Over the last decade, there has been a rising number of systematic reviews and meta-analyses of TCM. These reviews' themes frequently center on illnesses and medical goods (such as Chinese proprietary or single-herbal remedies) [39].

14.6.1 Efficacy of TCM Herbal Formulas in Oncology

Cancer patients' quality of life is significantly impacted by the side effects of chemotherapy alone. Chinese medicine has benefits in the treatment of cancer when taken with neoadjuvant chemotherapy. The meta-analysis of 12 articles on the use of TCM in the treatment of breast cancer was conducted by Shi et al. in 2021 [40]. Use of the performance status score's efficiency allowed tests for the 95% confidence interval, odds ratio, and heterogeneity. In terms of the incidence of gastrointestinal adverse reactions after chemotherapy, the objective curative effect, the TCM syndrome score change, and the performance status score effective rate, the meta-analysis verified that using TCM in conjunction with neoadjuvant chemotherapy to treat breast cancer has clear advantages over chemotherapy alone [40]. Additionally, a study comparing the effects of various TCM external therapy modalities paired with the three-step analgesic ladder approach on cancer pain was carried out using network meta-analysis [41]. There were 78 trials in all, with more than five thousand individuals and eight treatments. Ranking likelihood indicated that a four-week intervention period produced the highest clinically effective rate for manual acupuncture plus the three-step analgesic ladder approach. The influence of this type of treatment on the clinical result becomes more noticeable as treatment duration increases. The study noted that these results should be interpreted cautiously because of the moderate methodological eminence of the included RCTs. Furthermore, carefully planned research is required to validate these findings [41].

14.6.2 Insights from Meta-Analysis of COVID-19 Studies

It is still debatable whether TCM combined with Western medicine is safe and operative in mitigating COVID-19 in patients. Lei et al. (2022) conducted a meta-analysis to determine the efficacy of TCM [42]. The Cochrane Collaboration's technique was used to undertake a quality assessment of the included studies and assign ratings of low, high, and unclear risk of bias. Included were 13 RCTs with more than a thousand patients that were carried out in China. Cross-sectional data from multiple interventions showed that the TCM group outperformed the control group in terms of statistically higher rates of total efficacy, improvement in chest CT findings and cough, and a reduced frequency of conversion to severe cases.

Additionally, 7 RCTs were involved in the meta-analysis out of the 25 RCTs that involved more than two thousand participants and were chosen for the systematic review [43].

The combined evidence indicated a moderate degree of certainty that, when compared with routine treatment alone, TCM treatment plus routine treatment may improve chest imaging and lead to a clinical cure, while also lowering clinical deterioration, adverse event development, mechanical ventilation use, and mortality in COVID-19 patients. When compared with regular treatment alone, TCM treatment may not increase the negative result of the SARS-CoV-2 nucleic acid test. Patients with COVID-19 were proven to be safe during TCM treatment [43].

Treating COVID-19 with TCM in addition to Western therapy was more successful in symptom relief, encouraging patient recovery, and reducing the likelihood that patients would acquire serious diseases. To come to a more definitive conclusion, however, larger, better-designed trials with longer follow-up periods and sufficient power are needed.

14.6.3 Systematic Reviews of TCM for Cardiovascular Disorders

Evaluation, grading of recommendations, development, and assessment approach were utilized to acutely estimate whether published meta-analyses/systematic reviews of TCM for adults with ischemic stroke are of sufficient value [44]. A total of 83 studies were included. A measurement instrument used to assess systematic reviews revealed that about half of the items were reported; nevertheless, most of the reviews neglected to include information regarding funding, research design inclusion rationale, registration, and the list of omitted studies. Among included studies, over half had low or very low levels of evidence, according to the evidence evaluation. According to reports, the quality of abstracts from systematic reviews and meta-analyses on the use of TCM for ischemic stroke is low, making it difficult for clinical practitioners to obtain accurate information on time. Despite a medium level of methodological quality, there is a substantial risk of bias in individual research, which contributes to the lack of confidence in this evidence [44].

Additionally, a meta-analysis was conducted to assess TCM's therapeutic benefits for atrial fibrillation patients [45]. There were seven allusions in all. It was discovered that the effectiveness of TCM or Western medicine plus TCM combined therapy was higher and the incidence of adverse effects was lower than that of traditional Western medication. It was determined that the combined therapy showed superior therapeutic results to traditional Western medicine [45].

A summary of the meta-analyses and case studies of TCM clinical trials has been depicted in Figure 14.3.

14.7 STATUS OF CASE STUDY IN TCM

Clinical trials in TCM are frequently criticized for providing insufficient information regarding a patient's development following a therapy intervention. Treatments in TCM are frequently very individualized. Despite widespread recognition of the value of case studies, TCM case reporting is still of inadequate quality. The case report was essential to TCM history, as it was one of the main means of information transmission. Due to its distinctive focus on essential details for pattern classification without the efficacy evaluation criteria, it is deeply ingrained in Chinese culture. As a result, the modern medical system finds it difficult to accept and understand [39].

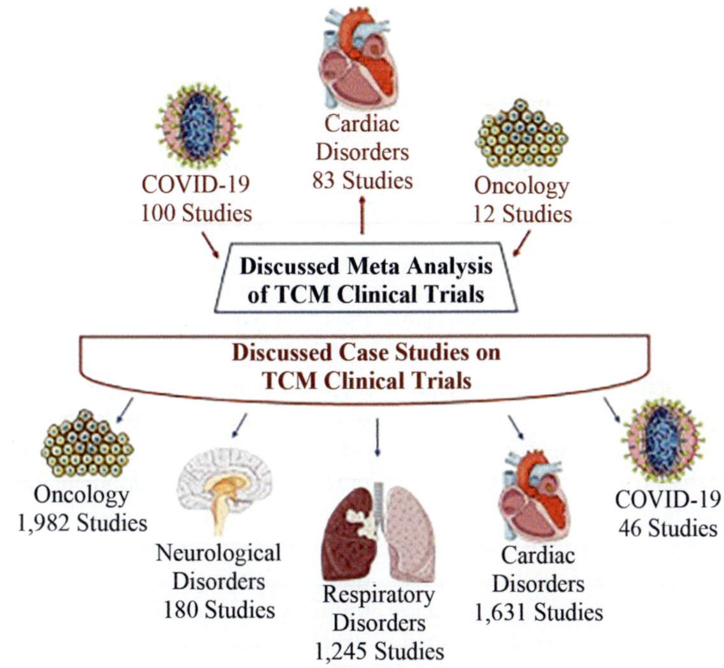

FIGURE 14.3 Number of case studies and meta-analyses of TCM clinical trials analyzed.

More focus is placed on documenting discussions about the diagnosis and management of difficult disorders in modern medical case reports. Each case report or record includes the entire medical file, including the diagnosis and effectiveness assessment. On the other hand, the majority of TCM case reports currently available do not satisfy the requirements of an ideal case study in order to offer strong evidence for additional research studies. Hence, in order to compare treatment efficacy and have evidence-based discussions, it is imperative that the protocol for detailing case studies in TCM practice be standardized and communicated in a common language that can be linked with modern conventional medicine [39].

Standardizing clinical terminology and record format, along with providing sufficient and conclusive information for diagnosis and efficacy evaluation, is currently the top focus. The use of case reports as an evidence-based method for evaluating the effectiveness of TCM treatment will be encouraged by further analysis using modern information processing technology of well-kept data in the TCM treatment of common diseases. This will help optimize the TCM treatment strategy and encourage the development of individualized treatment with TCM [39].

14.8 CASE STUDY ON REGULATORY AND ETHICAL ISSUES IN TCM CLINICAL TRIALS

The number of published clinical trials on TCM has increased recently, although it is unclear how clinical trial registration, ethical review, and informed consent are implemented in these studies. Zhao et al. investigated the status of informed consent, ethical review, and clinical trial registration in TCM clinical studies in 2018. A total of 659 clinical studies were examined. Clinical trial registration information was reported in only roughly 2% of trials. The percentages for informed consent and ethical review were close to 20% and 25%, respectively. Informed consent, protocol approval, and trial registration were not carried out properly [46].

There were other potential causes for these issues. Initially, the significance of informed consent, ethical review, and clinical trial registration was overlooked by the researchers. The majority of trial principal investigators in China are clinical practitioners who lack a firm understanding of the principles and procedures for managing and designing trials. Due to a lack of understanding of research procedures and clinical trial registration, many single-site trials did not properly complete informed consent, ethical review, or clinical trial registration. Second, the material about informed consent, ethical review, and clinical trial registration was not rigorously reviewed by the journals. Editors and peer reviewers are only permitted to evaluate the structure and content of clinical trial publications; they are not permitted to assess the registration status of RCTs. Third, in China, an institutional review board or ethics committee is present in every hospital that has received certification for conducting clinical trials from the Chinese Drug Administration; however, these committees were not established in other research institutions. It is imperative to approach these issues prudently and implement significant actions to that end. Researchers, practitioners, and journal editors should pay close attention to the registration and ethical assessment of clinical trials in TCM. Clinical trial technique training has to be improved [46].

In clinical studies using TCM, informed consent, ethical assessment, and clinical trial registration all performed significant but underappreciated roles. Therefore, the process of raising the caliber and openness of clinical trials in China ought to engage researchers and ethics committees. For accurate and trustworthy scientific outcomes, there should be more emphasis placed on clinical trial technique education and training.

14.9 INNOVATIVE TECHNOLOGIES AND APPROACHES IN TCM CLINICAL RESEARCH

To comply with international standards, TCM must be digitized, normalized, and standardized. Traditional TCM research mostly draws from literature reviews and personal experiences. In order to improve data analysis and experimental evidence, modern methodologies incorporate new technologies such as the Internet of Things, artificial intelligence, big data, and cloud computing. This change attempts to support creative and sustainable modernization while maintaining TCM culture [47]. Additionally, developments in information science and omics technology (metabonomics, genomics, proteomics) support TCM's holistic approach. These technologies enhance the precision and accuracy of TCM therapies by quantifying TCM syndrome-detection indicators.

Huang et al. (2018) proposed an architecture of digital Chinese medicine research that involves four aspects [47]. First, the epistemological approach makes use of computational thinking and systematic scientific big data to comprehend and analyze the intricate human body. In TCM, the second theoretical approach is centered on the expression of knowledge, information, and computation. Third, the technical approach entails gathering data on life status, differentiating between digital syndromes, and conducting a thorough assessment of human life status through the use of digital technology. Lastly, the application strategy builds a technical support system for scientific research, clinical practice, and information management by using clinical digitalization of TCM [47].

Technology integration is essential to TCM modernization. Collaborative operational models, interdisciplinary training, and an efficient organizational and management structure are necessary. Establishing a new digital TCM system will lay the fundamentals for intelligent digital Chinese medicine disciplines.

14.10 FUTURE DIRECTIONS AND OPPORTUNITIES FOR TCM IN GLOBAL HEALTHCARE

Unresolved concerns about the safety and effectiveness of TCM use in contemporary society impede the safe and efficient utilization of TCM's potential advantages.

Incorporating TCM pattern classification into research design can yield clinical evidence with additional value, demonstrating the effectiveness of TCM based on its own principles of diagnosis and therapy. Innovative RCT designs that integrate pattern information and disease diagnosis for group classification will undoubtedly uncover more significant contributions from TCM treatments and help to avoid the shortcomings of some study designs that use Western medical research methods to evaluate TCM efficacy without distinguishing between patterns that are distinctive features and the core of TCM diagnosis and treatment theory [39].

For TCM to be accepted by Western medical professionals and become a crucial part of the healthcare system, evidence-based medicine must be applied to the practice. When investigating TCM as a potentially cost-effective therapy option to address healthcare challenges such as functional disorders, chronic illnesses in the aging population, and growing healthcare expenses, this is an essential first step. Sufficient evidence in the form of RCTs evaluating both safety and efficacy is required to validate the appropriate use of TCM. But rather than being forcefully imposed on TCM, evidence-based medicine should be flexibly adjusted while taking into account the particularities of TCM practice [39].

The development of technology and research instruments has made it necessary to carry out extensive studies on the toxicities and potential pharmacological interactions of herbs, as well as to correlate laboratory results with clinical observations. This offers fresh perspectives on the herbs, enabling us to use them in the contemporary clinical context in a more sensible and secure manner. To protect TCM users from hazards, tight quality control in the production process, regulatory controls on TCM herbal products, and active safety surveillance to identify unfavorable events are essential [48].

Researchers should discover a novel method in the not too distant future for entering important patient and medical data into case study reports. Data for evaluation can be obtained by comparing the patient's or volunteer's TCM and conventional medical features. The evidence-based methodologies for evaluating the effectiveness of TCM treatment on individual patient cases can also benefit from the possibility of recording objective and subjective outcome indicators, such as biomarkers and TCM practitioners' treatment records, respectively. These integrative methods combine the best features of TCM recording methodology with contemporary case reports utilized in Western medicine. This could open the door to learning about TCM's holistic methods to illness diagnosis and treatment, as well as patient-specific treatment plans [39].

TCM has amassed vast amounts of data and rich experience with human use, both of which can be examined to produce empirical proof. Although incorporating real-world evidence into TCM regulation presents difficulties and constraints, efforts should be made to standardize TCM diagnosis and treatment, create reliable sources of real-world evidence, incorporate real-world evidence and RCTs, encourage cooperation among stakeholders, and offer education and training to raise awareness and comprehension of real-world evidence in TCM. By doing this, it will be possible to improve patient outcomes while using TCM therapies and to further promote the application of real-world evidence in TCM regulation [49].

14.11 CONCLUSION

The focus on TCM clinical studies will continue to grow on a global scale. In the clinical evaluation of TCM treatment efficacy and in clinical trials of TCM products that are generally in line with TCM principles and approved by regulatory bodies, researchers in both the basic and clinical sciences will need to take into account widely established standardized operating procedures. The aforementioned sections have covered a number of important topics, including the development of creative modifications to the standard clinical trial design, the acceptance and recognition on a global scale of both quantity and quality in TCM clinical studies, and the focus on particular research questions and suggested solutions found in TCM treatment principles. Sustained research and cooperation will be essential to maximizing TCM's potential in a way that is internationally recognized and supported by the scientific community.

REFERENCES

[1] Hyodo, I., Amano, N., Eguchi, K., et al. (2005). Nationwide survey on complementary and alternative medicine in cancer patients in Japan. *Journal of Clinical Oncology, 23*, 2545–2654.

[2] Youn, B. Y., Moon, S., Mok, K., Cheon, C., Ko, Y., Park, S., et al. (2022). Use of traditional, complementary, and alternative medicine in nine countries: A cross-sectional multinational survey. *Complementary Therapies in Medicine, 71*, 102889.

[3] Darby, S. B. (2009). Traditional Chinese medicine: A complement to conventional. *Nursing for Women's Health, 13*(3), 198–206.

[4] Chen, K., & Chen, H. (2020). Traditional Chinese medicine for combating COVID-19. *Frontiers in Medicine, 14*, 529–532.

[5] Zhou, Y., Yang, J., He, Y., Lv, Y., Wang, C., Deng, H., & Huang, J. (2022). Characteristic analysis of clinical trials for new traditional Chinese medicines in mainland China from 2013 to 2021. *Frontiers in Medicine, 9*, 1008683.

[6] Li, N., & Han, X. M. (2010). Tentative study on TCM standardization. *Journal of TCM Information, 27*(5), 1–3.

[7] Wu, W. W., Ji, X., Wang, H., Chen, F., Ding, Q., Zhang, G. D., et al. (2021). Pediatric clinical trials in mainland China over the past decade (from 2009 to 2020). *Frontiers in Medicine, 8*, 745676.

[8] Wu, D. W., Huang, H. Y., Tang, Y., Wang, H. X., Wang, J., Wang, S. H., et al. (2020). Progress on clinical trials of cancer drugs in China. *Chinese Journal of Oncology, 43*, 218–223.

[9] Zhong, Q., Tao, Y., Chen, H., Zhou, Y., Huang, L., Han, X., et al. (2021). The changing landscape of anti-lung cancer drug clinical trials in mainland China from 2005 to 2020. *Lancet Regional Health—Western Pacific, 11*, 100151.

[10] Xie, H., & Preast, V. (2007). *Xie's veterinary acupuncture* (1st ed.). Blackwell Publishing Ltd.

[11] Zhang, H. I. (2020). A possible representation of Yin Yang five elements (Yin Yang Wu Xing): Hypothesis of dual five-body-coordination. *Integrative Medicine Research, 9*(4), 100462.

[12] Cao, C., & Brown, B. (2019). Understanding Chinese medicine and Western medicine to reach the maximum treatment benefit. *Journal of Translational Science, 5*, 1–12.

[13] Wang, J., Guo, Y., & Li, G. L. (2016). Current status of standardization of traditional Chinese medicine in China. *Evidence-Based Complementary and Alternative Medicine, 2016*, 9123103.

[14] Wang, L. X., Wang, Z. G., & He, Y. (2011). Tentative study on the ways and methods for TCM standardization. *World Journal of Traditional Chinese Medicine, 6*(5), 375–378.

[15] Yi, X. F. (2016). *Seven industry standards of clinical trials of TCM.* (2nd ed.). China News of Traditional Chinese Medicine.

[16] Li, X. C., Dai, G. H., & Liu, X. C. (2016). Real-world study methods and its application in clinical efficacy evaluation of traditional Chinese medicine based on HIS. *Journal of Shandong University of Traditional Chinese Medicine, 40*, 415–418.

[17] Lu, X., Tian, L. J., & Chen, Z. L. (2013). Construction profile of TCM standardization. *TCM Journal, 54(10)*, 886–888.

[18] Marshall, A. C. (2020). Traditional Chinese medicine and clinical pharmacology. *Drug Discovery and Evaluation: Methods in Clinical Pharmacology*, 455–482.

[19] Li, C., Jia, W. W., Yang, J. L., et al. (2022). Multi-compound and drug-combination pharmacokinetic research on Chinese herbal medicines. *Acta Pharmacologica Sinica, 43*, 3080–3095.

[20] Busia, K. (2024). Herbal medicine dosage standardisation. *Journal of Herbal Medicine, 46*, 100889.

[21] Zhang, J., & Zhang, Z. M. (2021). The challenges of ethical review in clinical research of traditional Chinese medicine. *Evidence-Based Complementary and Alternative Medicine, 12*, 6754985.

[22] Ernst, E. (2004). Challenges for phytopharmacovigilance. *Postgraduate Medical Journal, 80*(943), 249–250.

[23] Zhang, T. J., Xu, J., Shen, X. P., & Han, Y. Q. (2016). Relation of "property-response-component" and action mechanism of Yuanhu Zhitong dropping pills based on quality marker (QMarker). *Chinese Traditional and Herbal Drugs, 47*, 2199–2211.

[24] Tian, G., Zhao, C., Zhang, X., et al. (2021). Evidence-based traditional Chinese medicine research: Two decades of development, its impact, and breakthrough. *Journal of Evidence-Based Medicine, 14*, 65–74.

[25] Wang, Y., & Huang, L. (2019). Taking a broad and long-term view to establish China Center for Evidence-Based Traditional Chinese Medicine (CCEBTCM). *Chinese Journal of Evidence-Based Medicine, 19*, 1131–1137.

[26] Sun, X., Li, L., Liu, Y., et al. (2021). Assessing clinical effects of traditional Chinese medicine interventions: Moving beyond randomized controlled trials. *Frontiers in Pharmacology, 12*, 713071.

[27] Shi, Q., Li, L., Li, S., & Sun, X. (2021). Statistical methods in pragmatic randomized controlled trials (I): Addressing non-compliance. *Chinese Journal of Evidence-Based Medicine, 21*, 117–124.

[28] Jianping, L., & Keji, C. (2015). Guide to clinical research methods in traditional Chinese medicine and integrative Chinese and Western medicine. *Chinese Journal of Integrative Medicine, 35*, 901–932.

[29] Qiu, R. J., Li, M., Hu, J. Y., & Chen, J. (2021). Methods for development of a core outcome set for clinical trials integrating traditional Chinese medicine and Western medicine. *Journal of Integrative Medicine, 19*(5), 389–394.

[30] Prinsen, C. A. C., Vohra, S., Rose, M. R., Boers, M., Tugwell, P., & Clarke, M. (2016). How to select outcome measurement instruments for outcomes included in a "core outcome set"—A practical guideline. *Trials, 17*(1), 449.

[31] Lv, D., Liu, Y., Tang, R., Fu, S., Kong, S., Liao, Q., et al. (2024). Analysis of clinical trials using anti-tumor traditional Chinese medicine monomers. *Drug Design, Development and Therapy, 18*, 1997–2020.

[32] Wei, J. J., Guo, R. J., Fu, G. J., et al. (2022). Registration of intervention trials of traditional Chinese medicine for four neurological diseases on Chinese clinical trial registry and ClinicalTrials.gov: A narrative review. *Journal of Traditional Chinese Medicine, 42*(1), 148–153.

[33] Ji, Z. C., Qiang, X. Y., Hu, H. Y., Chen, Z., Ou, Y., Feng, C. N., & Zhang, J. H. (2022). Clinical trials and evaluation of Chinese patent medicine for pneumonia. *Zhongguo Zhong Yao Za Zhi, 47*(9), 2343–2350.

[34] Sheng, X. D., Chen, Z., Ji, Z. C., et al. (2022). Clinical trials and evaluation of Chinese patent medicine for influenza. *Zhongguo Zhong Yao Za Zhi, 47*(9), 2338–2342.

[35] Wu, X. L., Qiang, X. Y., Ji, Z. C., Hu, H. Y., Feng, C. N., et al. (2022). Clinical trials and evaluation of Chinese patent medicine for heart failure. *Zhongguo Zhong Yao Za Zhi, 47*(9), 2322–2329.

[36] Nile, S. H., & Kai, G. (2021). Recent clinical trials on natural products and traditional Chinese medicine combating COVID-19. *Indian Journal of Microbiology, 61*(1), 10–15.

[37] Luo, H., Yang, M., Tang, Q. L., Hu, X. Y., Willcox, M. L., & Liu, J. P. (2021). Characteristics of registered clinical trials on traditional Chinese medicine for coronavirus disease 2019 (COVID-19): A scoping review. *European Journal of Integrative Medicine, 41*, 101251.

[38] Lai, S. L. (2000). Developing systematic review of clinical medical documents of TCM. *Zhongguo Zhong Yi Yao Xin Xi Za Zhi, 7*, 78.

[39] Jiang, M., Yang, J., Zhang, C., Liu, B., Chan, K., & Cao, H. (2010). Clinical studies with traditional Chinese medicine in the past decade and future research and development. *Planta Medica, 76*(17), 2048–2064.

[40] Shi, G., Yu, D., Wu, J., Liu, Y., & Huang, R. (2021). A systematic review and meta-analysis of traditional Chinese medicine with chemotherapy in breast cancer. *Gland Surgery, 10*(5), 1744–1755.

[41] Jin, H., Liang, J., Zhang, S., Ma, S., Qin, H., Zhang, D., Pang, X., & Zhang, M. (2024). External treatment of traditional Chinese medicine for cancer pain: A systematic

review and network meta-analysis. *Medicine, 103*(8), e37024.

[42] Lei, Y., Guan, H., Xin, W., & Yang, B. C. (2022). A meta-analysis of 13 randomized trials on traditional Chinese medicine as adjunctive therapy for COVID-19: Novel insights into Lianhua Qingwen. *BioMed Research International, 2022*, 4133610.

[43] Wang, H., Xu, B., Zhang, Y., et al. (2021). Efficacy and safety of traditional Chinese medicine in coronavirus disease 2019 (COVID-19): A systematic review and meta-analysis. *Frontiers in Pharmacology, 12*, 609213.

[44] Sun, S., Zhao, L., Zhou, X., et al. (2023). Methodological, reporting, and evidence quality of systematic reviews of traditional Chinese medicine for ischemic stroke. *Frontiers in Pharmacology, 14*, 1047650.

[45] Lei, Y., Tang, Y., Huang, L., & He, P. (2021). Systematic review and meta-analysis on efficacy of traditional Chinese medicine for atrial fibrillation through cluster analysis. *Annals of Palliative Medicine, 10*(8), 8982–8990.

[46] Zhao, H., Zhang, J., Yang, F., & Tan, L. (2018). Improving the ethical review of clinical trials on traditional medicine: A cross-sectional study of clinical trial registration, ethical review, and informed consent in clinical trials of traditional Chinese medicine. *Medicine (Baltimore), 97*(47), e13062.

[47] Huang, H. Y., Yan, J. F., & Hu, Z. X. (2018). Digital Chinese medicine research: An overview. *Digital Chinese Medicine, 1*, 1–8.

[48] Fung, F. Y., & Linn, Y. C. (2015). Developing traditional Chinese medicine in the era of evidence-based medicine: Current evidences and challenges. *Evidence-Based Complementary and Alternative Medicine, 2015*, 425037.

[49] Li, P., Wang, S., & Chen, Y. (2024). Use of real-world evidence in regulatory decisions for traditional Chinese medicine: Current status and future directions. *Therapeutic Innovation & Regulatory Science, 58*(1), 34–41.

15 Challenges and Future Perspectives of Herbal Medicines

Raja Chakraverty[1], *Prasanth Thumpati*[2], *Swarnali Goswami*[3], *Abhik Paul*[4], *and Tatini Debnath*[5]

[1]Department of Pharmacology, Haldia Institute of Pharmacy, ICARE Complex, Haldia, Purba Medinipur, West Bengal, India

[2]Virginia Commonwealth University, Richmond, Virginia, USA

[3]Manager, HEOR Complete Solutions, California, USA

[4]Doctoral Research Scholar, Department of Pharmaceutical Technology, Jadavpur University, Kolkata, West Bengal, India

[5]Doctoral Research Scholar, Department of Pharmaceutical Technology, Maulana Abul Kalam Azad University of Technology, West Bengal, India

15.1 INTRODUCTION

Herbal medicine, often referred to as phytomedicine, pertains to the utilization of various parts derived from medicinal plants, encompassing a broad spectrum of applications that extend beyond the confines of conventional medical practices. The discipline of herbalism is steeped in a rich historical tradition, showcasing its extensive application throughout human history, which has allowed it to flourish as a revered practice outside the modern medical arena. Over the course of the past several decades, herbal medicine has gradually transitioned into the mainstream medical narrative, propelled by significant advancements in both analytical techniques and quality control measures, alongside noteworthy progress in clinical research methodologies. According to the esteemed World Health Organization (WHO) [1, 2], the concept of phytomedicine, or herbal medicine, embodies a comprehensive amalgamation of knowledge, skills, and practices that are rooted in the cultural theories and experiences inherent to diverse indigenous populations, regardless of whether these practices can be explicitly articulated, and this holistic approach is employed for the purposes of maintaining overall health as well as for the prevention, diagnosis, enhancement, or treatment of both physical and mental ailments. Traditional herbal medicines are typically characterized as naturally occurring substances that are derived from plants, which necessitate minimal to no industrial processing, and have been historically employed as therapeutic interventions within localized or regional healing frameworks [3–8]. The application of herbal medicines has been an integral part of human health practices for thousands of years, long preceding the advent of the allopathic pharmaceutical system that dominates modern medicine today. Phytomedicines, originating from raw plant materials, can encompass a variety of forms, including tinctures, herbal teas, poultices, powdered extracts, and numerous other preparations derived from plants. The utilization of botanical resources for healing purposes is a practice that predates recorded human history, serving as a foundational element that has significantly contributed to the development of contemporary medicinal practices. Scholarly research and comprehensive studies focusing on clinical, pharmacological, and chemical investigations of these herbal remedies, which are predominantly sourced from plant material, have underscored their importance as the foundational basis for many early pharmaceutical compounds, such as aspirin, which is derived from willow bark; digitoxin, extracted from the foxglove plant; morphine, obtained from the opium poppy; quinine, sourced from the bark of the cinchona tree; and pilocarpine, which is derived from the jaborandi plant. The relevance of herbal medicines continues to be pronounced, particularly in the context of primary healthcare, where it is estimated that approximately 70% to 80% of healthcare treatments administered to individuals in developing countries are based on herbal solutions. This high percentage of reliance on herbal medicine within primary healthcare is largely attributable to the prevailing belief among populations that herbal remedies possess inherent therapeutic efficacy [9, 10].

15.2 BACKGROUND HISTORY OF HERBAL MEDICINE

The utilization of natural products derived from organic origins, particularly various parts of plants, has been an integral aspect of human civilization, serving as both sustenance and therapeutic agents over countless generations, particularly in the context of employing entire plants or specific plant components to treat and avert ailments. It is indeed a formidable challenge to ascertain the precise historical moment when humanity commenced the practice of

DOI: 10.1201/9781003522065-15

employing plants for medicinal purposes; however, several ancient literary works and other historical documents provide credible indications of its inception. The most archaic written documentation pertaining to the employment of medicinal flora for the formulation of pharmaceutical preparations has been unearthed on a Sumerian clay tablet originating from Nagpur, which is estimated to be around 5,000 years old, and this remarkable artifact encompasses 12 distinct recipes related to drug formulation while also alluding to over 250 different plant species, including notable alkaloids such as opium poppy, henbane, and mandrake. The revered Indian scriptures known as the Vedas explicitly delineate therapeutic interventions that incorporate the extensive variety of plants that are abundantly available throughout the Indian subcontinent. Furthermore, the ancient Chinese compendium titled *Pen T'Sao*, which is attributed to Emperor Shen Nung and dates back to approximately 2500 BC, meticulously documents the medicinal properties of 365 different botanical substances, many of which continue to be utilized in contemporary herbal medicine practices [11–13]. Theophrastus, who lived from 371 to 287 BC, is widely regarded as the pioneer of botanical science, a distinction that he earned through his seminal works titled *De Causis Plantarium*, which translates to *Plant Etiology*, and *De Historia Plantarium*, which translates to *Plant History*. Within the confines of these scholarly texts, he undertook the task of classifying around 500 medicinal plants that were known during his era. In his examination of plant toxicity, Theophrastus emphasized the critical importance of gradual acclimatization to these substances by humans through the incremental increase of dosage, thereby highlighting the necessity of careful administration. Each nation around the globe possesses its own historical records concerning the utilization of botanical and herbal remedies for the treatment and alleviation of various medical conditions, and these records often exhibit considerable variability across different cultures. In the context of India, Ayurveda stands out as the most ancient medical system, with its origins tracing back approximately 5,000 years [14]. The Ayurvedic texts are actually a component of the Atharvaveda, which dates back to around 1000 BC, and during earlier centuries, this knowledge was predominantly transmitted through oral traditions until the first comprehensive Ayurvedic textbook was meticulously documented and preserved in the ancient language of Sanskrit. Rather than merely combating diseases, the ancient principles of Ayurveda were fundamentally designed to promote overall health and well-being. The principal texts that are available concerning herbal medicine include the *Charak Samhita*, which dates to around 1000 BC, and the *Sushrut Samhita*, which was composed around 100 AD [15, 16]. The Ayurvedic literature provides exhaustive descriptions of more than 1,500 different herbs and an extensive array of 10,000 distinct formulations. Furthermore, the diagnostic criteria, accompanied by the signs and symptoms associated with over 5000 various diseases or disorders, are thoroughly articulated in the "Madhav Nidan," which dates back to approximately 800 AD.

15.3 CHALLENGES RELATED TO THE REGULATORY STATUS OF HERBAL MEDICINES

Clinical trials that focus on herbal medicines are confronted with a myriad of obstacles, which encompass critical aspects such as safety, quality assurance, and the necessity for standardization, all of which are essential for ensuring the reliability and efficacy of these natural products in clinical settings.

15.3.1 Safety

One major issue that impedes the evaluation of herbal medicines is the prevalent and often pronounced lack of scientific evidence that would allow for a comprehensive assessment of both their safety and therapeutic efficacy, which is crucial for gaining the trust of both practitioners and patients alike.

Furthermore, it is not uncommon for the safety of participants involved in clinical trials of herbal products to be significantly disregarded, thereby raising serious ethical concerns regarding the welfare of those who volunteer for such studies. Such negligence can result in erroneous risk assessment processes, which, in turn, can lead to dire and potentially life-threatening consequences for individuals who might suffer adverse reactions or lack of therapeutic benefit from these unregulated substances.

15.3.2 Quality

In addition to safety concerns, the quality of the herbal products being tested in clinical trials must undergo rigorous evaluation to ensure batch-to-batch uniformity regarding the active constituents, which is critical for achieving consistent therapeutic outcomes across different batches of the same product.

Moreover, the issues surrounding the standardization of herbal medicines present substantial challenges that must be addressed in order to facilitate their integration into mainstream medical practices and ensure the delivery of safe and effective treatments.

15.3.3 Standardization

The inherent variability in the composition of plant materials used in herbal formulations presents a significant barrier to achieving consistency in product quality and efficacy, making it exceedingly difficult to establish reliable dosing regimens for patients. Additionally, the complex interactions that occur within the various components of herbal formulations introduce further complications, as they can alter the pharmacological effects of the individual ingredients, thereby complicating the overall therapeutic profile of the product.

15.3.4 Regulation

The overarching regulation of herbal bioactive-based formulations is fraught with numerous challenges that encompass considerations of safety, pharmacovigilance, and quality assurance, all of which are necessary to protect consumers and ensure the responsible use of these alternative therapies. Lastly, there exists a substantial lack of comprehensive knowledge concerning herbal medicines among both healthcare professionals and the general populace, which ultimately hinders the advancement and acceptance of these natural remedies in a scientifically rigorous manner.

According to the formal definition established within the context of nutritional science and health policy, a dietary supplement is classified as a product that is specifically designed to be ingested by individuals, with the primary objective of augmenting or enhancing the nutritional value of the diet, and it must contain what is referred to as a "dietary ingredient." The dietary ingredients incorporated within these various products can encompass a wide array of essential components, such as diverse vitamins, essential minerals, a multitude of different herbs, or other botanical substances that are recognized as necessary for the proper functioning and overall health of the human body. Under the provisions of the Dietary Supplement Health and Education Act (DSHEA), it is noteworthy that any supplementary toxicity studies are typically not mandated if the herbal component in question was already available in the marketplace prior to the year 1994, thereby indicating a certain level of regulatory leniency for established products. In this context, it is the responsibility of the Food and Drug Administration (FDA) to bear the evidentiary burden of demonstrating that a particular herbal medicinal product or "dietary ingredient" poses a threat of toxicity or is otherwise deemed unsafe for human consumption. Furthermore, a significant challenge that persists in numerous countries is the issue related to the sharing of regulatory information pertaining to herbal medicines, which is often inadequately communicated between various regulatory authorities and the safety monitoring or pharmacovigilance centers that are responsible for ensuring public health and safety. This lack of information exchange can lead to gaps in the understanding of the safety profiles of these herbal products, ultimately complicating the regulatory landscape and posing potential risks to consumers who rely on these dietary supplements for their health needs [17–20].

15.4 NEED FOR SAFETY ASSESSMENT OF HERBAL DRUGS

It is an undeniable truth that the intricate requirements, alongside the multifaceted research protocols, rigorous standards, and diverse methodological approaches that are essential for the thorough evaluation of both the safety and efficacy of herbal medicines are significantly more elaborate and demanding when compared with the relatively straightforward criteria established for conventional or orthodox pharmaceuticals. A single herbal medicine or medicinal plant may possess a plethora of natural constituents, with the potential to contain well over a hundred unique compounds; when one considers a mixed herbal medicinal product, the number of natural constituents can be exponentially increased, potentially comprising several multiples of what is found in a single herb. Consequently, the task of conducting a comprehensive analysis of individual active constituents within such complex mixtures may become practically unfeasible, particularly in instances in which an herbal product is formulated from two or more distinct herbs, each contributing its own array of bioactive compounds and interactions [21, 22].

15.5 OBSTACLES ASSOCIATED WITH THE QUALITY CONTROL OF HERBAL MEDICINE

The overall quality and integrity of the raw materials employed in the manufacturing processes of herbal medicinal products significantly influence both their safety and therapeutic efficacy, creating a direct correlation between the initial ingredients and the final outcomes of these natural remedies. The caliber of the source materials or raw ingredients is not solely reliant on intrinsic genetic characteristics, which encompass inherent traits of the plants themselves, but is also significantly impacted by extrinsic factors such as the environmental conditions in which these plants are cultivated, as well as the implementation of effective agricultural practices and responsible collection techniques for medicinal flora, which together include critical aspects such as careful plant selection and appropriate cultivation methodologies [23–26]. It is this intricate interplay of numerous and diverse factors that complicates the execution of stringent quality control measures on the raw materials utilized in the production of herbal medicines, thereby presenting substantial challenges to manufacturers and regulators alike. In accordance with established Good Manufacturing Practices (GMP), the accurate identification of the various species of medicinal plants, along with their appropriate storage conditions and specialized cleaning protocols for different materials, emerge as a critical component in ensuring the quality control of the initial raw materials that serve as the foundation for these herbal products. However, the predominant obstacles are found in the quality control processes associated with the finished herbal medicinal products, particularly those that are composed of multiple herbal ingredients that may interact in complex ways. Consequently, the overarching requirements and methodologies employed for the quality assessment of completed herbal products tend to be markedly more intricate and nuanced than those applied to conventional pharmaceuticals, necessitating a more comprehensive approach to ensure compliance with safety standards. To safeguard the safety and effectiveness of herbal medicines, the WHO continues to advocate for the implementation of robust quality assurance and control measures, which include the establishment of National Quality Specifications and standardized criteria for herbal materials, adherence to GMP protocols, and the enforcement of labeling and licensing regulations governing manufacturing practices.

15.6 HERBAL MEDICINE AND SAFETY MONITORING

In recent years, there has been a notable and significant increase in the awareness and usage of herbal or natural medicines and products within developed nations, highlighting a trend that has captured the attention of researchers and healthcare professionals alike. Additionally, the reliance of numerous individuals residing in developing countries on plant-based sources for their medicinal needs, combined with the generally inadequate regulation of herbal medicines across many jurisdictions, has brought to the forefront serious safety concerns that necessitate a comprehensive examination of both the potential risks and benefits associated with these natural remedies [27–29]. The adverse effects that may arise from the consumption of herbal medicines can be attributed to a multitude of factors, including but not limited to the inappropriate selection of plant species, the adulteration of herbal products with harmful substances, the presence of undeclared pharmaceutical ingredients, contamination during processing, excessive dosages, and the improper utilization of these remedies by either healthcare practitioners or consumers themselves, as well as the concurrent use of herbal medicines with traditional pharmaceuticals. A significant gap in understanding persists regarding the critical importance of taxonomic botany and the meticulous documentation practices that are often overlooked by many manufacturers of herbal medicines, which creates distinctive challenges in the accurate identification and collection of the correct medicinal plants employed in the formulation of herbal remedies. In order to mitigate the confusion that arises from the use of various common names for these plants, it is imperative to adopt standardized binomial nomenclature that is widely recognized and utilized in scientific discourse, ensuring clarity and precision in communication. For instance, *Artemisia absinthium L.*, known by at least 11 different common names, possesses an active narcotic derivative that underscores the importance of accurate identification in relation to safety and efficacy. Consequently, the effective monitoring and regulation of herbal medicine require a robust and collaborative approach that brings together experts from diverse fields such as botany, phytochemistry, pharmacology, and various other stakeholders who play a pivotal role in this domain. Such interdisciplinary collaboration is essential for establishing rigorous standards and protocols that ensure the safety and efficacy of herbal medicines in both developed and developing countries, thereby fostering a more informed and responsible usage of these natural products. Ultimately, the intersection of traditional knowledge and contemporary scientific research holds tremendous potential for advancing our understanding and application of herbal medicines, while simultaneously addressing the pressing safety concerns that have emerged in recent years. It is through this synthesis of knowledge and expertise that we can hope to navigate the complexities of herbal medicine and harness its benefits in a manner that is both safe and effective for diverse populations around the globe [30–35].

15.7 FUTURE PERSPECTIVES OF HERBAL MEDICINE

Notwithstanding the increasingly widespread acknowledgment of the therapeutic efficacy inherent in herbal medicines, it is imperative to note that the domain of quality control associated with these products is confronted with numerous significant gaps and formidable challenges that necessitate comprehensive and meticulous examination. A particularly noteworthy challenge that emerges within this context is the substantial variability that exists in product composition and overall quality, which can be attributed to a myriad of factors including, but not limited to, geographical variations, varying methodologies employed in cultivation, and the diverse practices involved in post-harvest processing. This intrinsic diversity presents a distinctive obstacle when it comes to the establishment of consistent and reliable quality standards that can be uniformly applied across the spectrum of different herbal products available in the market. Although there have been rapid and remarkable advancements in analytical technologies, including but not limited to mass spectrometry and DNA barcoding, which have significantly altered the quality control landscape in various other industries, the application of these cutting-edge techniques within the framework of herbal medicine quality assessment has been sporadic, inconsistent, and, at times, insufficiently rigorous. In light of this, we advocate for the systematic incorporation of molecular fingerprinting techniques, such as DNA barcoding and metabolic profiling, as a means to authenticate the various herbal ingredients utilized in these products. These innovative methodologies not only facilitate precise species identification but also empower practitioners to detect potential adulterants or contaminants that may compromise product integrity. Additionally, we strongly recommend the integration of real-time monitoring systems and Internet of Things (IoT) devices throughout the entire herbal supply chain. By doing so, we can achieve continuous data collection regarding environmental conditions, which is essential for maintaining the botanical integrity and overall quality of the products. Furthermore, it is also crucial to recognize that the implementation of data-driven predictive modeling is an essential component that can significantly enhance quality control practices within the herbal industry, ultimately ensuring that consumers receive products that are both safe and effective. The anticipated trajectory of the future is characterized by a significant and burgeoning demand coupled with a rapidly expanding market for herbal medicines and various herbal healthcare products, which is evident in both nations classified as developing and those recognized as developed across the global landscape. This trend underscores the increasing recognition of the therapeutic potential and health benefits associated with herbal remedies, which are gaining prominence as viable alternatives to conventional pharmaceutical products in contemporary healthcare paradigms. As such, the convergence of consumer interest, regulatory support, and scientific validation is likely to fuel the growth of this sector, further demonstrating the pivotal role that herbal medicine is destined to play in the future of global health and wellness initiatives [36–40].

15.8 CONCLUSION

In the contemporary context, the worldwide acceptance and utilization of herbal medicines, along with an array of associated products, are experiencing a remarkable and exponential surge that cannot be overlooked. Furthermore, the issues related to adverse reactions that these herbal remedies may provoke have become increasingly prominent and apparent in recent times, resulting in a growing prevalence of reported incidents that are no longer subject to debate, especially given the prior misconceptions that erroneously categorized herbal medicinal products as inherently "safe" merely because they originate from "natural" sources. Consequently, there exists an urgent necessity for the regulatory policies governing herbal medicines to be not only standardized but also significantly strengthened on a global scale to ensure consumer safety and product efficacy. It is now incumbent upon regulatory bodies to take on the formidable responsibility of diligently monitoring the controlled and quality-assured distribution of herbal products, while also actively facilitating their progression toward the critical stages of clinical trials. In light of these developments, it becomes increasingly important for stakeholders in the herbal medicine sector to collaborate closely with regulatory authorities to ensure that appropriate safety measures are implemented and that the integrity of the herbal products is maintained throughout their lifecycle. Ultimately, the establishment of robust regulatory frameworks will play a pivotal role in fostering public confidence and promoting the safe integration of herbal medicines into mainstream healthcare practices worldwide.

REFERENCES

[1] Bhardwaj, S., Verma, R., & Gupta, J. (2018). Challenges and future prospects of herbal medicine. *International Research in Medical and Health Sciences*, *1*(1), 12–5.

[2] Kasilo, O. M., & Wambebe, C. (2021). *Traditional and complementary medicine in global health care. In Handbook of global health* (pp. 1459–1505). Cham: Springer International Publishing.

[3] Tilburt, J. C., & Kaptchuk, T. J. (2008). Herbal medicine research and global health: An ethical analysis. *Bulletin of the World Health Organization*, *86*, 594–599.

[4] Kamboj, A. (2012). Analytical evaluation of herbal drugs. *Drug Discovery Research in Pharmacognosy*, *3*, 23–55.

[5] Balunas, M. J., & Kinghorn, A. D. (2005). Drug discovery from medicinal plants. *Life Sciences*, *78*(5), 431–441.

[6] Butler, M. S. (2004). The role of natural product chemistry in drug discovery. *Journal of Natural Products*, *67*(12), 2141–2153.

[7] Pal, S. K., & Shukla, Y. (2003). Herbal medicine: current status and the future. *Asian Pacific Journal of Cancer Prevention*, *4*(4), 281–288.

[8] World Health Organization. (2004). Guidelines for the regulation of herbal medicines in the South-East Asia Region. In *Guidelines for the regulation of herbal medicines in the South-East Asia Region*.

[9] World Health Organization. (2005). National policy on traditional medicine and regulation of herbal medicines: report of WHO global survey. In *National policy on traditional medicine and regulation of herbal medicines: report of WHO global survey* (pp. 156–156).

[10] Conboy, L., Kaptchuk, T. J., Eisenberg, D. M., Gottlieb, B., & Acevedo-Garcia, D. (2007). The relationship between social factors and attitudes toward conventional and CAM practitioners. *Complementary Therapies in Clinical Practice*, *13*(3), 146–157.

[11] Kumar, V., & Kumar, V. (2009). An overview of herbal medicine. *International Journal of Pharmaceutical Sciences*, *1*(1), 1–20.

[12] Kelly, K. (2009). *History of medicine, facts on file*. New York, NY. 10001.

[13] Nair, D. (2020). *An Experimental Evaluation on "Shothaghna" (Anti-Inflammatory) Karma of Bakula Twak (Mimusops Elengi Linn.) In Wistar Albino Rats* (Doctoral dissertation, Rajiv Gandhi University of Health Sciences (India)).

[14] Wiart, C. (2006). *Ethnopharmacology of medicinal plants: Asia and the Pacific*. Totowa, NJ: Humana Press.

[15] Petrovska, B. B. (2012). Historical review of medicinal plants' usage. *Pharmacognosy Reviews*, *6*(11), 1.

[16] Katic, R. (1958). La medicine en Serbie au moyen age. *Beograd: Scientific Work*, 7–36.

[17] Bazala, V. (1943). The historical development of medicine in the Croatian lands. *Zagreb: Croation Publishing Bibliographic Institute*, 9–20.

[18] Nikolovski, B. (1995). Essays on the history of health culture in Macedonia. *Skopje: Macedonian Pharmaceutical Association*, 17–27.

[19] Goyal, M., Sasmal, D., & Nagori, B. P. (2012). Ayurveda the ancient science of healing: an insight. *Drug Discovery Research in Pharmacognosy. Croatia: InTech*, 1–10.

[20] Senior, K. R. editor. (2009). *The 100 most influential scientists of all time*. The Rosen Publishing Group, Inc.

[21] Israili, A. H. (1981). Humoral theory of Unani Tibb. *Indian Journal of History of Science*, *16*(1), 95–99.

[22] Smith, J. E. (2010). A brief history of herbal medicine. *Inspired Times*, *4*, 15.

[23] Dervendzi, V. (1992). *Contemporary treatment with medicinal plants*. Skopje: Tabernakul, 5–43.

[24] Mukherjee, P. K. (2002). *Quality control of herbal drugs: An approach to evaluation of botanicals*. Business horizons.
[25] Ernst, E., Schmidt, K., & Wider, B. (2005). CAM research in Britain: The last 10 years. *Complementary Therapies in Clinical Practice*, *11*(1), 17–20.
[26] Barnes, P. M., Bloom, B., & Nahin, R. L. (2007). *Complementary and alternative medicine use among adults and children: United States*.
[27] Verma, S., & Singh, S. P. (2008). Current and future status of herbal medicines. *Veterinary World*, *1*(11), 347.
[28] Moshi, M. (2005). Current and future prospects of integrating traditional and alternative medicine in the management of diseases in Tanzania. *Tanzania Journal of Health Research*, *7*(3), 159–167.
[29] Russo, M., Spagnuolo, C., Tedesco, I., & Russo, G. L. (2010). Phytochemicals in cancer prevention and therapy: Truth or dare? *Toxins*, *2*(4), 517–551.
[30] Purohit, C. K., & Gokhale, S. B. (2014). *A text book of pharmacognosy* (50th ed.). Pune: Nirali publication, 15–32.
[31] Bhattacharya, R., Reddy, K. R., & Mishra, A. K. (2014). Export strategy of Ayurvedic products from India. *International Journal of Ayurvedic Medicine*, *5*(1), 125–128.
[32] Bandaranayake, W. M. (2006). Quality control, screening, toxicity, and regulation of herbal drugs. *Modern Phytomedicine: Turning Medicinal Plants into Drugs*, *20*, 25–57.
[33] World Health Organization. (2005). *National policy on traditional medicine and regulation of herbal medicines: Report of a WHO global survey*. World Health Organization.
[34] Ekor, M. (2014). The growing use of herbal medicines: Issues relating to adverse reactions and challenges in monitoring safety. *Frontiers in Pharmacology*, *4*, 177.
[35] Ghosh, S., Bishal, A., Ghosh, S. K., Jana, K., Gayen, B., Sahu, S., & Debnath, B. (2023). Herbal medicines: A potent approach to human diseases, their chief compounds, formulations, present status, and future aspects. *International Journal of Membrane Science and Technology*, *10*, 442–464.
[36] Zhou, J., Ouedraogo, M., Qu, F., & Duez, P. (2013). Potential genotoxicity of traditional Chinese medicinal plants and phytochemicals: An overview. *Phytotherapy Research*, *27*(12), 1745–1755.
[37] World Health Organization. (2004). *WHO guidelines on safety monitoring of herbal medicines in pharmacovigilance systems*. World Health Organization.
[38] Rodrigues, E., & Barnes, J. (2013). Pharmacovigilance of herbal medicines: The potential contributions of ethnobotanical and ethnopharmacological studies. *Drug Safety*, *36*(1), 1–2.
[39] Farah, M. H., Edwards, R., Lindquist, M., Leon, C., & Shaw, D. (2000). International monitoring of adverse health effects associated with herbal medicines. *Pharmacoepidemiology and Drug Safety*, *9*(2), 105–112.
[40] Ghosh, S., Bishal, A., Ghosh, S. K., Jana, K., Gayen, B., Sahu, S., & Debnath, B. (2023). Herbal medicines: A potent approach to human diseases, their chief compounds, formulations, present status, and future aspects. *International Journal of Membrane Science and Technology*, *10*, 442–464.

16 Future Trends, Emerging Technologies, and Interdisciplinary Collaborations for Clinical Trials of Herbal Drugs

Richi Goel, Versha Chandila, and Sweta Bawari
Amity Institute of Pharmacy, Amity University Campus, Noida, Gautam Buddha Nagar, Uttar Pradesh, India

ABBREVIATIONS

AI	Artificial intelligence
ANNs	Artificial neural networks
ARIA	Automatic retinal image analysis
AUC	Area under the curve
AUEC	Area under the effect curve
BMI	Body mass index
CA19-9	Carbohydrate antigen 19–9
CCMQ	Constitution in Chinese Medicine Questionnaire
CEA	Carcinoembryonic Antigen
CFS	Chalder Fatigue Scale
CHM	Chinese herbal medicine
CNNs	Convolutional neural networks
CSBM	Complete spontaneous bowel movements
DED	Dry eye disease
DNA	Deoxyribonucleic acid
ED	Erectile dysfunction
EMA	European Medicines Agency
FD	Functional dyspepsia
FDA	Food and Drug Administration
FSRS	Framingham Stroke Risk Score
FSS	Fatigue Severity Scale
GJD	Gongjin-Dan
HJIG	Hongjing I Granules
HLP-II	Health-Promoting Lifestyle Profile II
HPLC	High Performance Liquid Chromatography
IFN-γ	Interferon-gamma
IL-6	Interleukin-6
IL-16	Interleukin-16
IND	Investigational New Drug
ISI	Insomnia Severity Index
LDL	Low-density lipoprotein
NGS	Next-generation sequencing
OTE	Overall treatment effect
PBPK	Physiologically based pharmacokinetic modeling
PLSS	Professional Life Stress Scale
RCT	Randomized clinical trial
RF	Random forests
SBM	Spontaneous bowel movements
SCC	Squamous Cell Carcinoma Antigen
SDS	Single Dyspepsia Symptom Scale
SF-36	Short-Form 36 Health Survey
SHT	Ssanghwa-Tang
SOC	Standard of care
SOL	Shehlingcao oral liquid
SVM	Support vector machines
TCM	Traditional Chinese medicine
TDS	Total Dyspepsia Symptom Scale
t_{max}	Time of maximum
TNF-α	Tumor necrosis factor
TSH	Thyroid stimulating hormone
VBE	Virtual bioequivalence
XBD	Xuanfei Baidu Decoction
XBJ	Xuebijing injection

16.1 INTRODUCTION

India is recognized as a major hub for clinical trials of herbal medicine, driven largely by its traditional medicinal systems, particularly Ayurveda. The Drugs and Cosmetics Act, 1940, governs the use and marketing of herbal medicines in the country, and their approval for safety and efficacy data, adherence to quality control standards, compliance with labeling and packaging, manufacturing practices, and clinical trial protocols require adherence to the latest scientific standards. The scientific validation of traditional knowledge is one of the main goals of herbal medicine clinical trials in India. Even though many plant-based medicines have been used for the treatment of a plethora of diseases since time immemorial, their systematic investigation is the need of the hour to reduce the gap between traditional and conventional medicines and modern methods, as researchers have indicated that many plants may have potential adverse effects that are not evident from their traditional use; therefore, scientific investigation is necessary to close the gap between conventional medicine and modern methods [1].

Promising results have been reported in clinical trials assessing the effectiveness of herbal medicines such as Ginseng for chronic conditions like diabetes and cardiovascular ailments. When compared to conventional therapy, research on Ayurvedic formulations for diabetes management has

DOI: 10.1201/9781003522065-16

shown notable improvements in metabolic indicators and glycemic control. The use of herbal remedies in treatment regimens is supported by clinical evidence, which also has a beneficial effect on how people view their usefulness and efficacy in treating chronic ailments. These results highlight the potential of herbal remedies in contemporary healthcare to complement or replace current pharmacotherapies. Positive clinical findings linked to St. John's Wort for depression have significantly changed how the general public and the medical community view herbal remedies. The number of clinical trials exploring herbal medications has significantly increased; this is supported by a systematic analysis of clinical trials using herbal medicine and focusing on a variety of disorders, published between 2019 and 2022, which found that 1,517 publications were eligible for inclusion out of 5,918 records. Another bibliometric analysis examining trends in herbal medicine for pain from 1990 to 2019 found that 2,986 original articles were published, with China and the United States contributing the most, indicating collaborative global research in herbal medicine. Clinical studies that concentrate on the effectiveness and safety of popular herbal supplements, such as echinacea and ginkgo biloba, have been significant in shifting views. By providing scientific data rather than anecdotal information, the evidence obtained from these trials regarding the safety profiles and therapeutic benefits of these supplements supports their integration into patient treatment [2]. Additional validations have been provided by clinical research examining the use of particular standardized herbal medicines in the treatment of cancer. Significant impacts on tumor growth and post-treatment patient quality of life have been observed in trials evaluating the anticancer potential of natural compounds such as curcumin and epigallocatechin-3-gallate. These results strengthen the notion that herbal remedies can make a significant contribution to cancer treatment, increasing their acceptance by both patients and medical professionals. Because of the outcomes of these clinical trials, herbal medicines are now seen as scientifically verified treatment solutions rather than traditional therapies. By promoting the adoption of herbal remedies and pushing more healthcare policies toward their inclusion in clinical guidelines, this change promotes a more integrated approach to healthcare [3].

16.2 DIVERGENCE IN CLINICAL TRIAL FRAMEWORKS: HERBAL MEDICINE VERSUS CONVENTIONAL MEDICINE

The regulatory frameworks, procedures, standardization, and intrinsic complexity of herbal formulations are the main areas where clinical trials for herbal medicine diverge greatly from those for conventional medications. Herbal medications are frequently exempt from the FDA's requirement for an IND application; these include dietary supplements, functional foods, and cosmetics marketed without any health claims and subject to fewer rigorous regulations. Conventional medications, on the other hand, must pass extensive clinical trial phases to prove their safety and efficacy before they can be released.

Given the inherent heterogeneity of plant materials, standardization of herbal medications often proves problematic. Moreover, these are generally composed of multiple active constituents, which can vary widely in their effects and interactions. This makes it difficult to determine the relative contributions of each component, which ultimately makes evaluating their therapeutic outcomes in clinical trials more difficult. Conventional medications, on the other hand, typically concentrate on a single active ingredient, making it easier to attribute the observed effects to that substance [4].

The consistency and efficacy of herbal formulations might also differ greatly from batch to batch, in contrast to synthetic medications, which have a predetermined composition. Therefore, clinical trials for traditional medicines often use a standardized active component to guarantee uniform dosage and hence ensure meaningful results on safety and efficacy from the trials [5].

While both conventional and herbal medicine frequently employ randomized controlled trials (RCTs), the approaches can vary. This is because herbal products have unique characteristics, such as taste and odor, and double-blinding can be difficult to apply in the context of herbal treatment. Furthermore, herbal trials frequently rely on subjective reports that can introduce variability because of patients' perceptions and cultural backgrounds and impact their evaluation of the treatment's effect, whereas conventional medications usually use objective outcome measures.

16.3 CURRENT STATUS OF CLINICAL TRIALS OF HERBAL DRUGS

Due to the frequent adverse effects encountered with synthetic medications and increasing microbial and cancer cell resistance toward a multitude of conventional drugs, current therapies for various illnesses stand inefficient, specifically those that are chronic in nature. Due to this, there has been a significant revisiting of the use of herbal products in recent years. It is believed that traditional or folk medicine accounted for around 75% of the plant-based medicinal substances employed globally. About 70% of contemporary medications in India are derived from natural resources, and several synthetic counterparts have been created using plant-derived prototype chemicals [6]. With the aid of modern scientific knowledge, technology, and ideas, multinational pharmaceutical corporations and researchers have begun to rediscover medicinal plants as an alternative source of novel medication prospects according to conventional knowledge. Thus, the number of clinical trials of herbal drugs has increased in the past decade. Lately, numerous herbal formulations, like Hongjing I granule, Xuanfei Baidu decoction, Xuebijing injection (XBJ), and Shenlingcao Oral Liquid (SOL), made up of various phytoconstituents, have been subjected to clinical trials for the treatment of a plethora of diseases, and some of them have been discussed in Table 16.1.

TABLE 16.1
Current Status of Clinical Trials for Herbal Drugs

Herbal Formulation/ Phytoconstituents	Disease Targeted	Experimental Design	Phases of CT	Observations in the Ongoing Trials/ Outcome of Completed Trials	CT Number	References
AYUSH-64	COVID-19	Multicentric, controlled, RCT	Phase II	134 patients out of 140 patients qualified for efficacy, with complete recovery and enhanced quality of life; few adverse effects were reported	CTRI/2020/06/025557	[27]
BV-4051	SARS-CoV-2	Double-blind, placebo-controlled, RCT	Phase III	Reduced duration of illness, reduced severity scores, and decreased IL-6	CTRI/2020/09/027817	[18]
Powder mixture of cinnamon, garlic, ginger, fenugreek, and arjuna	Metabolic syndrome	Parallel, RCT	Phase I	Observation of systolic blood pressure, triglyceride level, diastolic blood pressure, glycosylated hemoglobin, high-density lipoprotein level, and waist-to-hip ratio	NCT06515652	[8]
Leaf and root extract of *Withania somnifera*	Aging	Crossover and double-blind, RCT	Early phase I	Measurements of plasma concentration of withanolides after administration of herb	NCT06171724	[9]
Xuanfei Baidu decoction	COVID-19	Pilot, RCT	Phase II	Reduced fever, coughing, and exhaustion	ChiCTR2000034795	[20]
SCD-2101	Functional constipation	Single-arm, open-label pilot clinical trial	NA	Changes in CSBM and SBM, and observation of adverse effects after administration of the drug	NCT06534996	[7]
Brucea javanica oil emulsion	Advanced colorectal cancer	Prospective, multicenter, controlled RCT	Phase IV	Progression-free survival and quality of life	NCT05897749	[10]
Kratom	Herbal interaction	Non-randomized, crossover clinical trial	Early phase 1	AUC ratio in the presence to absence of kratom	NCT05846451	[11]
Xuebijing injection (XBJ)	Sepsis	Multicenter, double-blind, placebo-controlled RCT	Phase IV	Lower mortality rate and reduced ICU mortality	NCT03238742	[22]
Chinese herbal medicine teabag	Stroke risk reduction	Parallel, RCT	Phase 1	Score change of stroke risk estimated with ARIA	NCT05801575	[12]
Curcumin Oral Capsule	Lupus Nephritis	Parallel, RCT	Phase II	Examination of urine and serum samples within week 12, examination of blood samples for ESR level and CRP level	NCT05714670	[13]
Lavender teabag	Anxiety, depression	Single-blind RCT	Phase I	Reduced anxiety and depression	IRCT20090304001742N4	[21]

(Continued)

TABLE 16.1 ***(Continued)***
Current Status of Clinical Trials for Herbal Drugs

Herbal Formulation/ Phytoconstituents	Disease Targeted	Experimental Design	Phases of CT	Observations in the Ongoing Trials/ Outcome of Completed Trials	CT Number	References
Shenlingcao Oral Liquid (SOL)	Long COVID	Parallel, RCT	Phase II	Change in scores on the Chalder Fatigue Scale (0–33 points), quality of life, changes in behavior, muscle strength	NCT05684952	[14]
Costus root paste	Antibacterial effects	Parallel, RCT	Phase III	Bacterial count in children	NCT05657132	[28]
Xian-Hua-Cha	Hyperlipidemia, obesity	Crossover, RCT	Phase III	Changes in serum lipid profile involving serum triglycerides, total cholesterol, changes in body weight	NCT05632068	[15]
Wei's Qi Ju Gan Lu Formula	Dry eye	Parallel, RCT	Early phase I	Change in SPEED score to assess the improvement of dry eye symptoms	NCT05319041	[16]
Jiangtang Tiaozhi recipe	Type 2 diabetes dyslipidemia	Parallel, RCT	Phase I/phase II	Changes in HbA1c, blood glucose at fasting, low-density lipoprotein (LDL), BMI, total cholesterol, gut microbiota	NCT04623567	[17]
Ayush-SR	Occupational stress	Double-blind, placebo-controlled, prospective, RCT	Phase II	Professional Life Stress Scale (PLSS), Manasa Bhava (psychobehavioral parameters)	CTRI/2020/07/026276	[19]
Yukgunja-tang	Functional dyspepsia (FD)	Multicenter, three-arm, assessor-blind, parallel-group RCT	Phase III	Total dyspepsia symptom scale (TDS), overall treatment effect (OTE)	ChiCTR-IIR-15007567	[23]
Hongjing I granule	Erectile dysfunction	Double-blind, placebo-controlled multicenter RCT	Phase I	Total sexual intercourse, changes in scores of sexual desire, and observation of adverse effects	ChiCTR2000041127	[24]
Raw cinnamon powder	Postprandial hyperglycemia	Crossover RCT	Phase I	Diminished hyperglycemic effects	ReBEC: RBR-98tx28b	[25]
Gongjin-Dan and Ssanghwa-Tang	Chronic fatigue	Placebo-controlled, double-blinded RCT	Phase I	Safe, effective, reduced fatigue, improved quality of life	KCT0007515	[26]

In the last five years, every country has fought against the COVID-19 pandemic, which sparked the search for drug repurposing in complementary and alternative systems of medicine. An interdisciplinary AYUSH Research and Development team was established by the Ministry of AYUSH, which focused mainly on managing and curing COVID-19. AYUSH-64, a polyherbal Ayurvedic formulation, was co-administered with a standard of care (SOC) to treat mild and moderate COVID-19 in a controlled, multicentric, and RCT (CTRI/2020/06/025557) (phase II). The formulation of AYUSH-64 includes the bark of *Alstonia scholaris, the* rhizome of *Picrorhiza kurroa,* the whole plant of *Swertia chirata,* and some amount of seed powder of *Caesalpinia crista,* which was administered twice a day. Primary outcomes include clinical recovery and the percentage of patients who showed clinical recovery within 28 days, and observation of body temperature, absence of cough, and symptoms related to COVID-19. Additionally, AYUSH has substantially improved the number of secondary outcomes and improved quality of life without any mortality. Some adverse effects were reported, including hyperglycemia, while some patients suffered from comorbid diseases, which could be risk factors for various other fatal diseases.

SCD-2101, composed of four herbs, is a formula of Chinese herbal medicine (CHM) that has been utilized to treat functional constipation and is administered twice a day, although the formulation has not been revealed yet (NCT06534996). It is a single-arm, open-label pilot study conducted on elderly patients diagnosed with functional constipation with not more than two complete spontaneous bowel movements (CSBM) in one week. Outcomes include improvement in CSBM and spontaneous bowel movements (SBM), and adverse effects were observed [7].

A randomized, parallel comparative study was conducted between herbal medicine and allopathic medicines used to treat metabolic syndrome, majorly including dyslipidemia (NCT06515652). Herbal medicine included a powder mixture of cinnamon, garlic, ginger, fenugreek, and arjuna, whereas the allopathic drugs included rosuvastatin, telmisartan, and sitagliptin, administered continuously for three months. Outcomes of herbal medicine include observation of glycosylated hemoglobin, systolic blood pressure, diastolic blood pressure, triglyceride level, and waist-to-hip ratio at day 0, day 30, day 60, and day 90 [8].

Essential components extracted from the leaves and roots of *Withania somnifera,* available commercially as Shoden ®, were administered to healthy old seniors *at* two different concentrations (240 mg and 480 mg) to investigate the pharmacokinetics and oral bioavailability of NCT06171724. It is a crossover and double-blind RCT with outcomes involving measurement of plasma concentrations of all the withanolides from blood samples within a 48-hour period to evaluate pharmacokinetic parameters, including maximum concentration, area under the curve (AUC) (0 to infinity), and AUC (0 to t). Secondary outcomes include time of maximum concentration (t_{max}), half-life (t½), electrocardiography, liver function, thyroid-stimulating hormone (TSH), testosterone levels, and blood cells counts [9].

The safety and efficacy of *Brucea javanica* oil emulsion injection were assessed in patients suffering from metastatic colorectal cancer who were denied receiving multiline treatment; a multicenter, prospective RCT was conducted (NCT05897749). The primary outcome was progression-free survival, while the secondary outcome involved improvement in quality of life [10].

Herbal interaction of a botanical substance, namely *kratam*, and an opioid medicine, namely oxycodone, was investigated in a nonrandomized, crossover clinical trial explaining the effect of *kratam* (administered in the form of tea) on the metabolism and pharmacological effect of oxycodone (NCT05846451). The primary outcome includes the AUC ratio in the presence versus absence of *kratam.* Secondary outcomes include the area under the effect curve (AUEC), $C_{max,}$ and t½ [11].

Safety and clinical effectiveness of the CHM teabag were chosen for reducing the risk of stroke in the older population in a stepped-wedge group RCT design (NCT05801575). The CHM teabag includes five herbs: rood and rhizome of dried Notoginseng; Mu Dan Pi, which is the root bark of the *Paeonia suffruticosa* tree; root of *Puerariae lobatae*; dried bud of *Rosa rugosa;* and *Citri randis* Exocarpium. The stroke risk, which is calculated by automatic retinal image analysis (ARIA), was the main outcome measure, represented by a score from 0 to 1. The Framingham Stroke Risk Score (FSRS), the Constitution in Chinese Medicine Questionnaire (CCMQ), the Health-Promoting Lifestyle Profile II (HLP-II), and observation of adverse events constitute secondary outcomes [12].

Oral curcumin capsule is investigated to study its effects on pediatric patients suffering from lupus nephritis in children is a serious and frequent pediatric problem; thus, a parallel RCT is conducted (NCT05714670). Primary outcome includes examination of urine samples for estimation of creatinine and albumin levels and examination of serum samples from baseline and week 12. Secondary outcome includes examination of a blood sample for erythrocyte sedimentation rate and C-reactive protein [13].

Efficacy and safety of Shehlingcao oral liquid (SOL), which is a CHM, were used to treat fatigue associated with long COVID in a double-blinded, placebo-controlled RCT (NCT05684952). *Ganoderma lucidum* and powder of *Cordyceps sinensis* as minister medicine, American Ginseng as the main ingredient, and rose as an adjuvant are used in the formulation of SOL. The primary outcomes include a questionnaire measuring the level of tiredness, namely, the Chalder Fatigue Scale (CFS) at four weeks, whereas the secondary outcomes include the Insomnia Severity Index (ISI), quality of life, changes in behavior, muscle strength, safety, and analysis of gut microbiota [14].

Xian-Hua-Cha, constituting *Citrus reticulata* Blanco, *Zizyphus jujuba Mill., Rosa rugosa* Thunb, *Nelumbo nucifera* Gaertn, *Cassia obtusifolia* L., *Stevioside,* and resistant starch mixed with water, was used as an experimental drug for lipid and body fat issues, which are a major global health concern, particularly in an obese population. A crossover RCT was conducted, which aimed to enroll patients with active hyperlipidemia and a body mass index (BMI) of more than 27 (NCT05632068). This was accomplished by analyzing the changes in body fat, patient body weight, and other metabolic parameters involving blood sugar and cholesterol at the conclusion of the study [15].

Wei's Qi Ju Gan Lu herbal formula was used to treat dry eye disease (DED), which is more common among post-menopausal women than in the general population, in a parallel RCT recruiting peri- and postmenopausal women with an age range of 40 to 79 years (NCT05319041). The primary outcome included a questionnaire assessing improvement in dry eye symptoms ranging from 0 (no dry eye symptoms) to 64 (most symptomatic) within three months, whereas the secondary outcomes included changes in sleep score and amelioration in depression score within three months [16].

The effectiveness of CHM Jiangtang Tiaozhi Recipe was evaluated in a parallel-controlled RCT on patients with active type 2 diabetes mellitus along with dyslipidemia (NCT04623567). The composition of Jiangtang Tiaozhi Recipe involves various herbs like *Anemarrhena asphodeloides* Bunge, *Salvia miltiorrhiza* Bunge, *Zingiber officinale* Roscoe, *Momordica charantia* L., *Schisandra chinensis* (Turcz.) Bail., *Coptis chinensis* Franch., and *Aloe vera* (L.) Burm.f. Changes in HbA1c, fasting blood glucose, low-density lipoprotein (LDL), BMI, total cholesterol, gut microbiota, and fecal metabolites are evaluated at different intervals [17].

A double-blind, placebo-controlled clinical study using a polyherbal drug, namely BV-4051, was conducted (CTRI/2020/09/027817) to treat patients with COVID-19, as an elevation in vital immune mediators such as interleukin-6 (IL-6), tumor necrosis factor (TNF-α), and interferon-gamma (IFN-γ) causes immunological hyperactivation, which results in shock, respiratory failure, or multi-organ dysfunction. Therefore, improvement in inflammatory biomarkers could be a key to improving outcomes in patients with active COVID-19. This Ayurvedic polyherbal formulation has significantly lowered IL-6 and symptoms that were associated with SARS-CoV-2 and was administered to hospitalized patients. The composition of BV-4051 (ARTOVID-20®) includes Shallaki *(Boswellia serrata)*, Turmeric (*Curcuma longa*), Ginger (*Zingiber officinale)*, and Ashwagandha (*Withania somnifera);* additionally, these phytoconstituents are known for modulating the immune system and inflammatory biomarkers. The primary and secondary endpoints include mean severity scores for headache, smell, taste, sore throat, and quality of life, whose progressive results indicate that this polyherbal formulation is quite effective [18].

Another study was conducted to evaluate the efficacy and safety of Ashwagandha (*Withania somnifera* (L.) Dunal), a plant with neuroprotective and antidepressant properties, which was used in an Ayurvedic formulation—"Ayush-SR"—as the main component (CTRI/2020/07/026276) to relieve this occupational stress in a prospective, single-centered RCT. Since nursing is one of the most demanding professions, it requires efficiency and stability as well as the capacity to deal with numerous professional circumstances associated with poorer health and high occupational stress. According to Ayurveda, the equilibrium of *Sareerika Doshas* (bodily humors) and *Manodoshas* (humors of the mind) determines whether complex and interconnected psychophysiological traits are harmful or health-promoting. The endpoints include the Professional Life Stress Scale (PLSS), improvement in quality of life, and Manasa Bhava (psychobehavioral parameters) [19].

Patients suffering from COVID-19 were again selected to assess the clinical effectiveness of Xuanfei Baidu decoction (XBD), either in combination with a traditional drug or alone, in an RCT (ChiCTR2000034795). XBD consists of 13 Chinese herbs, including the rhizome of *Phragmitis* and *Atractylodis*; the rhizome and root of *Polygoni Cuspidati* and *Glycyrrhizae*; various herbs such as *Ephedrae, Gypsum Fibrosum, Pogostemonis, Artemisiae Annuae,* and *Verbenae;* Coix seed; *Descurainiae Semen Lepidii Semen*; *Armeniacae Semen Amarum*; and *Citri Grandis* Exocarpium. Outcomes include the absence of COVID-19 symptoms, while secondary outcomes include the absence of symptoms along with amendments in the immune system. The results demonstrated that XBD, along with traditional medicine, could reduce clinical symptoms such as fever, coughing, and exhaustion [20].

It is believed that the aroma of herbal tea made up of lavender flowers stimulates the limbic system by activating olfactory nerve cells, which release different types of neurotransmitters, including serotonin, encephalin, and others. Thus, lavender (*Lavandula officinalis*) can be used to uplift the mood of an individual and thereafter can be used to treat anxiety and depression. Based on this hypothesis, a lavender tea bag was prepared, and its effect on anxiety and depression was evaluated in a single-blinded RCT (IRCT20090304001742N4). Some questionnaires, including the Spielberger Anxiety Inventory and the Beck Depression Inventory, were used as endpoints to assess the outcomes. The results of this study demonstrated that the depression and anxiety intervention group had decreased scores, suggesting the effectiveness of lavender [21].

Sepsis is a dreadful health condition but lacks proper treatment. In relevance to this, a Chinese herbal formula-based injection has been prepared, which has the potential to modulate the septic response. Based on these results, the clinical effectiveness of Xuebijing injection (XBJ) against placebo on mortality among patients with active sepsis on 28 days was evaluated in a multicenter, double-blind, placebo RCT (NCT03238742). XBJ is composed of five herbs, including the roots of *Paeonia lactiflora*, *Salvia miltiorrhiza*, and *Angelica sinensis*; the rhizome of *Ligusticum chuanxiong*; and the flowers of *Carthamus tinctorius*. The major end point was mortality after randomization at 28 days. Results concluded that patients with sepsis treated with XBJ had a considerably decreased mortality rate at 28 days compared to those treated with a placebo. Additionally, when compared with placebo, XBJ was linked with relief from ventilator use and ICU stay in patients with active sepsis during the 28 days, suggesting that it was also able to lower ICU mortality [22].

Herbal medicines have great potential to treat functional dyspepsia (FD) since they are characteristic multitarget, multicomponent drugs. Yukgunja-tang, a polyherbal formulation, is composed of *Citri unshii pericarpium, Hoelen, Ginseng radix alba, Glycyrrhizae radix, Pinelliae tuber, Zingiberis rhizoma crudus, Atractylodis rhizoma alba, and Zizyphi fructus and has* been used to treat FD in a multicenter, parallel-group RCT (ChiCTR-IIR-15007567). The primary outcome includes a questionnaire, namely the Total Dyspepsia Symptom Scale (TDS), with scoring, whereas secondary outcomes include questionnaires including the Single Dyspepsia Symptom Scale (SDS), Overall Treatment Effect (OTE), HADS, and many others [23].

Erectile dysfunction (ED) is one of the most frequently occurring sexual dysfunctions affecting patients both mentally and physically. Early findings indicate that Chinese herbal formulas, such as Hongjing I granules (HJIG), have emerged as alternative methods for treating ED. To evaluate the safety and clinical efficacy, a double-blind, multicenter, placebo-controlled RCT was conducted (ChiCTR2000041127). The composition of HJIG consists of *Rhodiola rosea, Radix Astragali, Codonopsis pilosula, Angelica sinensis, Salvia miltiorrhiza, Radix Paeoniae Alba, Lycium chinense Miller, Epimedium brevicornum, Chuan-Niu-Xi,* and *Cyathula officinalis.* End points include total sexual intercourse, changes in scores of sexual desire, and observation of adverse effects [24].

A recent study investigated to determine the acute impact of a single oral dose of raw cinnamon (*Cinnamomum zeylanicum* Nees) powder, either as a beverage dissolved in water or as regular hard gelatin capsules, on postprandial hyperglycemia in patients with type 2 diabetes that was induced by a carbohydrate-rich balanced breakfast (ReBEC: RBR-98tx28b). Measurements of blood glucose were made at fasting and at 0.25, 0.5, 0.75, 1, 1.5, and 2 hours after a meal. Following each breakfast, the palatability scores for taste intensity (bitterness, creaminess, saltiness, sourness, and sweetness), as well as visual appeal, scent, and pleasantness of taste were evaluated. Regardless of dosage, consuming raw cinnamon diluted in water reduced the hyperglycemic blood glucose peak and the meal-induced big glucose spike. Cinnamon diminished the hyperglycemic properties greatly when taken as capsules. Consuming raw cinnamon flattened the glycemic curve (lower iAUC) without altering the shape [25].

The inability to identify the exact cause of persistent fatigue hinders the formulation of effective treatment plans. Therefore, substantial research is ongoing with limited treatment options. There are two widely available herbal prescriptions in China and Korea, namely Gongjin-Dan (GJD) and Ssanghwa-Tang (SHT), used for relieving the symptoms associated with fatigue. A study protocol was designed to evaluate the clinical safety and effectiveness of these two herbal formulations in a placebo-controlled, double-blinded RCT (KCT0007515). GJD is a black-brown pill constituting *Cervus nippon* Temminck, *Panax ginseng* C.A. Meyer, *Angelica gigas* Nakai, *Cornus officinalis* Seibold & Zucc., *Moschus moschifrus* L., and *Rehmannia glutinosa* (Gaertn.) DC., whereas SHT is a brown-colored solution constituting *Rehmannia glutinosa* (Gaertn.) DC., *Zizyphus jujuba* Miller var. hoonensis T. B. Lee, *Paenoia lactiflora* Pall., *Zingiber officinale* Rosc., *Cnidium officinale* Mak., *Cinnamomum cassia* J. Presl, *Glycyrrhizia uralensis* Fisch., *Angelica gigas* Nakai, and *Astragalus membranous* Bunge. The trial concluded that these two herbal formulations are effective and safe, decreasing fatigue symptoms measured by the Fatigue Severity Scale (FSS), CFS, and Short Form 36 Health Survey (SF-36) scores, and improving quality of life without any adverse effects, indicating that they can be used for treating chronic physical and mental fatigue [26].

16.4 INTERDISCIPLINARY COLLABORATIONS FOR CLINICAL TRIAL OF HERBAL DRUGS

The development and evaluation of traditional knowledge into evidence-based medicine for clinical use require a multidisciplinary approach to ensure safety, efficacy, and regulatory compliance.

The field of ethnobotany, which studies the interactions between people and plants, resides at the interface of the social and biological sciences. According to data from the World Health Organization (WHO), 80% of people worldwide rely mostly on traditional medicine [27]. Traditional and folklore medicine have provided the rationale for various clinical investigations, and the traditional knowledge needs to be validated; the first step in achieving that is the botanical identification for creating a standardized database. Pharmacognosy is the study of medicinal drugs derived from plants and other natural sources. It encompasses the identification, isolation, and characterization of bioactive compounds found in plants, which are responsible for therapeutic effects, including that of curcumin from *Curcuma longa* L. (turmeric) for cardiac ischemia and reperfusion in the trial CTRI/2012/02/002442, and for nonalcoholic liver disorders in the IRCT2014110511763N18 clinical investigation. Complementary to pharmacognosy, pharmacology involves studying the effects of substances on biological systems, which is essential in evaluating the safety and efficacy of herbal drugs. It aids in understanding the mechanism of action of curcumin; these insights strengthen the rationale for curcumin's inclusion in clinical trials, guiding dosage selection and assessing possible drug interactions. Moreover, the identified bioactive moieties were isolated and purified from the plant source for further analysis and clinical investigation under a number of trials, for example, R-(-)-Gossypol Acetic Acid (NCT00773955, NCT00397293, and NCT00666666), beta-lapachone (NCT00524524, NCT00099190, and NCT00358930), and Lovastatin (NCT00462280, NCT00584012, and NCT00902668) [28], chemistry plays a crucial role in enhancing curcumin's stability and bioavailability, driving research toward the integration of pharmaceutical formulations to develop advanced delivery systems, such as

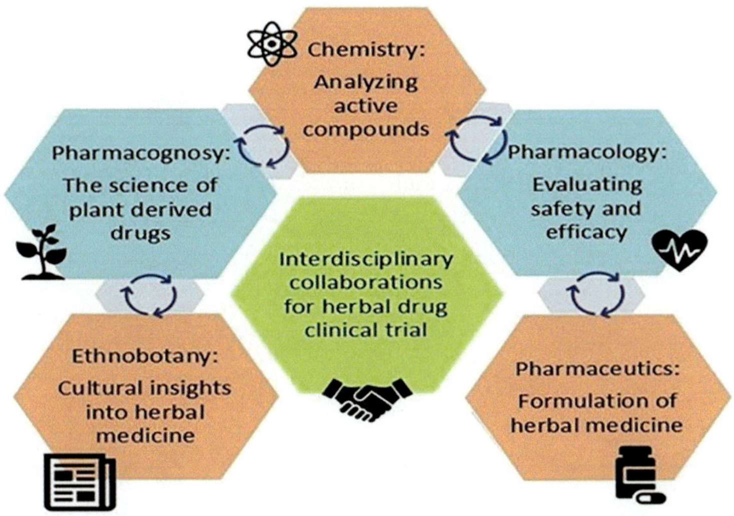

FIGURE 16.1 Interdisciplinary collaboration for the clinical trials of herbal drugs.

nanoparticles and liposomal formulations, which are essential for maximizing curcumin's therapeutic effects while addressing its limitations, thereby ensuring successful application in clinical trials. Numerous clinical investigations have been carried out to assess the effectiveness of *Lavandula angustifolia,* and one such example includes the inhalational formulation of lavender oil for the treatment of anxiety in trial number NCT03722771. *Beta vulgaris* (beetroot) roots and leaves have been attributed in protecting the cardiovascular system in trial number NCT04094233 and assisting in detoxification processes, *Allium sativum* L. (garlic) is well known for having antioxidant, antibacterial, and anti-inflammatory effects [29].

Betalains, mainly betacyanins and betaxanthins, are the primary bioactive compounds found in beetroot and are of particular interest due to their non-toxic nature and potential health benefits against cancer and cardiovascular conditions, among a number of other ailments. Among the many bioactive constituents found in garlic, organosulfur compounds such as ajoene, allicin, and alliin—in particular allicin—exhibit antibacterial and antioxidant properties [30]. The field of pharmacology helps to understand the possible mechanism of action by employing various *in vitro* and *in vivo* models; the biological activity of the aforementioned plants was studied, and the isolated curcumin, betalains, among others, showed promise as therapeutic agents for various diseases, including cancer and neurodegenerative disorders. Clinical studies have proven that the bioactive compound betanin acts as an antioxidant, anti-inflammatory, and anticancer agent and has cardiovascular benefits as well. It acts by reducing reactive oxygen species as well as enhancing the expression of endogenous antioxidant enzymes; by inhibition of proinflammatory cytokines such as TNF-α and IL-16; and by induction of apoptosis and inhibition of tumor growth, thereby contributing to improved endothelial function along with reduced oxidation of low-density lipoproteins. Pharmaceutics focuses on formulation, manufacture, and delivery of herbal medicine, ensuring that they are prepared in a manner that maximizes their therapeutic effect and minimizes adverse effects. Curcumin comes in a variety of pharmacological formulations such as ointments, tablets, and capsules, among many, and has been investigated under different clinical settings, a few of which are IRCT201505301165N4 and NCT01052597, delivering accurate and efficient dosage. Beetroot extracts are developed in various forms, including powders, juices, and capsules, making them suitable for dietary supplements and functional foods, working toward enrichment of the cardiovascular system [31].

These fields work together to provide a thorough framework for the development and clinical assessment of herbal medications (Figure 16.1). Pharmacognosy guarantees quality and standardized raw materials, while ethnobotanical insights provide the real-life traditional claims for the rationale of clinical trials of the herbal preparations. Pharmacology clarifies the mechanism of action, pharmaceutics facilitates efficient dose and administration, and chemistry aids in understanding the relationship between the structure and its interaction with different biological systems in clinical investigations. The complete success of a clinical trial is improved by this interdisciplinary approach, which eventually results in safer and more efficient herbal therapies in modern healthcare.

16.5 EMERGING TECHNOLOGIES FOR THE APPLICATION IN CLINICAL TRIALS

Clinical trials are changing as a result of new advancements in technology, especially in the herbal medicine sector. A more sophisticated comprehension of the complex relationships between herbal substances and biological systems is being made possible by advancements in

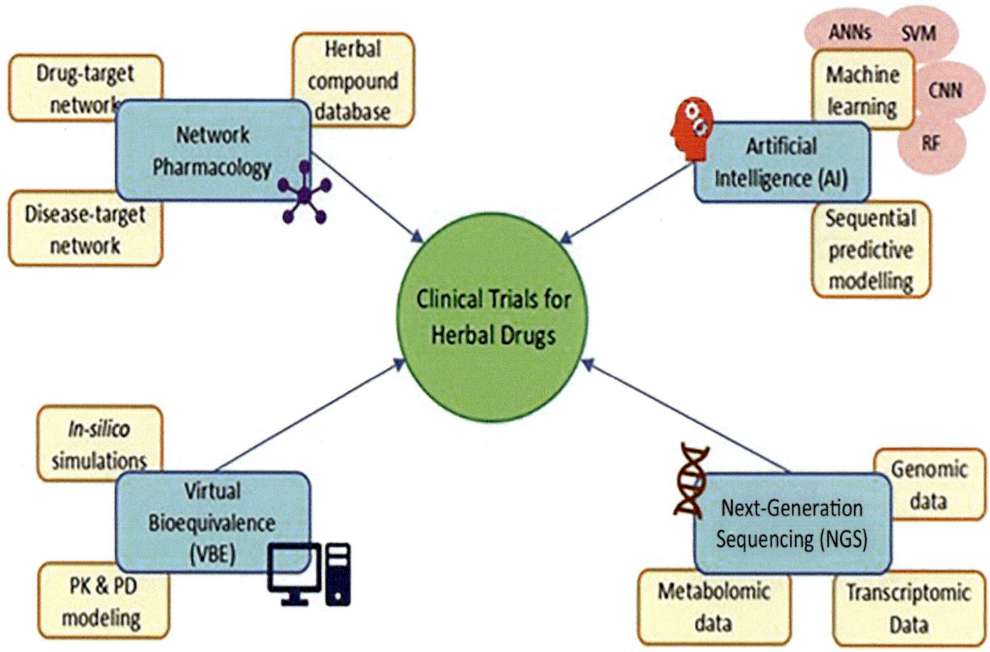

FIGURE 16.2 Emerging technologies in clinical trials for herbal drugs: ANNs (artificial neural networks); CNNs (convolutional neural networks); RF (random forests); SVM (support vector machines).

network pharmacology, virtual bioequivalence (VBE), next-generation sequencing (NGS), proteomics, and genomics. These technologies promise to improve medication development and discovery accuracy as they come together, making it possible to successfully incorporate herbal medicines into modern healthcare (Figure 16.2).

16.5.1 Network Pharmacology

The disciplines of systems biology and traditional pharmacology are brought together in network pharmacology to unravel the complex relationships between biological targets and herbal compounds. Traditional pharmacology primarily focuses on a linear approach, typically emphasizing the interactions between a single drug and a specific biological target. In contrast, network pharmacology acknowledges the complexities of drug interactions through its multitarget, multi-pathway approach, which makes it a useful tool in the design of clinical trials, as it not only increases therapeutic efficacy but also lowers the possibility of side effects. Following its introduction in 2007, network pharmacology rapidly gained traction, particularly in the context of traditional Chinese medicine (TCM), where multitarget interactions are a fundamental characteristic. By connecting the pharmacological effects of TCM formulations to certain biological processes, it helps validate assumptions about their effectiveness and raises the legitimacy of clinical trials employing these treatments. The field of clinical trials is greatly enhanced by network pharmacology when it comes to mechanism prediction, drug target identification, and the incorporation of traditional medical practices. A more sophisticated understanding of how drugs work is anticipated through this comprehensive approach, which also creates opportunities for novel treatments [32].

16.5.2 Virtual Bioequivalence

In VBE, the pharmacokinetic characteristics of drug formulations are predicted and simulated using computational models instead of actual clinical trials. The implementation of VBE in clinical trials of herbal drugs typically includes several key methodologies: physiologically based pharmacokinetic modeling (PBPK), *in silico* simulations, integration of *in vitro* data, and available clinical data. Despite its advantages, the application of VBE in herbal drug trials faces challenges, including the need for high-quality *in vitro* and *in vivo* data to inform model accuracy [33].

The consistency of herbal products' active ingredients can be challenging to achieve, which affects the credibility of trial outcomes. The design of control groups is complicated by the nearly impossible task of producing placebo formulations that closely resemble the flavor, color, and odor of herbal medications. In order to overcome these difficulties, VBE uses virtual simulations of the pharmacokinetic characteristics of the herbal medication and its intended placebo, eliminating the necessity for physical placebo formulations. VBE is now being accepted by regulatory frameworks as a legitimate technique for proving bioequivalence, especially when it comes to herbal substances. This approach is getting popularity for its capacity to deliver solid evidence of these formulations' efficacy and safety while abiding by legal requirements.

By offering additional knowledge about how individual variability and environmental factors affect drug absorption, VBE can be used in conjunction with conventional pharmacokinetic models. Researchers can make more accurate predictions about the performance of herbal medicines in real-world settings [34].

16.5.3 Next-Generation Sequencing

When compared to conventional techniques, NGS refers to a collection of advanced DNA sequencing technologies that allow for significantly quicker and affordable sequencing of entire genomes or specific DNA sections. One of the principal applications of NGS in herbal drug development is its role in authentication and quality assessment by analyzing the DNA sequence of medicinal plants. The application of NGS facilitates pharmacogenomic studies that can elucidate how genetic variations among individuals influence their responses to herbal medicines. By analyzing transcriptomic and metabolomic data, researchers can determine the expression levels of relevant genes in response to specific herbal treatments. The genome of *Gardenia jasminoides*, a widely used medicinal plant, was decoded utilizing NGS technology, and the sequencing of the *Artemisia annua* genome by NGS led to the identification and characterization of its biosynthetic pathway. The analysis of various herbal plants through NGS has resulted in the optimization of their active constituents for enhanced efficacy [35].

Combining NGS with other omics technologies, such as genomic, transcriptomic, and metabolomics data, maximizes its potential and allows for a more comprehensive study of the mechanism of action of herbal drugs and their possible effects on different patient populations.

16.5.4 Artificial Intelligence and Machine Learning

A fast-emerging new technical science is artificial intelligence or simply AI. Its technical applications, theories, and procedures mimic, increase, and broaden human intelligence. AI employs a number of machine learning (ML) and deep learning methods, such as SVM, Bayesian networks, CNNs, RF, and ANNs. AI tools like sequential predictive modeling and hierarchical interaction networks use past trial data to forecast the chances of success for upcoming studies based on factors like target diseases, drug molecule characteristics, and patient eligibility requirements.

AI can speed up the drug discovery process by automating labor-intensive procedures and analyzing massive datasets that have historically required a lot of human effort. Researchers can find possible therapeutic targets and forecast their efficacy by using ML algorithms to evaluate genomic data. By forecasting the pharmacological characteristics of lead compounds from their chemical structures, AI plays a critical role in drug design. It might be daunting to deal with the enormous volumes of data produced during herbal medication clinical trials. Effective management is made possible by AI techniques such as data mining and natural language processing, which automate the process of organizing and extracting pertinent information from unstructured databases [36].

One of the most crucial parts of clinical trials is still finding new patients, which frequently takes up a large amount of the study's time. Researchers can effectively identify potential participants and optimize eligibility criteria with AI-driven tools like Trial Pathfinder and Criteria2Query. These tools enable quicker recruitment by examining clinical databases and electronic health records, guaranteeing that trials include the most appropriate patient populations. Furthermore, through tailored interventions, AI can help sustain participant engagement and adherence, lowering dropout rates.

16.6 FUTURE TRENDS

With the improved pharmacological properties of medicinal plants in the 21st century, herbal therapy has emerged as a potentially effective future treatment option. A worldwide trend known as "Return to Nature" can be said to represent a shift from synthetic to herbal therapy. However, numerous challenges have been observed, causing the decline of plant-based herbal remedies, which are a source of drug discovery. With widespread wildcrafting and unsustainable harvesting methods, plant populations are frequently threatened when a plant is commercialized as an herbal remedy or when one of its constituents begins to be used as a pharmaceutical drug. The alleged "Taxol supply crisis" served as the traditional instance of this compound supply issue. This was more troublesome because the entire production process included laborious bark harvesting, drying, extraction, and purification, which took a long time. The quality and mix of the available plant material can vary, which might make it difficult to evaluate its potential medicinal benefits. In addition to species identity and harvest time, other factors that affect the chemical composition include soil composition, altitude, the real climate, manufacturing, and storage conditions. Additionally, chemicals may undergo change and degradation during extraction and isolation procedures. This method makes it more difficult to meet clinical trial inclusion and exclusion requirements [37].

According to reports, almost 60% of cancer medications currently on the market or under study are derived from natural materials. Nowadays, plants constitute the source of almost 80% of antibacterial, immunosuppressive, cardiovascular, and anticancer medications. Of the 177 licensed anticancer medications, about 70% are derived from natural products or mimics. There are approximately 121 prescription medication entities in use worldwide, and about 25% of them are derived from plant sources. Over 100 medications based on natural products are undergoing clinical trials, and 13 medications of natural origin were approved in the United States between 2005 and 2007. Additionally, it was calculated that 11% of the 252 medications listed in the WHO's Essential Medicine List are solely derived from plants [38].

The emergence of new technology being used in clinical trials for herbal drugs has enhanced patient outcomes significantly. Various chronic diseases and orphan diseases have gained recognition due to the possible treatment of them with the help of herbal medicines. Synergistic therapies of conventional medicine and herbal medicine have increased efficacy and reduced toxicity, particularly in oncology [39]. The introduction of advanced genomic technologies has shifted the focus toward personalized medicine, where herbal medicine plays a major role in improving treatment outcomes. Technological advancements are also playing a pivotal role in the evolution of herbal medicine. For instance, innovations such as nanotechnology are being explored to improve the bioavailability of herbal drugs, making them more effective in treating various conditions. Additionally, ongoing research in the field of phytochemistry seeks to isolate active compounds from herbs, potentially leading to the development of new therapeutic agents with reduced side effects [40].

In recent years, there has been an increasing focus on the research and standardization of herbal medicines. These trials aim to address concerns regarding the potential toxicity of herbal products and their interactions with conventional pharmaceutical drugs. In addition to safety studies, regulatory frameworks are being strengthened, with organizations like the WHO promoting guidelines for the quality and standardization of herbal products. Such measures are crucial for the successful integration of herbal medicine into mainstream healthcare systems. However, there is also concern about the potential for adverse interactions between herbal and conventional treatments, which indicates the need for cautious and well-informed integration of these remedies into clinical practice.

Despite many challenges, various interdisciplinary groups, including pharmacognosists, pharmacologists, and botanists, have shown their interest in studying medicinal plants for novel phytoconstituents, leading to the development of medications for various diseases, which is proven by the massive publication of research papers and an increased registry of herbal clinical trials. In reference to this, it can be predicted that herbal medicinal practices will soon be included in conventional medications due to recent advancements in quality control and regulatory measures in many nations worldwide [41].

16.7 CONCLUSION

Emerging technologies and interdisciplinary cooperation hold great promise for the future of clinical trials of herbal medication. Major challenges with herbal drugs, including standardization and scientific validation due to variation in active ingredients and uneven quality, will be lessened with the aid of these developed technologies. While omics technologies like genomics and proteomics are boosting our understanding of the molecular mechanisms in reference to herbal medicines, AI and ML have revolutionized the analysis of data, the design of data, and the prediction of outcomes. Clinical trials are becoming more adaptable, precise, and patient-centered because of digital health platforms like wearable smartwatches and smartphone apps that allow real-time patient monitoring and data collection. Furthermore, blockchain technology has improved trial data transparency and credibility while providing a creative method to integrate data. The complexities of conducting clinical trials of herbal drugs have been made easier with the collaboration of various disciplines. The combination of Ayurvedic knowledge and conventional methods ensures that herbal medicines are thoroughly investigated for safety and effectiveness. Overall, the future of clinical trials for herbal drugs has promising results driven by technology, resulting in efficient therapies and better incorporation of herbal medicine into medical care based on scientific evidence.

REFERENCES

[1] Pelkonen, O., Xu, Q., & Fan, T.-P. (2014). Why is research on herbal medicinal products important and how can we improve its quality? *Journal of Traditional and Complementary Medicine*, 4(1), 1–7, 1. https://doi.org/10.4103/2225-4110.124323

[2] Salm, S., Rutz, J., van den Akker, M., Blaheta, R. A., & Bachmeier, B. E. (2023). Current state of research on the clinical benefits of herbal medicines for non-life-threatening ailments. *Frontiers in Pharmacology*, 1–14, 1234701. https://doi.org/10.3389/fphar.2023.1234701

[3] Sen, S., & Chakraborty, R. (2017). Revival, modernization and integration of Indian traditional herbal medicine in clinical practice: Importance, challenges and future. *Journal of Traditional and Complementary Medicine*, 7(2), 234–244. https://doi.org/10.1016/j.jtcme.2016.05.006

[4] Paczkowska-Walendowska, M., Sip, S., Staszewski, R., & Cielecka-Piontek, J. (2022). Single-pill combination to improve hypertension treatment: Pharmaceutical industry development. *International Journal of Environmental Research and Public Health*, 19(7), 4156. Article 7. https://doi.org/10.3390/ijerph19074156

[5] David, S., & Kim, P. Y. (2024). Drug Trials. In *StatPearls*. StatPearls Publishing. http://www.ncbi.nlm.nih.gov/books/NBK546595/

[6] Sen, S., & Chakraborty, R. (2015). Toward the integration and advancement of herbal medicine: A focus on traditional Indian medicine. *Botanics: Targets and Therapy*, 5, 33–44. https://doi.org/10.2147/BTAT.S66308

[7] Bian, Z. (2024). A Single Arm Pilot Study of the Chinese Herbal Medicine Formula (SCD-2101) for the Functional Constipation in the Elderly (Clinical Trial Registration No. NCT06534996). https://clinicaltrials.gov/study/NCT06534996

[8] Zaidi, S. A. (2024). To Compare the Efficacy and Safety of Herbal Medicine With Allopathic Therapy for Metabolic Syndrome: A Randomized, Multi-center, Clinical Study (Clinical Trial Registration No. NCT06515652). https://clinicaltrials.gov/study/NCT06515652

[9] Speers, A. (2024). Safety, Tolerability, and Pharmacokinetics of an Oral Withania Somnifera Product in Older Adults (Clinical Trial Registration No. NCT06171724). https://clinicaltrials.gov/study/NCT06171724
[10] Zhang, Y. (2023). Clinical Study on the Effect of Brucea Javanica Oil Emulsion Injection on the Survival of Patients With Advanced Colorectal Cancer Who Failed to Receive Multi-line Treatment (Clinical Trial Registration No. NCT05897749). https://clinicaltrials.gov/study/NCT05897749
[11] Paine, M. (2023). Evaluating a Potential Pharmacokinetic Kratom-oxycodone Interaction Concurrent With Clinical Endpoints (Clinical Trial Registration No. NCT05846451). https://clinicaltrials.gov/study/NCT05846451
[12] Bian, Z. (2023a). Efficacy of the CHM Teabag in Decreasing Stroke Risk Among Elderly People in Hong Kong: A Stepped Wedge Cluster Randomized Trial (Clinical Trial Registration No. NCT05801575). https://clinicaltrials.gov/study/NCT05801575
[13] Hosam, Sujoud. (2024). The Effect of Curcumin on the Clinical Outcome of Pediatric Patients With Active Lupus Nephritis (Clinical Trial Registration No. NCT05714670). https://clinicaltrials.gov/study/NCT05714670
[14] Bian, Z. (2023b). The Efficacy and Safety of a Chinese Herbal Medicine (Shenlingcao Oral Liquid) for Treating Long COVID Associated Fatigue: A Double-blinded, Placebo-controlled, Randomized Clinical Trial (Clinical Trial Registration No. NCT05684952). https://clinicaltrials.gov/study/NCT05684952
[15] Chang Gung Memorial Hospital. (2024). The Efficacy of Xian-Hua-Cha for Improving Hyperlipidemia Among Obese Population: A Crossover Trial (Clinical Trial Registration No. NCT05632068). https://clinicaltrials.gov/study/NCT05632068
[16] Li, Jie. (2022). "Qizhu Yuling Prescription in the Prevention and Treatment of Postoperative Metastasis and Recurrence of Esophagus Cancer: A Randomized, Double-Blind, Controlled and Multi-Center Clinical Study." Clinical trial registration NCT05626309.
[17] Tian, J. (2023). Efficacy Assessment of Chinese Herbal Medicine Jiangtang Tiaozhi Recipe Treating Participants With Glucose and Lipid Metabolism Disturbances and Associations Study Between Glucose and Lipid Metabolism Disturbances and Cardiovascular Risk Factors (Clinical Trial Registration No. NCT04623567). https://clinicaltrials.gov/study/NCT04623567
[18] Chitre, D., Nadkarni, S., Jagtap, N., Tulle, R., Gitte, A., Rahate, P., Chaskar, S., & Dey, D. (2023). Phase III randomized clinical trial of BV-4051, an Ayurvedic polyherbal formulation in moderate SARS-CoV-2 infections and its impact on inflammatory biomarkers. Phytotherapy Research, 37(4), 1232–1241. https://doi.org/10.1002/ptr.7683
[19] Chiluveri, S. K., Deshpande, S. N., Bhatia, T., Yadav, B., Gupta, B., Jameela, S., Rana, R. K., Singhal, R., Chandrasekhara Rao, B., & Srikanth, N. (2024). Study protocol of a randomized, double-blind, placebo-controlled clinical trial on the efficacy and safety of an Ayurveda formulation (Ayush-SR) in the management of occupational stress among nurses. *Journal of Research in Ayurvedic Sciences*, 8(2), 85–97. https://doi.org/10.4103/jras.jras_294_23
[20] Xiong, W., Wang, G., Du, J., & Ai, W. (2020). Efficacy of herbal medicine (Xuanfei Baidu decoction) combined with conventional drug in treating COVID-19:A pilot randomized clinical trial. *Integrative Medicine Research*, 9(3), 100489. https://doi.org/10.1016/j.imr.2020.100489
[21] Bazrafshan, M.-R., Jokar, M., Shokrpour, N., & Delam, H. (2020). The effect of lavender herbal tea on the anxiety and depression of the elderly: A randomized clinical trial. *Complementary Therapies in Medicine*, 50, 102393. https://doi.org/10.1016/j.ctim.2020.102393
[22] Liu, S., Yao, C., Xie, J., Liu, H., Wang, H., Lin, Z., Qin, B., Wang, D., Lu, W., Ma, X., Liu, Y., Liu, L., Zhang, C., Xu, L., Zheng, R., Zhou, F., Liu, Z., Zhang, G., Zhou, L., EXIT-SEP Investigators. (2023). Effect of an herbal-based injection on 28-day mortality in patients with sepsis: The EXIT-SEP randomized clinical trial. *JAMA Internal Medicine*, 183(7), 647–655. https://doi.org/10.1001/jamainternmed.2023.0780
[23] Lee, B., Ha, N.-Y., Park, H.-J., Kim, A.-R., Kwon, O.-J., Cho, J.-H., Shin, S. M., Kim, J., & Yang, C. (2023). Herbal medicine yukgunja-tang for functional dyspepsia: A protocol for a randomized, controlled, multicenter clinical trial. *Healthcare*, 11(10), Article 10. https://doi.org/10.3390/healthcare11101456
[24] Xu, R.-N., Ma, J.-X., Zhang, X., Liao, Z.-D., Fu, Y.-J., & Lv, B.-D. (2023). Efficacy of Chinese herbal medicine formula in the treatment of mild to moderate erectile dysfunction: Study protocol for a multi-center, randomized, double-blinded, placebo-controlled clinical trial. *International Journal of General Medicine*, 16, 5501–5513. https://doi.org/10.2147/IJGM.S436347
[25] Moreira, F. D., Reis, C. E. G., Gallassi, A. D., Moreira, D. C., & Welker, A. F. (2024). Suppression of the postprandial hyperglycemia in patients with type 2 diabetes by a raw medicinal herb powder is weakened when consumed in ordinary hard gelatin capsules: A randomized crossover clinical trial. *PLoS One*, 19(10), e0311501. https://doi.org/10.1371/journal.pone.0311501
[26] Choi, J.-Y., Choi, B., Kwon, O., Seo, C.-S., Kim, A.-R., Shin, H., & Kim, K. (2024). Efficacy and safety of herbal medicine Gongjin-Dan and Ssanghwa-Tang in patients with chronic fatigue: A randomized, double-blind, placebo-controlled, clinical trial. *Integrative Medicine Research*, 13(1), 101025. https://doi.org/10.1016/j.imr.2024.101025
[27] Azaizeh, H., Fulder, S., Khalil, K., & Said, O. (2003). Ethnobotanical knowledge of local Arab practitioners in the Middle Eastern region. *Fitoterapia*, 74(1), 98–108. https://doi.org/10.1016/S0367-326X(02)00285-X
[28] Dhupal, M., & Chowdhury, D. (2020). Phytochemical-based nanomedicine for advanced cancer theranostics: Perspectives on clinical trials to clinical use. *International Journal of Nanomedicine*, 15, 9125–9157. https://doi.org/10.2147/IJN.S259628
[29] Tesfaye, A. (2021). Revealing the therapeutic uses of garlic (Allium sativum) and its potential for drug discovery. *The Scientific World Journal*, 2021(1), 8817288. https://doi.org/10.1155/2021/8817288
[30] Melguizo-Rodríguez, L., García-Recio, E., Ruiz, C., Luna-Bertos, E. D., Illescas-Montes, R., & J. Costela-Ruiz, V. (2022). Biological properties and therapeutic applications of garlic and its components. *Food & Function*, 13(5), 2415–2426. https://doi.org/10.1039/D1FO03180E
[31] Pinheiro, V. dos S., Junior, O. J. F. R., Ortmann, C. F., Pande, A., Conte-Junior, C. A., & Alvares, T. S. (2024). Evaluation of 12-week standardized beetroot extract

supplementation in older participants: A preliminary study of human health safety. *Nutrients*, 16(12), Article 12, 1942. https://doi.org/10.3390/nu16121942

[32] Zheng, S., Xue, T., Wang, B., Guo, H., & Liu, Q. (2022). Application of network pharmacology in the study of mechanism of Chinese medicine in the treatment of ulcerative colitis: A review. *Frontiers in Bioinformatics*, 2, 928116. https://doi.org/10.3389/fbinf.2022.928116

[33] Hussaarts, L., Mühlebach, S., Shah, V. P., McNeil, S., Borchard, G., Flühmann, B., Weinstein, V., Neervannan, S., Griffiths, E., Jiang, W., Wolff-Holz, E., Crommelin, D. J. A., & de Vlieger, J. S. B. (2017). Equivalence of complex drug products: Advances in and challenges for current regulatory frameworks. *Annals of the New York Academy of Sciences*, 1407(1), 39–49. https://doi.org/10.1111/nyas.13347

[34] Yang, R., Lin, Y., Chen, K., Huang, J., Yang, S., Yao, A., Yang, X., Lei, D., Xiao, J., Yang, G., & Pei, Q. (2024). Establishing virtual bioequivalence and clinically relevant specifications for omeprazole enteric-coated capsules by incorporating dissolution data in PBPK modeling. *The AAPS Journal*, 26(4), 82–94. https://doi.org/10.1208/s12248-024-00956-0

[35] Delgado-Tejedor, A., Leekitcharoenphon, P., Aarestrup, F. M., & Otani, S. (2021). Evaluating the usefulness of next-generation sequencing for herb authentication. *Food Chemistry: Molecular Sciences*, 3, 100044. https://doi.org/10.1016/j.fochms.2021.100044

[36] Birjandi, S. M., & Khasteh, S. H. (2021). A survey on data mining techniques used in medicine. *Journal of Diabetes & Metabolic Disorders*, 20(2), 2055–2071. https://doi.org/10.1007/s40200-021-00884-2

[37] Bucar, F., Wube, A., & Schmid, M. (2013). Natural product isolation—how to get from biological material to pure compounds. *Natural Product Reports*, 30(4), 525–45. https://doi.org/10.1039/C3NP20106F

[38] Pan, S.-Y., Zhou, S. F., Gao, S. H., Yu, Z. L., Zhang, S. F., Tang, M., Sun, J. N. et al. (2013). New perspectives on how to discover drugs from herbal medicines: CAM's outstanding contribution to modern therapeutics. Evidence-Based Complementary and Alternative Medicine: eCAM 2013:627375. https://doi.org/10.1155/2013/627375

[39] Hemaiswarya, S., Prabhakar, P. K., & Doble, M. (2022). Clinical trials on synergistic herb–drug interactions. In S. Hemaiswarya, P. K. Prabhakar, & M. Doble (Eds.), Herb–Drug Combinations: A New Complementary Therapeutic Strategy (pp. 251–263). *Springer Nature*. https://doi.org/10.1007/978-981-19-5125-1_17

[40] Gao. (2023b). Application of third-generation sequencing to herbal genomics. *Frontiers in Plant Science* 1–14 (March). https://doi.org/10.3389/fpls.2023.1124536

[41] Chikezie, P. C. (2015). Herbal medicine: Yesterday, today and tomorrow. *Alternative & Integrative Medicine*, 4(3), 2327–5162. https://doi.org/10.4172/2327-5162.1000195

Index

A

active comparator, 106, 137
acute toxicity, 63, 94
agricultural practices, 64, 92, 131, 134, 196
animal studies, 11, 39, 135, 137
anticancer activity, 17, 20, 44, 186, 201, 207, 209
anti-inflammatory activity, 17–19, 40, 42, 54, 140, 143, 163, 172, 207
antimicrobial activity, 17, 42, 61, 143
antioxidant activity, 17–18, 39–41, 43, 143, 163, 207
artificial intelligence, 102, 153, 190, 209
ayurveda, 6, 16, 26, 37, 38, 40, 43, 45, 49, 53, 83, 102, 106, 117–118, 124, 135, 147, 149, 151, 152, 156, 157, 167–169, 171, 173, 177, 195, 200, 205
AYUSH, 6, 28, 43, 53, 59, 106, 117, 135, 149, 151–153, 156, 202–204

B

batch consistency, 8, 45, 62, 95, 97, 99–101
batch-to-batch consistency, 45, 62, 95
batch-to-batch variation, 12, 37, 60, 99
Bayesian, 142, 209
bias control, 122
bioactive compounds, 17, 20, 32, 44, 63, 98, 100–101, 106, 170, 196, 206–207
bioavailability, 17–20, 33, 42, 64, 71, 108, 155, 169, 204–205, 206, 210
biodiversity conservation, 53–54, 82–83
bioethics, 76, 86
biomarkers, 18, 46, 62, 70, 137, 191, 205
biopiracy, 36, 75, 78–79, 82–83, 86–87, 151, 152
botanical authentication, 7–8, 18, 61
botanical characterization, 94, 95, 99
botanical drugs, 8, 59, 101
botanical identity, 6–7, 135

C

carcinogenicity evaluation, 69
case reports, 2, 11, 105, 114, 190, 191
case-control studies, 5, 170
causality assessment, 9
chemical fingerprinting, 6, 68–69, 139, 158
chemical profiling, 17, 38–40, 63, 105, 134
chemometric analysis, 160, 162
clinical development, 98, 123
clinical efficacy, 11, 36, 42, 68, 72, 120, 135, 150, 173, 187, 206
clinical protocol, 7, 95–96
clinical trial protocol, 8, 10, 15, 106–107, 112, 114, 118, 123, 152, 200
clinical trial registration, 107, 113, 120, 173, 187, 190
cohort studies, 4, 5, 169, 171, 185
combination therapy, 19, 20, 184, 185
community engagement, 84, 127
complementary medicine, 6–7, 25, 58–59, 107, 110, 124, 150
CONSORT, 33, 54, 120, 131, 142
consumer safety, 27, 58, 61, 198
contaminant testing, 6, 7
cross-sectional studies, 7, 170, 186
cultural sensitivity, 75, 82–83, 85, 138, 171, 184
current good manufacturing practice, 26, 93
cytochrome P450, 19, 131, 134

D

data confidentiality, 78
data integrity, 77, 78, 151, 152, 170
data management, 115, 141, 177
data sharing, 53, 54, 151
declaration of Helsinki, 76–78, 114, 118, 138
design of experiments, 45
dietary supplements, 26, 28, 50, 52, 57–59, 77, 91–93, 100, 101, 105–107, 109, 119, 150, 152, 196, 201, 207
DNA barcoding, 61, 62, 67, 80, 140, 158, 159, 197
DNA fingerprinting, 159
dose determination, 131, 137
dose escalation, 19, 137
dose-response relationship, 19
double-blind studies, 32, 71
drug development, 8, 9, 12, 16, 17, 19, 45, 61, 70, 77, 95, 97, 100, 102, 118, 121, 209
drug discovery, 10, 17, 36, 70, 147, 163, 209
drug efficacy, 110, 151
drug safety, 8, 119
drug-herb interactions, 151, 153
Drugs and Cosmetics Act, 1940, 9, 59, 106, 148, 149, 200
Drugs and Cosmetics Rules, 1945, 148

E

efficacy assessment, 13, 111, 132
electronic health records, 102, 152, 209
ethical approval, 1, 85, 116, 127
ethical challenges, 75, 77–80, 84, 138
ethical guidelines, 6, 10, 52, 76, 82, 84, 111, 117, 124, 139, 171
ethics committees, 6, 7, 9, 72, 85, 86, 113, 116, 124, 131, 138, 139, 149, 190
ethnobotany, 206
ethnomedicine, 167
ethnopharmacology, 8, 72
European medicines agency, 6, 7, 15, 50, 56, 106, 150, 200
extraction techniques, 17, 18, 32, 33, 136

F

FAIR principles, 54
first-in-human trials, 133
fixed dose combinations, 118
flavonoids, 17, 18, 20, 136, 158, 161
FTIR fingerprinting, 161

G

GC-MS, 56, 61–64, 70, 161–162
genomics, 45, 54, 72, 102, 190, 208, 210
genotoxicity testing, 69
global healthcare, 25, 33, 182, 189
global herbal medicine market, 25, 29
good agricultural and collection practices, 6–7, 9, 80, 106, 131, 133
good clinical practice, 2, 10, 15, 76–77, 97–98, 106, 113, 116–117, 124, 131, 149, 152, 170, 182
good manufacturing practice, 7, 26, 49, 56, 64, 78–79, 91, 93, 99, 106, 131, 134, 150, 156, 196
government initiatives, 148, 152

H

heavy metal contamination, 11, 134
herbal bioactive molecules, 16, 17
herbal clinical trials, 1, 6, 7, 9, 19, 210
herbal cosmetics, 155
herbal dietary supplements, 26
herbal drug standardization, 60–62
herbal efficacy, 105
herbal extracts, 141
herbal formulations, 6, 7, 8, 33, 37, 38, 40, 49, 51, 57, 62, 105, 148, 149, 152, 157, 169–171, 173, 175–177, 184, 195, 201, 206
herbal healthcare products, 197
herbal pharmacology, 103

I

ICH guidelines, 69
IND application, 92–96, 201
indigenous knowledge, 75, 78–80, 82, 83, 86, 151, 152
institutional review board, 8, 76, 110, 131, 132, 190
investigational new drug, 8, 56, 91, 92, 95, 119, 129, 200

J

Jamu, 116–117

K

Kampo medicine, 8, 105, 120

L

LC-MS, 18, 44, 56, 62, 63, 66–67, 161–162

M

machine learning (ML), 152, 153, 209
marker compounds, 7, 61–63, 95, 125, 140
meta-analysis, 54, 121, 170, 188, 189
metabolomics, 45, 61–63, 68, 70, 72, 102, 140, 209
molecular mechanisms, 70, 210, 171
multidisciplinary approach, 85, 206

N

Nagoya protocol, 79, 82, 151
nanocarriers, 20, 33

nanotechnology, 17, 20, 33, 210
nanotechnology in herbal medicine, 20
National Center for Complementary and Integrative Health (NCCIH), 50, 53, 93, 102
natural products, 26, 36, 58–59, 61, 69–70, 119, 123, 125, 155, 177, 188, 194–195, 197, 209
network pharmacology, 44, 208
neurotoxicity, 70
next-generation sequencing, 61, 72, 140, 200, 208, 209
non-randomized clinical trials, 3

O

observational studies, 2–4, 108, 169
omics technologies, 70, 72, 209, 211
organ toxicity, 72, 94

P

participant autonomy, 138
participant protection, 9, 15, 76, 111, 114, 124
participant rights, 15, 113, 184
patient safety, 1, 7, 10, 53, 58, 62, 72, 98, 119, 148, 151
personalized healthcare, 25, 173
personalized medicine, 19, 45, 102, 210
pharmacodynamics, 10, 20, 54, 96, 120, 123, 126, 134, 177
pharmacogenetics, 19, 172
pharmacogenomics, 19, 54, 172
polyherbal formulations, 7, 37, 39, 40, 44, 69, 106, 151
post-marketing surveillance, 8, 19, 68, 96, 97, 99, 102
predictive modeling, 197, 209
principal component analysis, 57, 62, 159, 161

Q

quality assurance, 7, 11, 28, 49–51, 57, 60–63, 99, 106, 114, 125, 150, 163, 176, 195–197
quality by design, 45, 78
quality management systems, 72
quality standards, 6–9, 54, 62–63, 71, 79, 91, 99, 121, 140, 150, 197

R

randomization, 3, 4, 83, 118, 122, 137, 169, 171, 185, 205
randomized clinical trials, 3, 37, 71, 105, 131, 134, 142, 188
reproducibility, 1, 7–9, 17, 18, 32, 37, 38, 49, 61, 62, 67, 94, 95, 105, 125, 133, 141, 153, 171, 177
research integrity, 76, 86, 128
retrospective studies, 2
reverse pharmacology, 36, 37, 128
risk-benefit analysis, 80, 124, 132

S

safety monitoring, 8, 9, 54, 68, 72, 96, 102, 112, 120, 127, 196, 197
safety profiling, 136
scientific rigor, 3, 8, 84, 85, 148, 152, 176, 177, 184
selection bias, 4, 5, 169, 171
serious adverse events, 7, 91, 98, 112, 114
siddha medicine, 167–169, 173, 174

T

therapeutic efficacy, 20, 28, 58, 62, 99, 117, 169, 185, 194–197, 208
toxicity assessment, 63
traditional healthcare systems, 155
transcriptomics, 45, 70, 140
transparency in clinical trials

U

UHPLC, 57, 69, 70, 160
Unani medicine, 17, 117, 147, 149, 156, 167, 169, 175–176

V

voucher specimens, 8
vulnerable populations, 82, 84, 96, 117

W

WHO guidelines, 1, 9, 64, 71, 78, 106–107, 113–114
WHO monographs, 9

For Product Safety Concerns and Information please contact our
EU representative GPSR@taylorandfrancis.com Taylor & Francis
Verlag GmbH, Kaufingerstraße 24, 80331 München, Germany